Recent Advances in Water Management: Saving, Treatment and Reuse

Recent Advances in Water Management: Saving, Treatment and Reuse

Special Issue Editors

José Alberto Herrera-Melián
José Alejandro Ortega Méndez

MDPI • Basel • Beijing • Wuhan • Barcelona • Belgrade

MDPI

Special Issue Editors

José Alberto Herrera-Melián
Campus Universitario de Tafira
Spain

José Alejandro Ortega Méndez
University of Las Palmas de Gran Canaria
Spain

Editorial Office
MDPI
St. Alban-Anlage 66
Basel, Switzerland

This is a reprint of articles from the Special Issue published online in the open access journal *Water* (ISSN 2073-4441) from 2017 to 2018 (available at: http://www.mdpi.com/journal/water/special_issues/Recent-Advances-Water-Management)

For citation purposes, cite each article independently as indicated on the article page online and as indicated below:

LastName, A.A.; LastName, B.B.; LastName, C.C. Article Title. *Journal Name* **Year**, *Article Number*, Page Range.

ISBN 978-3-03897-031-6 (Pbk)
ISBN 978-3-03897-032-3 (PDF)

Cover image courtesy of José Alberto Herrera-Melián.

Contents

About the Special Issue Editors

José Alberto Herrera-Melián studied a 5-year course of Marine Sciences in the University of Las Palmas de Gran Canaria (ULPGC). His doctoral thesis was devoted to study the chemical speciation of Ni and Co in seawater by using high speed voltammetric methods. During this period, he visited Woods Hole Oceanographic Institution (USA) and the University of Liverpool (UK) as a researcher. In 1995, he was given a position as a teacher of chemistry, chemical oceanography and marine pollution in the Department of Chemistry of the ULPGC. Then, he started to be interested in wastewater treatment, particularly in advanced oxidation technologies (Fenton and TiO2-photocatalysis) and natural wastewater treatment methods (ponds and constructed wetlands). He has published about 120 papers, book chapters, congress proceedings, posters and oral presentations on water analysis and treatment. His h-index is 25 (Google Academic).

José Alejandro Ortega Méndez studied a 5-year course of Marine Sciences in the University of Las Palmas de Gran Canaria (ULPGC) and a 3-year course of industrial technical engineering with specialization in industrial chemistry. His doctoral thesis was devoted to study of detoxification of hazardous waste through biological and photocatalytic treatments and their combination. In 2014, he worked at the Escuela Superior Politécnica del Litoral (Ecuador) as a researcher. Then, he started to be interested in wastewater treatment, particularly in advanced oxidation technologies (Fenton and TiO2-photocatalysis) and natural wastewater treatment methods (ponds and constructed wetlands). He has published about 30 papers, congress proceedings, posters and oral presentations on water analysis and treatment. His h-index is 9 (Google Academic). He is currently an Assistant Professor at the University of Las Palmas de Gran Canaria and he collaborates with the Institute of Environmental Studies and Natural Resources (i-UNAT-ULPGC).

Preface to "Recent Advances in Water Management: Saving, Treatment and Reuse"

Water has always determined the development of peoples and civilizations. Historically, the human being has settled on the edge of rivers that could provide water for consumption and help to get rid of waste. In addition to this, water courses have also served to exchange wealth, raw materials and manufactured products, but above all have served as a way for the dissemination of knowledge and culture.

It is evident that water is an essential component for life. What is not so evident is that in an ever growing population, we can still guarantee access to quality water, due to increasingly diminishing natural resources, including: deforestation, with the consequences of loss of fertile soil erosion, reduction of infiltration and replacement of aquifers; eutrophication and nitrification of lakes, rivers and coastal waters; the appearance and increment of emerging pollutants, mainly pesticides, PCBs, PAHs, personal care products, flame retardants, UV filters, etc., and their toxic, both acute and chronic effects, but also carcinogenic, teratogenic, endocrine disruptive effects, on the biota and the human population.

To all this, we must add the threat of climate change, whose real impact is yet to be determined since it will depend on the world's ability to control its greenhouse gas emissions. In any case, an even greater radicalization of the climate is to be expected, with an increase in the number of extreme events of drought and floods. This, in turn, is leading to migrations of populations from the most affected areas—presumably from the poorest as they have the least money to combat climate change—to the richest countries, particularly Europe and North America.

It is in this scenario that a water management system must be implemented, whose objective should be to guarantee access to quality water for the entire population while minimizing the environmental impact. To achieve this ambitious objective, it will be necessary to implement measures of a diverse nature. Broadly speaking, we can divide them into two types: administrative measures, that is, of a socio-economic, educational and political nature, and scientific and technological measures, related to the increase in the efficiency of the use of water, in order to minimize the environmental impact of the extraction, use, treatment and discharge of water back to Nature, in optimal conditions.

The authors of this book have offered their talent, daily effort and commitment, to achieve, perhaps without being fully aware of it, a healthier and fairer world. In short, a better planet for all of us.

José Alberto Herrera-Melián, José Alejandro Ortega Méndez
Special Issue Editors

water

MDPI

Article

Spanish Agriculture and Water: Educational Implications of Water Culture and Consumption from the Farmers' Perspective

Juan-Carlos Tójar-Hurtado *, Esther Mena-Rodríguez and Miguel-Ángel Fernández-Jiménez

Department of Research Methods and Evaluation in Education, University of Malaga, ES-29071 Malaga, Spain; emena@uma.es (E.M.-R.); mafjimenez@uma.es (M.-Á.F.-J.)
* Correspondence: jctojar@uma.es; Tel.: +34-952-132-543

Received: 14 October 2017; Accepted: 6 December 2017; Published: 11 December 2017

Abstract: The responsible management and consumption of water is a challenge that involves all segments of society. Having access to sufficient quality and quantity of water is not only a technological issue, but requires that the adopted measures and programmes take into account the dimensions of society and education. Spanish agriculture, as in other areas of the world, is a major consumer of water and more so than other sectors, including household consumption. Within the field of environmental education, this study covered the water culture and consumption of Andalusian farmers, based on their own perceptions. For this purpose, a questionnaire was created and validated, and included a sample of 1030 farmers selected with pseudorandom number sampling. An analysis of the data showed relevant results with respect to the values and notions supporting the justification for farmer behaviours, both from a cognitive-representative viewpoint and from an affective-expressive stance, as well as assertions made by the irrigators about other key sectors concerning the responsible management of water usage and water consumption. The findings of this study may assist in the design of environmental education programmes addressing this sector, which could also include other similar populations.

Keywords: foreign countries; agricultural occupations; water; environmental education; surveys; sustainability

1. Introduction

The responsible management and consumption of water is a challenge that involves all segments of society. Having access to sufficient quality and quantity of water is not only a technological issue, but requires that the adopted measures and programmes consider the dimensions of society and education. Awareness and environmental education programmes addressed to the population have a positive effect on the rationing and reduction of water consumption. Nevertheless, for large consumers, these extensive education programmes must be more focused and address their specific needs and behavioural patterns [1]. Spanish agriculture, as in other areas of the world, requires vast amounts of water, more than the industrial sector and domestic consumption. The proportion of water used in Spanish agriculture has increased steadily, from 62.00% in 1987 to 68.19%, in 2012, based on the latest published data. During the same period, the extraction of water for household consumption has increased from 12.00% to 14.21% [2].

Table 1, summarizing the data collected from the AQUASTAT information system [2], depicts the extraction of water according to sector—agriculture, industry, and municipal—and the total per capita. This table helps to compare water usage in Spain, using 2012 data, with other surrounding countries and countries around the world. It reflects the relative significance of the agricultural water usage compared to both industry and municipal usage. Apart from agriculture generally consuming greater volumes, some appreciable data also exists, such as for those countries that use minimal water in agricultural practices, for example, the Central African Republic or Seychelles. In some countries, agriculture consumes high volumes of water, for example, China and the United States. The total water consumption per capita reveals telling data, such as the high consumptions in countries such as Azerbaijan, Chile, New Zealand, United States and Turkmenistan.

Table 1. Water withdrawal by sector and country.

Country	Agriculture [a]	Industry [a]	Municipal [a]	Total [a]	Total per Capita [b]
Argentina	27.93	4.00	5.85	37.78	897.50
Australia	10.59	2.77	4.01	17.37	724.70
Azerbaijan	10.10	2.36	0.52	11.97	1279.00
Brazil	44.90	12.72	17.21	74.83	369.70
Canada	4.75	33.12	5.88	38.80	1113.00
Central African Republic	0.00	0.01	0.06	0.07	17.25
Chile	29.42	4.74	1.27	35.43	2152.00
China	392.20	140.60	75.01	607.80	431.90
Comoros	0.00	0.00	0.00	0.01	17.38
Egypt	67.00	2.00	9.00	78.00	910.60
France	3.14	21.61	5.48	30.23	475.60
Germany	0.21	32.60	5.41	33.04	410.50
Greece	7.92	0.33	1.29	9.63	865.20
Iraq	52.00	9.70	4.30	66.00	2646.00
Israel	1.02	0.11	0.71	1.95	282.30
Italy	12.89	16.29	9.45	53.75	899.80
Japan	54.43	11.61	15.41	81.45	640.60
Lesotho	0.00	0.02	0.02	0.04	23.24
Maldives	0.00	0.00	0.01	0.01	17.11
Mexico	61.58	7.28	11.44	80.30	657.80
Morocco	9.16	0.21	1.06	10.43	316.20
Portugal	8.77	1.50	0.91	9.15	867.30
New Zealand	3.21	1.18	0.81	5.20	1172.00
Saudi Arabia	20.83	0.71	2.13	23.67	907.50
Seychelles	0.00	0.00	0.01	0.01	150.80
Spain	*25.47*	*6.57*	*5.31*	*37.35*	*800.90*
Turkmenistan	26.36	0.84	0.75	27.95	5753.00
United Kingdom	1.05	1.19	5.87	8.21	129.20
United States of America	175.10	248.40	62.09	485.60	1543.00

Notes: [a] 10^9 m^3/year; [b] m^3/inhabitant/year. Adapted from AQUASTAT [2].

As shown in Table 1, Spain's situation is unique in Europe. Water consumption per capita is among the highest in Europe (800.9 m^3/inhabitant), much higher than in the United Kingdom (129.2), Germany (410.5) and France (475.6); but similar although somewhat lower than Greece (865.2), Portugal (867.3), and Italy (899.8). In absolute terms, Spain leads consumption in agriculture (25.47 $\times$ 10^9 m^3/year). Regarding water consumption in industry and by citizens, water consumption in Spain (68.19%) is only exceeded by Greece (82.24%), and Portugal (95.85%), which are Mediterranean countries like Spain that have very little industrial water consumption, at 0.33 and 1.5, respectively.

Farmers, a key component in the consumption of water and in various aspects concerning the quality and quantity of water, are far too often overlooked in terms of scientific research. Generally, this is a sector of the population that is difficult to access and has its own culture and traditions that are dependent on local contexts, which are seldom addressed or understood by other associated populations [3,4]. A review of the international literature shows that not many studies have addressed this issue. Research in the field of agriculture and environmental education is scarce. In the following paragraphs an analysis of the existing literature is made, highlighting the aspects that are the focus of this research.

In Oberkircher and Hornidge [3], a study was conducted with farmers from Khorezm, Uzbekistan. The unsustainable use of water for irrigation has created a major crisis in the Aral Sea. This study analysed farmer perceptions of water and its management, as well as how certain practices could promote water conservation and savings. Another study in Papua New Guinea [4] showed how little "indigenous knowledge" is acknowledged regarding environmental and agricultural education. This knowledge, a fundamental aspect of indigenous culture, is essential for the management and responsible consumption of water. Also, the results of an educational outreach programme on water resource management, and their effects on the beliefs and attitudes of local farmers in the Upper Taieri River Catchment, New Zealand [5], were analysed. Moreover, a review was undertaken in Iran using 36 studies with farmers [6], which showed the importance of education in improving sustainable behaviours.

Despite these examples, most of the studies on water management and consumption issues were conducted with the general population or with educational populations in mind [7–10]. In Thompson and Serna [11], a study was conducted revealing that 94.00% of the students who participated in an educational programme on water conservation had broadened their knowledge base and increased their commitment. For this reason, an examination of the behaviour of water management and consumption in specific sectors of the population, such as farmers, is pertinent and relevant from a researcher's perspective.

The Autonomous Community of Andalusia, Spain, was chosen as the area of study. Andalusia is the most populated autonomous community in Spain. It covers an area of 87,268 km^2, of which 45.74% is arable land. According to official data [12], noting that groundwater and treated wastewater were not included, Andalusia is the region in Spain where agriculture annually consumes the most water, 28.20% of the total, amounting to 4,216,350,000 m^3.

Accordingly, we conducted a study on water consumption and culture of farmers, based on their own points of view from an environmental education perspective. The specific objectives of the study were (1) to determine the understanding of farmers, their attitudes and moods concerning water management and consumption; and (2) to determine their position in terms of proposals for change and possible improvements in that subject; additional specific objectives include (3) verifying if any differences or correlations existed between the information, attitudes, and moods of farmers, and other variables such as age, gender, employment situation, cultivated surface area, and production.

2. Materials and Methods

A descriptive study was completed in a pre-research phase [13]. In that study, a sample of 24 participants, selected by theoretical sampling, was interviewed in depth. In the theoretical sampling, the participants are selected because they fulfil a series of characteristics according to the objectives of the research [14]. The participants belonged to several sectors with a relevant role related to the management and responsible use of water, including employees or members of water companies, administration, conservation associations, and environmental education and specialised media companies. The interview script included three main categories: (1) how they perceive and the importance they attribute to problems related to water; (2) the responsibility the entity assumes in this problem; and (3) solutions that it considers suitable for the problems related to the consumption and management of the water.

From the information gathered during the interviews, a 30-element questionnaire was designed, using a Likert scale from 1 to 5, with 1 meaning "fully disagree" to 5 meaning "fully agree". The questionnaire was formulated with the purpose of determining various aspects relating to water use and consumption, along with understanding farmer values and culture. The structure of the questionnaire consisted of three dimensions. The dimensions were based on Jakobson's model of language functions [15]: (1) representative, or referential, to gather information on various relevant facets of water management, with a total of 6 elements; (2) emotive, or expressive, to gather information on farmer feelings, attitudes and moods, with a total of 17 elements; and (3) appellative, or conative, to determine any appraisals regarding proposals for change and improvement directed at various sectors, with a total of 7 elements.

Furthermore, a number of questions related to classification variables, such a gender, age, employment situation, surface area, crop type and production, were included to achieve a better understanding of the selected sample and to conduct differential analyses.

Before starting the interviews, an expert validation occurred. Seven research methodology and environment experts reviewed and assessed the adequacy of the elements and dimensions of the questionnaire. After considering the experts' suggestions, a second version of the questionnaire was drafted. Using this second version, a pilot application of the questionnaire was conducted using a sample of 105 participants.

A reliability study, through internal consistency using Cronbach's alpha, and structural validity, through factorial analysis of principal components, were performed on the data collected during the pilot application. The reliability study provided a Cronbach's alpha of 0.79, which is considered acceptable [16]. A factorial analysis allowed for a model of nine components to be elaborated, which accounted for 68.45% of the total variance. The components of the model were fully consistent with the dimensional structure of the questionnaire.

After several adjustments had been made to the questionnaire based on the pilot application, a second application of the questionnaire was conducted on a pseudorandom and non-probabilistic sample of 1030 participants. The sample consisted of both men (53.00%) and woman (47.00%), between the ages of 17 and 77, with a mean age of 36 and standard deviation of 11.13. Other data that define the sample are the cultivation area, with a mean of 18.13 hectares and standard deviation of 8.62, the type of crop (olive grove 47.54%, cereals 23.16%, industrial crops 10.67%, fruit trees 9%, and other 9.63%), and production, with a mean of $53,915.10/year. A post evaluation study on the representativeness of the sample, by comparisons of distributions across χ^2, showed how the variables of age, gender, surface area of cultivation, type of crop, production and geographical areas were represented in similar proportions as in the source population.

As for the data gathered after the second application, descriptive analyses (measures of central tendency and dispersion), nonparametric tests of χ^2 (comparing observed and expected frequencies), analyses using the Pearson correlation coefficient (between classification variables such as age, surface area of cultivation, and productivity and the remaining elements on the questionnaire) and multivariate analysis of variance (provinces and employment situation with the rest of the questionnaire elements) were conducted. All analyses were performed using the SPSS v.22 statistical package.

3. Results

First, the descriptive results of the questionnaire are presented along with a brief analysis of the frequency distribution observed regarding the expected frequencies, including Pearson's χ^2 test. Second, the results of the bivariate, correlation coefficients, and multivariate analyses of variance are presented.

3.1. Descriptive Results

Tables 2–4 present the most relevant results from the questionnaire (Table S1 contains all the results). The most frequent options, the mean, and standard deviation are summarized. Non-parametric tests using χ^2 demonstrated significant differences ($p < 0.0005$) for all observed frequency distributions compared with the expected value, and for each element on the questionnaire. Table 2 displays some of the most significant results in terms of percentages, corresponding to the elements associated with the representative function (objective 1). Based on this function, we thought that information would be obtained for some relevant aspects of water usage and consumption from the farmer perspective.

Table 2. Results expressed in terms of a percentage of the respondents of the representative function.

Element	5	4	3	2	1	Me	SD	χ^2 *
1. When it comes to consumption, the agricultural sector should have more say in political decisions on water management	48.20	27.30	20.40	2.60	1.50	4.18	0.94	769.25
2. Water management would be better if the situation of farmers was considered	41.40	29.30	23.50	4.00	1.80	4.04	0.98	589.13
5. Water is not a problem for the general population, instead, it is a problem for farmers	7.00	10.70	17.80	12.20	52.30	2.07	1.32	702.69
6. It is a pity that all this water is lost at the river mouth	46.50	16.50	20.60	7.80	8.60	3.84	1.32	511.73

Note: * χ^2 Pearson Test, with df = 4, all significant with $p < 0.0005$.

A large majority of the respondents considered that the agriculture sector should have more of a say in political decisions on water management, with 48.20% fully agreeing and 27.30% agreeing to a certain extent, and that it would be better if water management considered farmers' circumstances. The average of both these elements was high, with means of 4.18 and 4.04, respectively, with a low dispersion of opinions, with standard deviations of 0.94 and 0.98, respectively.

Farmers, although they belong to the sector that consumes more water, do not think that the water problem is exclusively theirs. On the contrary, they do not agree that water is not a problem for the general population, with 52.30% totally disagreeing and 12.20% partially disagreeing. Nevertheless, most believe that the water "lost" at the river mouth is a pity, with 46.50% totally agreeing and another 16.50% partially agreeing. For both cases, the dispersion of opinions is not low (1.32), however, a marked tendency stretched in both directions.

Table 3 includes the most important elements corresponding to the emotive function. This function was intended to obtain an approximate notion of the feelings, attitudes and moods of farmers regarding water consumption (objective 2).

Table 3. Results of the emotive function.

Element	5	4	3	2	1	Me	SD	χ^2 *
8. If the infrastructure were improved, there would be a larger irrigated area	48.10	28.10	18.10	4.00	1.70	4.17	0.97	746.60
10. Using fertilisers above the recommended rates of application improves production	6.10	9.00	14.70	12.20	58.00	1.93	1.27	951.28
15. A social criterion should be utilised for the distribution of water (crops that generate more employment)	33.90	27.50	27.10	7.10	4.40	3.79	1.12	365.46
17. Development and growth cannot slow down due to a lack of water	30.70	21.40	27.80	10.00	10.00	3.53	1.29	194.30
18. Fertilisers are responsible for soil and water pollution	33.50	18.20	29.10	11.10	8.10	3.58	1.27	251.89
19. Improvements to infrastructure would allow for more irrigation	46.20	26.40	18.70	5.50	3.20	4.07	1.08	629.11
20. Investing in more efficient irrigation techniques would make it possible to endure times of drought	57.00	22.60	15.70	3.80	0.90	4.31	0.93	1041.07
21. Low quality or recaptured water could be used for agriculture	44.40	26.10	18.70	6.40	4.40	3.99	1.31	547.23

Note: * χ^2 Pearson Test, with df = 4, all significant with $p < 0.0005$.

Farmers support the idea of infrastructure improvements to achieve a larger irrigated area with 48.10% fully agreeing and 28.10% partially agreeing, whereas the average was high at 4.17. A large majority, 58.00%, of respondents disagreed with using more than the recommended rates of fertilisers to enhance production. Nevertheless, a high dispersion was seen for this case (1.27), denoting an opposing opinion of those favouring the use of rates greater than those recommended by some irrigators.

Although the opinions were dispersed around a mean of 3.47, a vast majority of respondents admitted that more water should be made available for crops that help maintain populations in the local area, with 22.90% totally agreeing and 24.70% partially agreeing. The social criterion for the distribution of water towards crops that generate further employment was supported by most of the respondents with 33.90% totally agreeing and 27.50% partially agreeing.

Most respondents stated that development and growth cannot be slowed down due to a lack of water (30.70% totally agree, with an average of 3.53), although the opinions were dispersed (SD = 1.29). Most farmers that answered the questionnaire, at 33.50%, admitted that fertilisers are responsible for soil and water pollution. Even more prominent was the opinion that improvements made to infrastructure would allow for more irrigation (46.20% totally agree). In this case, the statement was generic and it was not entirely clear if the farmers were referring to a larger irrigated area or to higher volumes per unit surface, or perhaps both.

Most agreed that investing in more efficient irrigation techniques would allow for times of drought to be endured (57.00% totally agree). The same occurred with the idea that reused water could be used in agriculture (44.40% fully agree).

Table 4 shows several of the results of the elements relating to the appellative function, the opinions and appreciations of the farmers partaking in the questionnaire regarding proposals for change and improvements targeting various sectors (continuing with objective 2).

Table 4. Results of the appellative function.

Element	5	4	3	2	1	Me	SD	χ^2 *
26. Other sectors, such as industry and tourism, manage water more poorly than agriculture	31.20	24.20	27.30	10.30	7.00	3.62	1.21	236.62
27. Domestic water consumption conceals unjustified water costs	35.30	26.50	25.50	7.70	4.90	3.80	1.15	353.23
28. There are many non-farmers who use a lot of water to cultivate their plots of land	42.40	23.70	22.50	7.20	4.20	3.93	1.15	481.62
29. Management should pay more attention to the opinion of farmers	39.80	30.20	22.50	4.90	2.60	3.99	1.03	532.05
30. Technological modernisation saves more water than advertising campaigns	42.80	25.90	24.10	4.90	2.30	4.02	1.03	597.64

Note: * χ^2 Pearson Test, with df = 4, all significant with $p < 0.0005$.

A slight trend was seen for assuming that other sectors, such as industry and tourism, manage water more poorly than agriculture, with a mean of 3.62 and SD of 1.21. Farmers participating in the questionnaire presumed that household water consumption concealed unjustified water costs, as 35.30% fully agreed and 26.50% partially agreed. Even more resounding was the view that many non-professional farmers producing furtive crops consume a lot of water to cultivate their plots of land with 42.40% totally agreeing and 23.70% partially agreeing.

The respondents believed that the administration should listen more to the opinions of farmers (39.80% fully agree, 30.20% partially agree). Along the same lines was the view that technological modernisation saves more water than advertising campaigns, as 42.80% fully agreed and 25.90% partially agreed.

3.2. Further Results

The analyses performed to meet the additional specific objectives showed a correlation between age, cultivated surface, and production, and the elements of the questionnaire (objective 3). As age increased, farmers were more in agreement with "When it comes to consumption, the agricultural sector should have more say in water management" ($r_s = 0.24$, $p < 0.0005$). Moreover, those with a larger cultivated surface area and/or higher production held the view that "more irrigation for rainfed crops would increase efficiency" ($r_s = 0.20$, $r_s = 0.27$, respectively, and both $p < 0.0005$). Less agreement existed for those who had a small cultivated surface area and/or reduced production. Finally, irrigators with higher production levels believed that more water should be provided for crops that help retain more people in the local area. Meanwhile, those who had a lower production level did not agree with this opinion ($r_s = 0.22$, $p < 0.0005$).

The multivariate analysis of variance determined that significant correlations existed between various elements of the questionnaire and the variables of gender, province, and current employment situation.

Specifically, male farmers, with a mean of 3.63, were more in agreement than female farmers, with a mean of 3.34, in thinking more water should be given to crops that encourage people to stay in the local area ($p < 0.0005$). A significant difference ($p = 0.03$) existed between the viewpoints of female farmers (mean of 3.38), who agree more than male farmers (mean of 3.21) in terms of the main use of river water being for agriculture. Likewise, women (mean of 4.10) had a significantly different opinion ($p = 0.001$) from men (mean 3.85), in thinking that many people who are not farmers use a lot of water to cultivate their plots of land.

The current employment situation (employed, self-employed, member of a cooperative or unemployed) provided some significant results. The self-employed, with a mean of 3.85, were less concerned with paying more to have access to more water than employed workers, with a mean of 2.62 ($p = 0.006$) or the unemployed (mean of 2.43, $p = 0.003$). The unemployed (mean of 3.53), also believed that more water should be provided to the larger cultivated areas than the employed workers (average of 3.53 and $p = 0.033$).

The multivariate analysis of the variance provided significant results with interesting nuances depending on if the crop area was drier or wetter. For example, respondents in drier areas, with a mean of 4.35 and p-value of 0.027, were more in agreement with the idea that "the water issue would be resolved by transferring water from catchment areas with a surplus to those in deficit" than those from the wetter areas, with a mean of 3.40. The results showed that all farmers agree with the water transfers. This result indicates how, in the drier areas of cultivation, the transfers are valued more positively as a solution. Similarly, farmers in drier areas (mean 3.88, $p = 0.05$) agreed with the opinion that "if the infrastructures were improved, there would be a larger irrigated area", more so than those from coastal and wetter areas (average 3.98). These results agreed with the previous results. All farmers hope to increase the irrigated area by improving infrastructures, but those in drier areas more strongly supported this idea ($p = 0.05$) than those in wetter areas.

Farmers in wetter areas (mean 4.52, $p = 0.032$) believe that "water of a lower quality, or recaptured, could be used for agriculture", more so than those in drier areas (mean 4.06). Although all farmers positively valued the use of low quality or recaptured water, those in more humid areas valued it more ($p = 0.032$). Respondents from drier areas (mean of 4.06) were more in agreement with "domestic water consumption concealed unjustified water costs" than those in more humid areas (mean of 3.51, $p = 0.025$). Similarly, all farmers thought that the water consumption of the citizens that conceals the waste of water is not justified. In this sense, farmers in the driest areas were those who were significantly more concerned ($p = 0.025$) with this issue.

4. Discussion

As in other studies [1,3,4], this research has shown the importance of cultural referents and the values of farmers for determining their water consumption behaviours. This culture, defined by a set of concrete traits, can determine farmers' behaviour towards developing sustainable water management practices (objective 1). Huan and Lamm [1] verified how large consumers of water are less inclined to participate in water saving programmes. This study depicts a similar situation. As the cultivation area increases, farmers are less likely to save water. Farmers participating in the questionnaire preferred to save water by opting for technological modernisation instead of participation in campaigns and educational programmes. A close correlation exists between the cultural values of farmers and the setting in which they live and work. For the Aral Sea in Uzbekistan, Oberkircher and Hornidge [3] examined the effects of religious values and the risk of being fined in encouraging water savings. These farmers believed that the state is responsible for water management and their perceived water needs were beyond their own geographical reality. A similar situation occurred in this study. In Spain, farmers remarked that the growing demand for water should be satisfied by public investment aimed at building hydraulic infrastructures, to provide more efficient technologies, and to manage drought and water scarcity. For this to happen, the farmers proposed that the administration should listen to them more often and that their opinion should have more weight (objective 2).

However, some of the farmer conceptions about water were erroneous, such as the idea that water entering the mouth of rivers is wasted water, but these ideas define them and must be considered when developing educational programmes. Other notions cannot be classified as erroneous, but they determine a particular mindset that is not conducive to saving water. An example of this is when the farmers indicated that development cannot be slowed due to a lack of water. As in Radcliffe et al. [4], new crops were found to be determined more by market and less by local uses and traditions, which are more respectful in terms of sustainable water use. Thus, Spanish farmers are prepared to abandon traditional rainfed crops in favour of irrigated crops, which require more water consumption. The same occurs with the possibility of introducing more "marketable" crops to generate further employment, even if they consume more water. Despite this, as observed by Tyson et al. [5], crop choice, the development of water allocation schemes, management, and addressing water shortage and quality problems could be approached from a communicative and educational process (continuing with objective 2).

As confirmed by Vaninee et al. [6], there is an important correlation between understanding and sustainable behaviours in agriculture, where environmental education can foster this sustainable behaviour so that substantial water savings may be achieved [3]. Understanding the demands of the agricultural sector, as demonstrated by Huan and Lamm [1] elsewhere in the world, allows us to identify the specific needs and behaviour patterns of key groups regarding water management and consumption for the general population.

5. Conclusions

The analysis of the data elicited the opinions and conceptions of farmers in Spain, where the consumption of water is significant. The attitudes and moods of these farmers were analysed, along with proposals for change and possible improvements suggested for various aspects related to water usage and consumption (research objectives 1 and 2). Farmers feel that their sector should have a louder voice when it comes to water management and that management would improve if their opinions were considered. Although they admit that agricultural practices produce waste water, they say that water shortage is an issue that is due to the general population rather than agriculture.

A large majority of farmers support improvements to water infrastructure that would allow for more land to be irrigated and consider that water should not be "let go to waste" at the mouths of rivers. This erroneous belief is deeply rooted among farmers and a large portion of the Spanish population. Moreover, farmers are supportive of a growth model that supports further irrigation. Whereas the state claims it is investing more in water infrastructure and efficient technologies to counteract the effects of climate change, famers are also of the opinion that development should never be halted because of a water shortage. Concepts such as sustainability in water management seem to be subject to economic development and growth. Along these lines, farmers agreed with "social criteria" to replace traditional crops with more commercial crops that are more desirable in the marketplace and to encourage crops that allow people to stay in the area, so that rural areas remain populated, despite the fact that these new crops would require water consumption.

Several relevant and statistically significant differences were unveiled in the opinions of the respondents, and in the variables including age, gender, employment situation, surface area of cultivation, and production. Accordingly, the specific objectives of the study were accomplished (objective 3).

Following the analysis of the data, we concluded that significant results were obtained about the mindsets and values behind the rationalisation of farmer behaviour, both from a cognitive-representational viewpoint and from an affective-expressive perspective. Assertions that farmers have raised against other core economic sectors, along with the administration, that use and manage water were included, based on their own perspectives.

The findings of this study contain a wealth of information for the preparation of environmental education programmes. Having an understanding of the preconceptions and cultural behaviours of Spanish farmers may assist in the development of specific programmes that further understanding, education on values, and training in attitudes and behaviours that are more respectful towards water usage and sustainable management.

Supplementary Materials: The following is available online at www.mdpi.com/2073-4441/9/12/964/s1, Table S1: Results expressed in terms of a percentage of the respondents of the total questionnaire.

Acknowledgments: The funds for the realization of this research were contributed by the Andalusian Plan of Research, Development and Innovation, Ministry of Economy and Knowledge, Junta de Andalucía (Andalusia, Spain). The funds to cover the costs of publishing in open access have been provided by Universidad de Málaga (Spain).

Author Contributions: J.-C.T.-H. conceived and designed the study, conducted the field analysis and drafted the manuscript. E.M.-R. performed the sample collection, the statistical analysis and helped in the data interpretation. M.-Á.F.-J. interpreted the statistical analysis and participated in drafting the manuscript.

Conflicts of Interest: The authors declare no conflicts of interest.

References

1. Huan, P.; Lamm, A.J. Informing Extension Program Development through Audience Segmentation: Targeting High Water Users. *J. Agric. Educ.* **2016**, *57*, 60–74. [CrossRef]
2. AQUASTAT-FAO's Information System on Water and Agriculture. Available online: http://www.fao.org/nr/water/aquastat/water_use/index.stm (accessed on 3 October 2017).
3. Oberkircher, L.; Hornidge, A.K. "Water Is Life"—Farmer Rationales and Water Saving in Khorezm, Uzbekistan: A Lifeworld Analysis. *Rural Sociol.* **2011**, *76*, 394–421. [CrossRef]
4. Radcliffe, C.; Parissi, C.; Raman, A. Valuing Indigenous Knowledge in the Highlands of Papua New Guinea: A Model for Agricultural and Environmental Education. *Aust. J. Environ. Educ.* **2016**, *32*, 243–289. [CrossRef]
5. Tyson, B.; Edgar, N.; Robertson, G. Facilitating Collaborative Efforts to Redesign Community Managed Water Systems. *Appl. Environ. Educ. Commun.* **2011**, *10*, 211–218. [CrossRef]
6. Vaninee, H.S.; Veisi, H.; Gorbani, S.; Falsafi, P.; Liaghati, H. The Status of Literacy of Sustainable Agriculture in Iran: A Systematic Review. *Appl. Environ. Educ. Commun.* **2016**, *15*, 150–170. [CrossRef]
7. Bajzelj, B.; Fenner, R.; Curmi, E.; Richards, K. Teaching sustainable and integrated resource management using an interactive nexus model. *Int. J. Sustain. High. Educ.* **2016**, *17*, 2–15. [CrossRef]
8. McBroom, M.; Bullard, S.; Kulhavy, D.; Unger, D. Implementation of Collaborative Learning as a High-Impact Practice in a Natural Resources Management Section of Freshman Seminar. *Int. J. High. Educ.* **2015**, *4*. [CrossRef]
9. Seehamat, L.; Sanrattana, U.; Tungkasamit, A. The Developing on Awareness of Water Resources Management of Grade 6 Students in Namphong Sub-Basin. *Int. Educ. Stud.* **2016**, *9*, 156. [CrossRef]
10. Chanse, V.; Mohamed, A.; Wilson, S.; Dalemarre, L.; Leisnham, P.; Rockler, A.; Shirmohammadi, A.; Montas, H. New approaches to facilitate learning from youth: Exploring the use of Photovoice in identifying local watershed issues. *J. Environ. Educ.* **2016**, *48*, 109–120. [CrossRef]
11. Thompson, R.; Serna, V. Empirical evidence in support of a research-informed water conservation education program. *Appl. Environ. Educ. Commun.* **2016**, *15*, 30–44. [CrossRef]

12. INEbase/Agricultura y Medio Ambiente/Agua/Estadísticas Sobre el uso del Agua/Últimos Datos. Available online: http://www.ine.es/dyngs/INEbase/es/operacion.htm?c=Estadistica_C&cid=1254736176839&menu=ultiDatos&idp=1254735976602 (accessed on 3 October 2017).

13. Matas-Terrón, A.; Estrada-Vidal, L.; Martín-Jaime, J. Perspectiva de los agentes institucionales ante la gestión del agua. In *VII Congreso Ibérico Sobre Gestión y Planificación del Agua "Ríos Ibéricos +10. Mirando Al Futuro Tras 10 Años de DMA*; Nueva Cultura del Agua: Talavera de la Reina, Spain, 2011; pp. 1–6.

14. Rovio-Johansson, A. Students' knowledge progression: Sustainable learning in Higher Education. *Int. J. Teach. Learn. High. Educ.* **2016**, *28*, 427–439.

15. Brown, J.W. Communicative competence vs. communicative cognizance: Jakobson's Model revisited. *Can. Mod. Lang. Rev.* **1984**, *40*, 600–615.

16. Jisu, H.; Delorme, D.; Reid, L. Perceived Third-Person Effects and Consumer Attitudes on Preventing and Banning DTC Advertising. *J. Consum. Aff.* **2006**, *40*, 90–116. [CrossRef]

water

Article

Water Use and Conservation on a Free-Stall Dairy Farm

Etienne L. Le Riche [1,2], **Andrew C. VanderZaag** [2,*], **Stephen Burtt** [2], **David R. Lapen** [2] and **Robert Gordon** [3]

[1]   School of Environmental Sciences, University of Guelph, Guelph, ON N1G 2W1, Canada;
      eleriche@hotmail.com
[2]   Science and Technology Branch, Agriculture and Agri-Food Canada, Ottawa, ON K1A 0C6, Canada;
      stephen.burtt@agr.gc.ca (S.B.); david.lapen@agr.gc.ca (D.R.L.)
[3]   Department of Geography & Environmental Studies, Wilfrid Laurier University,
      Waterloo, ON N2L 3C5, Canada; rogordon@wlu.ca
*   Correspondence: andrew.vanderzaag@agr.gc.ca; Tel.: +613-759-1254

Received: 13 October 2017; Accepted: 8 December 2017; Published: 15 December 2017

Abstract: Livestock watering can represent as much as 20% of total agricultural water use in areas with intensive dairy farming. Due to an increased emphasis on water conservation for the agricultural sector, it is important to understand the current patterns of on-farm water use. This study utilized in situ water meters to measure the year-round on-farm pumped water (i.e., blue water) on a ~419 lactating cow confined dairy operation in Eastern Ontario, Canada. The average total water use for the farm was 90,253 $\pm$ 15,203 L day^{-1} and 33,032 m^3 annually. Water use was divided into nutritional water (68%), parlour cleaning and operation (14%), milk pre-cooling (15%), barn cleaning, misters and other uses (3%). There was a positive correlation between total monthly water consumption (i.e., nutritional water) and average monthly temperature for lactating cows, heifers, and calves ($R^2 = 0.69$, 0.84, and 0.85, respectively). The blue water footprint scaled by milk production was 6.19 L kg^{-1} milk or 6.41 L kg^{-1} fat-and-protein corrected milk (FPCM) including contributions from all animal groups and 5.34 L kg^{-1} milk (5.54 L kg^{-1} FPCM) when excluding the water consumption of non-lactating animals. By applying theoretical water conservation scenarios we show that a combination of strategies (air temperature reduction, complete recycling of milk-cooling water, and modified cow preparation protocol) could achieve a savings of 6229 m^3 annually, a ~19% reduction in the total annual water use.

Keywords: milk production; water; footprint; water recycling; conservation; partitioning; efficiency

1. Introduction

In the past 100 years, agricultural production has accounted for as much as 80% of global freshwater consumption [1]. While green water can be made scarce and is important for global water resource allocation, blue water is more relevant from the point of view of industrial environmental impact assessments [2]. This is partially because natural vegetation consumes green water in much the same way as rain-fed agricultural land [3], whereas blue water withdrawals are almost entirely anthropogenic, and, in cases of fossil groundwater, non-renewable [4].

Total agricultural blue water (fresh surface/groundwater) use in Canada is estimated to be between 1.7 and 2.3 billion m^3 year^{-1}. While irrigation represents the bulk of this agricultural water use, livestock watering makes up between 5% to 10% of the total, which in turn represents up to 230 million m^3 of blue water annually [5,6]. In Canadian provinces where rain-fed agriculture predominates and there is intensive dairy production, such as Ontario and Quebec, livestock watering approaches 20% of the provincial totals [6].

Non-irrigation blue water use on dairy farms typically includes water consumption, milking equipment, parlour, and pipeline cleaning, washing down of the holding area, milk cooling, and temperature control [7]. In a European study, water meters read monthly by farmers determined a milk production water footprint (WF) of between 1.2 to 9.7 L kg^{-1} of fat-and-protein-corrected milk (FPCM) [8]. Capper et al. [9] found that water consumption on American dairies has decreased from 10.8 L kg^{-1} milk to 3.8 L kg^{-1} milk between 1944 and 2007. Drastig et al. [10] calculated that the mean blue water (fresh surface/ground water) consumption required to produce 1 kg of milk was 3.94 $\pm$ 0.29 L. Drastig et al. [10] reported that the majority of water use was for cow consumption (82%), whereas milk processing (cow preparation, bulk tank cleaning and line flushing) contributed 11% of the water use and the remainder (7%) was for barn cleaning and disinfection. However, some these figures were derived from models that may not include the water requirements of on-farm replacement animals. Moreover, a detailed understanding of dairy farm water uses and temporal dynamics is required to understand how farmers can adjust management practices to conserve water.

Water is the most important foodstuff of lactating cows [11,12] and daily water consumption of lactating dairy cows in Ontario can be as much as 155 L day^{-1}, up to triple that of dry cows [13]. In order to achieve optimal milk production in dairy cows, sufficient amounts of water, energy, protein and minerals are necessary [14]. Cardot et al. [15] identified several factors that affect free water intake, namely dry matter intake (DMI), milk yield, and to a lesser extent minimum temperature and rainfall. Links between production and heat stress have been demonstrated previously [16]. Both the consumption of dry matter (DM) and milk production decrease when the temperature humidity index (THI) was >60 [17]. Furthermore, water consumption increases linearly under mild heat stress when THI exceeds 30 [17] and hence daily water use fluctuations are typically greater in summer months [18]. Heat stress mitigation, such as cow showers, can decrease cow body temperature by 0.2 °C and showered cows spend half as much time near water bowls [19]. Lin et al. [20] showed that misters can decrease average daily air temperature by ~2 °C using 16.7 L cow^{-1} day^{-1} and ~4 °C using 44.2 L cow^{-1} day^{-1}.

To improve understanding of the current patterns of on-farm water use and potential avenues for water conservation, this study intended to:

1. Determine the total annual pumped groundwater (on-farm blue water) and blue water footprint of a dairy farm.
2. Partition the groundwater flow by type of use.
3. Identify areas for blue water conservation and provide estimates of potential savings.

2. Materials and Methods

2.1. Dairy Farm Site

The one-year monitoring period was from 1 October 2015 to 30 September 2016 for a total of 366 consecutive days. The trial was conducted on a confined dairy operation located in Eastern Ontario (44.981804°, −75.366390°). Herd information was collected from detailed monthly farm records obtained from the dairy herd management service (CanWest DHI) and the farmers. The operation included ~973 Holstein cows. During the monitoring period, the herd averaged 419 $\pm$ 13 lactating cows and 54 $\pm$ 6 dry cows (~11% of herd). In addition, it was estimated based on quarterly observations (counts) that there were ~60 transition cows (pre-fresh, fresh). The replacement animal populations fluctuated from month to month but were typically ~240 heifers and ~200 calves.

2.1.1. Animal Housing

The cow, heifer and calf animal groups were each housed in separate barns on the farm. The free-stall main barn housed lactating cows, transition cows (pre-fresh, fresh), and dry cows. A second free-stall barn housed the heifers, and a third barn housed the calves in 21 pens (~10 calves per pen). The main barn was cooled using 16 box fans evenly distributed throughout the building (four per quadrant). The calf barn was cooled using five high-volume low-speed fans and air circulation was aided by two positive pressure ventilation ducts. The heifer barn relied on passive ventilation from the roof, open sides (controlled with curtains), and ends of the building.

2.1.2. Animal Diets

Lactating cows were fed 25.2 kg day^{-1} dry matter (DM) as a total mixed ration (TMR) comprised of corn silage, ensiled field peas, high moisture corn, and supplements. Feed was analysed using the following methods: AOAC 930.15 for DM, Dumas combustion method for crude protein, and ICP-OES for nutrients. A dietary analysis of the feed given to the main animal categories is presented in Table 1. Pre-weaned calves (3–72 day) were fed milk replacer delivered by CF1000+ calf feeders (DeLaval Canada, Peterborough, ON, Canada).

Table 1. Typical feed constitution for each animal type (heifers and cows). Each analyte was measured in duplicate from feed laid out for each animal type. Values are mean ± SD.

Parameter	Heifers	Cows
Dry Matter (%)	45.7 ± 1.00	49.2 ± 3.43
Crude Protein (%DM)	13.3 ± 0.54	14.9 ± 1.29
Ca (%DM)	1.32 ± 0.01	0.96 ± 0.04
P (%DM)	0.34 ± 0.01	0.36 ± 0.03
K (%DM)	1.37 ± 0.06	1.02 ± 0.07
Mg (%DM)	0.33 ± 0.01	0.40 ± 0.01
Na (%DM)	0.44 ± 0.01	0.34 ± 0.14
Ca:P ratio	3.91 ± 0.06	2.70 ± 0.57

2.1.3. Milk Production

The milkhouse holding area and milking parlour (12 × 2 parallel) was perpendicularly connected to the main barn. The dairy cows, which were housed in the barn year-round, were milked 3× daily at 0300 h, 1100 h, and 1900 h with each milking event taking ~4–5 h. The bulk tank (31,593 L capacity) was emptied every 1–2 days depending on milk pick-up.

Average daily milk production was extrapolated from test day production data and herd size data corresponding to the monitoring period, which were obtained from CanWest DHI (Guelph, ON). FPCM was calculated using the following equation:

$$FPCM = M_{raw} \times \left(0.337 + 0.116 \times M_{fat} + 0.06 \times M_{pr}\right), \tag{1}$$

where *FPCM* is fat-and-protein-corrected milk, in kg, and M_{raw} is the average daily milk production, in kg. M_{fat} and M_{pr} are the respective average fat and protein contents of the milk, expressed as a percentage [21].

The average daily milk production based on monthly farm records for the monitoring period was 34.8 ± 0.8 kg cow^{-1} day^{-1} with a fat content of 3.8% and a protein content of 3.2%. Corrected to 4.0% fat and 3.3% protein, the milk production averaged 33.6 kg cow^{-1} day^{-1} FPCM.

2.2. Water Use Overview

Water was used in various aspects of the farm management, specifically, drinking water for each group of cattle (lactating cow, dry cow, heifer, and calf), milk parlour sanitization, milk pre-cooling (i.e., plate cooler), cow misting and general farm cleaning (i.e., barn floor and farm equipment wash-down). All on-farm water was drawn from two wells located on the property (Total Dissolved Solids 1039 mg L^{-1}, pH 7.5, nitrate-N 10.5 mg L^{-1}, $p < 1$ mg L^{-1}, Na 186 mg L^{-1}, sulphate 95.7 mg L^{-1}). These figures are all within the range of the acceptable guidelines, where applicable [22]. Water was analysed using the following methods: electrical conductivity (EC) for total dissolved solids, ion-selective electrode meter (ISE) for NO_3–N, and ICP–OES for nutrients.

Drinking water was stored in a 5678 L plastic reservoir with inlets controlled by float valves (Figure 1, "primary reservoir"). In addition, the milk pre-cooling water was freely discharged into this reservoir (without float valve control). Any overflow from this reservoir was diverted to an overflow reservoir (Figure 1, "overflow"). This reservoir was always full when inspected on site visits. All water that went into overflow was considered wasted, although attempts were made to use some of it for milkhouse floor cleaning. Overflow from this reservoir flowed to the manure pit.

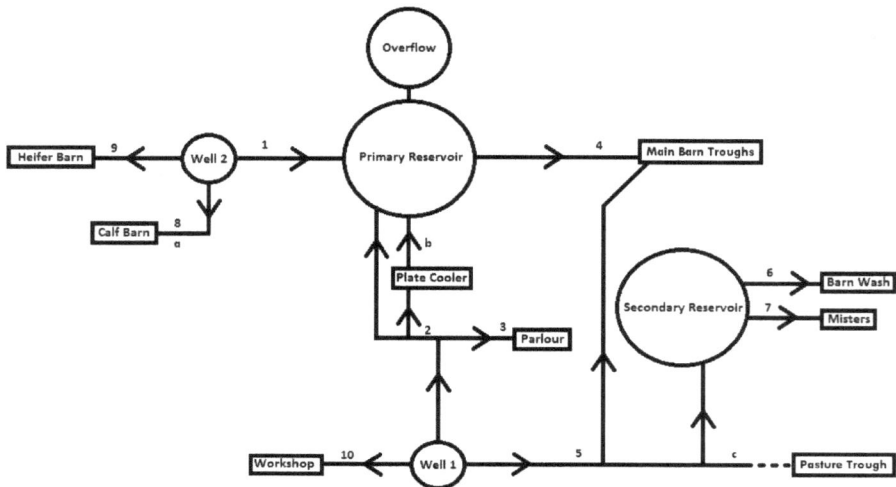

Figure 1. Simplified water flow diagram outlining the location of the 10 in-line flow meters (1–10) and placement of the transit-time flow meters (TTFMs) used to measure water to the calf barn (**a**), from the plate cooler (**b**), and to the pasture trough (**c**). Not to scale.

2.2.1. Nutritional Water

Cows in the main barn had free access to drinking water by means of 11 automatically replenishing 227 L troughs. Furthermore, water was added daily to the Total Mixed Rations (TMR). The heifer barn was equipped with seven automatically replenishing 250 L tip tank troughs. During a construction period from 15 June 2016 to 8 October 2016, the dry cows were moved to a nearby pasture equipped with a single large water trough. Calves received water delivered with the milk replacer described in the previous section and also had access to eight automatically replenishing ~20 L water bowls.

2.2.2. Milk Pre-Cooling

Milk was pre-cooled before entering the bulk tank using an in-line plate cooler system (Fabdec Limited, Ellesmere, UK). Water used by the plate cooler was discharged into the primary reservoir (Figure 1).

2.2.3. Parlour Sanitizing and General Cleaning

Sanitizing, rinsing, detergent washing and acid rinsing of the milk pipelines was conducted after each milking and the milkhouse floor was cleaned daily (parlour sanitizing). According to the sanitization protocol, each pipeline cleaning event used ~720 L of water for a total of 2160 L daily. The bulk tank was cleaned routinely after milk was removed for transport. This used ~400 L of water per wash according to the prescribed protocol, and a portion of this was reused for floor cleaning. After each milking, the standing area in the main barn was hosed down with a high-volume hose pumping from a ~500 L basin that was gradually filled by a low-volume hose with a float valve (general cleaning). General cleaning also included occasional farm equipment cleaning.

2.2.4. Cow Misting

The main barn was equipped with high-pressure misters located above the feed bunks and arranged in four zones for cooling the cows. These misters were automatically activated when the in-barn temperature reached or exceeded 21 and 24 °C, as per the following automated two-step program:

1. 21 °C, 24 s in each zone successively followed by a 10-min rest period.
2. 24 °C, 36 s in each zone successively followed by a 7-min rest period.

2.3. Flow Measurements

The farm owners and the farm's plumber were interviewed to understand the sources and pathways of water throughout the farm. In addition, water pipes were visually inspected and surveyed with a portable transit time flow meter (TTFM) (Greyline Instruments Inc., Long Sault, ON, Canada) to confirm the information. Ten in-line model 1000JLPRS multi-jet propeller flow meters with pulse outputs (Carlon Meter, Grand Haven, MI, USA) were installed between 1 August 2015 and 22 September 2015 in strategic locations to monitor and partition whole-farm water use (Figure 1). Seven were dispersed in the main barn to measure: (1, 2) inflow from the two wells; (3) flow to the parlour; (4) flow to the main barn troughs from the primary reservoir; (5) flow from well 1 to the main barn bowls and secondary reservoir; (6) flow used for washing the main barn floor; and (7) flow to the misters. The other three meters measured flow to: (8) calf barn; (9) heifer barn; and (10) farm workshop (Figure 1). Due to a plumbing change, the flow to the calf barn was measured using a TTFM from 26 October 2015 to 14 June 2016. Data from six meters were stored on data loggers (CR200X, CR800; Campbell Scientific, Logan, UT, USA) and the other four meters were stored on USB storage devices (USB-505, Measurement Computing, Norton, MA, USA) as 10 min, 1 h, and 1 d averages. Due to a partial instrument failure with the meter on the mister line, daily mister water use for the entire period was estimated using an equation generated from periods of successful data acquisition. Plate cooler waste was visually observed overflowing from the primary reservoir. This waste flow was determined by subtracting the difference between measured inflow (Meters 1 and 2) and outflow (Meter 4) from the primary reservoir.

For further partitioning water use, a follow-up measurement campaign was conducted using a TTFM to measure flow of the plate cooler water return from 30 June to 6 July 2016. Another TTFM was installed on the line supplying the dry cow pasture water trough from 15 June to 24 June 2016. Gaps in the dry cow pasture drinking water time series before the TTFM was installed were filled using a water intake vs. temperature response equation developed from lactating cow data. The pasture trough was visually observed to be overflowing due to the trough not being level. This waste flow was determined by measuring flow into the trough when no cows were drinking during site visits, and verified each day by flow measured in the middle of the night when cows were inactive.

2.4. Environmental Measurements

In-barn air temperature was measured using a shielded thermistor every 10 s and recorded as 10 min, 1 h, and 1 d averages on a CR200X datalogger (Campbell Scientific, Logan, UT, USA). In-barn humidity was measured using a CS215 temperature RH probe (Campbell Scientific); however, the sensor failed in the midst of the study, therefore gaps were filled using average daily relative humidity (RH) recorded at the Ottawa Central Experimental Farm Weather Station (45.383262°, −75.714079°). With these data, THI was calculated according the following equation [23]:

$$THI_{avg} = (1.8 \times T_a + 32) - (0.55 - 0.0055 \times RH) \times (1.8 \times T_a - 26), \tag{2}$$

where THI_{avg} is the average daily THI, T_a is the average daily air temperature (°C), and RH is the average daily relative humidity (%).

3. Results

3.1. Environmental Conditions

The average RH and air temperature (T_a) for the monitoring period was 69 ± 15% and 12.5 ± 7.3 °C, respectively. The resulting average THI was 57 ± 11. The average monthly temperatures and THI are presented in Figure 2, illustrating the seasonal changes with high values occurring from May to Aug. The number of days in which daily average T_a exceeded 25 °C was 11, 5, 3, and 4 for May, June, July, and August, respectively. Likewise, the number of days in which THI exceeded 75 was 8, 3, 3, and 1 for May, June, July, and August, respectively.

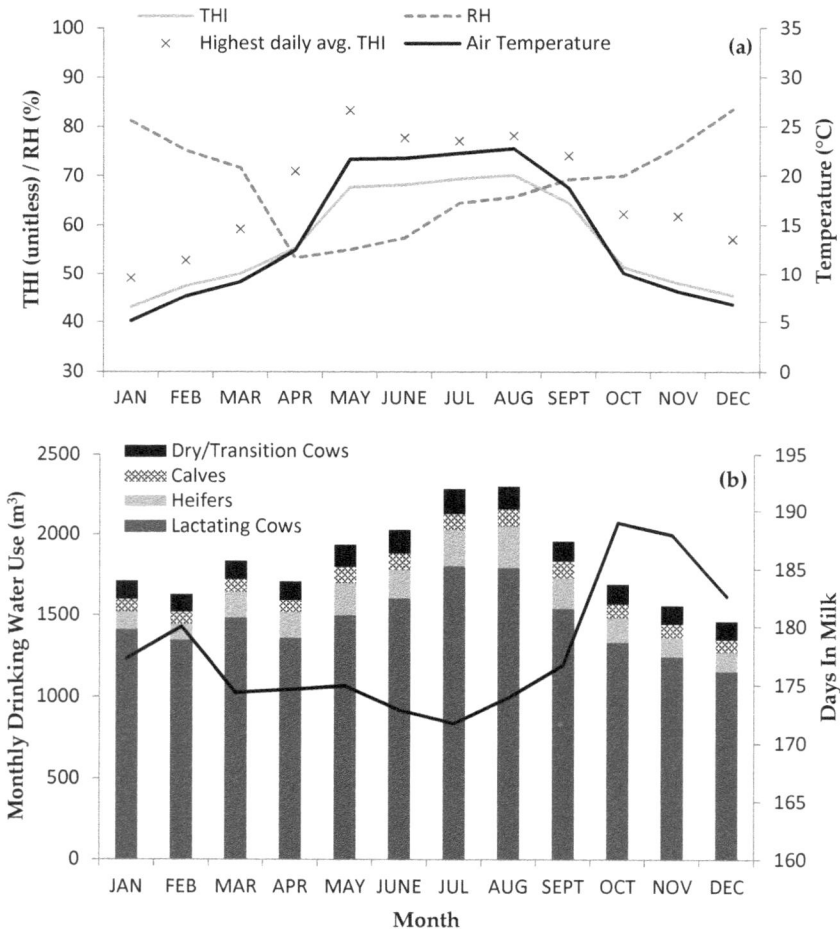

Figure 2. (**a**) Average monthly THI and air temperature (°C). (**b**) Total monthly drinking water consumption (m³) broken down by animal category (lactating cows, dry/transition cows, calves and heifers). The solid line is the average monthly days in milk (DIM).

3.2. Total Farm Water Use

The average total daily water use (1 October 2015 to 30 September 2016) for the farm was 90,253 L $\pm$ 15,203 L and the annual water use was 33,032 m^3 (Table 2). The majority of the on-farm water use was for nutritional water (68%), while milking parlour cleaning and operation contributed 14%, waste represented 15% (including unrecovered plate cooler return water and pasture trough overflow), and barn cleaning, misters and other water use (misters, cleaning) represented 3% (Figure 3).

Misters were operational between May and October and were estimated to have had a cumulative water use of 480.5 m^3 for this period (Table 2). The cumulative value was based on measured and gap-filled data. Gaps were filled using the following equation, which was developed by regression of measured air temperature and water use for misting:

$$MIST_{daily} = 658.79 \times (T_a) - 11,250, \tag{3}$$

where $MIST_{daily}$ is the total daily water demand of the mister system (L day^{-1}), and T_a is the average daily barn air temperature ($^\circ$C) (RMSE = 712, R^2 = 0.84, $p < 0.001$).

Table 2. Allocation of total on-farm water uses.

Component	Annual Water Use (m^3 year^{-1})	Daily Water Use (m^3 d^{-1})
Drinking Water	22,101	60.4 $\pm$ 8.8
Plate Cooler Waste	4649	12.7 $\pm$ 7.9
Milk Parlour	4451	12.2 $\pm$ 1.7
Barn Cleaning	702	1.9 $\pm$ 0.89
Misters	481	1.3 $\pm$ 2.1
TMR	474	1.3 $\pm$ 0.81
Pasture Waste *	175	0.48 $\pm$ 0.82
Total	33,032	90.3 $\pm$ 15.2

Note: * Overflow in the pasture water trough occurred during a portion of the summer, but for consistency of calculation was assigned a daily value based on the entire year.

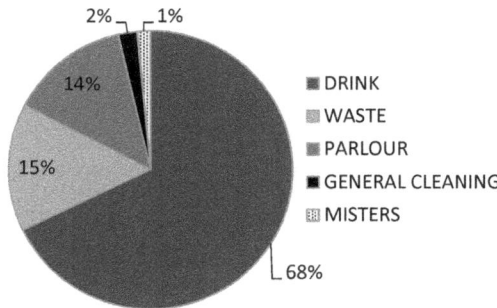

Figure 3. Breakdown of total farm water use (%) including drinking, waste, parlour (foot baths, parlour floor cleaning, cow cleaning, line sanitization), general cleaning (i.e., barn floor and farm equipment), and mister water use. Waste includes water that was not recovered from the plate cooler return and water spilled from the pasture bowl.

3.3. Drinking Water

The majority of the drinking water (80%) was used to service the lactating cows, whereas heifers, dry/transition cows, and calves made up the remaining 9%, 7%, and 4%, respectively (Figure 4). The average daily water consumption per animal for the lactating cows (excluding TMR water addition) was 114 $\pm$ 13 L day^{-1}, for dry cows was 36 $\pm$ 5.2 L day^{-1}, for heifers was 22 $\pm$ 8.2 L day^{-1}, and for calves was 12 ± 2.9 L day^{-1}. These water consumption values are generally in the ranges identified in local government documents [13] (Table 3). Note that dry cow drinking water for the entire monitoring period was estimated using an equation developed from the period where they were pastured separately in combination with the drinking water temperature response of lactating cows:

$$DC_{drink} = 0.636 \times T_a + 27.03, \qquad (4)$$

where DC_{drink} is the daily water consumption per dry cow (L cow^{-1} day^{-1}) and T_a is the daily average barn air temperature ($°C$) (RMSE = 3.0, R^2 = 0.48, $p < 0.001$).

Table 3. Measured and published water consumption per animal category (L day^{-1}) showing the mean $\pm$ SD of measured daily values as well as the published range of water consumption.

	Measured Water Consumption (L day^{-1})	Published Water Consumption [†] (L day^{-1})
Lactating Cows	114 ± 13	110–132 [‡]
Dry Cows	36 ± 4.7	34–49
Heifers	22 ± 8.2	14.4–36.3
Calves	12 ± 2.9	4.9–13.2

Note: [†] [13]; [‡] Adjusted for Holstein dairy cows producing 34.8 kg day^{-1} of milk.

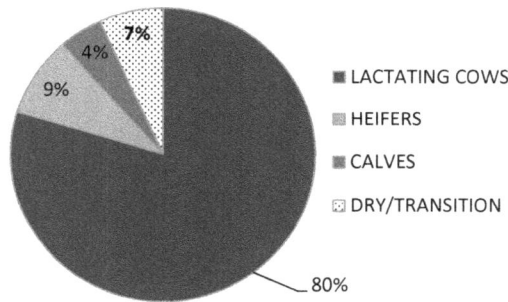

Figure 4. Breakdown of drinking water use (%) by animal category (lactating cows, dry/transition cows, calves and heifers). The dry and transition cow water was modelled based on a period when the dry cows were placed in pasture on a separate water supply.

Water consumption was greater in warm weather months compared to cool months and this was observed for all animal categories (Figure 2). The relationship between each month's average daily water consumption and average monthly temperature had a positive correlation for lactating cows, heifers, and calves ($R^2 = 0.69$, $p < 0.001$; $R^2 = 0.84$, $p < 0.001$, $R^2 = 0.85$, $p < 0.001$; respectively) (Figure 5a). The heifer barn was not equipped with cooling equipment (i.e., fans, misters) and this may explain the steeper slope ($\sim 3\times$) of the water consumption response of this animal group compared to lactating cows and calves. The THI was also positively correlated to water consumption but did not provide better correlation than simply using air temperature as a predictor. For example, using daily data, both THI and T_a had similar fits ($R^2 = 0.60$, $p < 0.001$) with the total drinking water use (Figure 5b). The results were no different if only considering the drinking water supplied to lactating cows. In a long trial such as this it appears that temperature was the major driver of THI, as exemplified by the fact that average daily THI and average daily air temperature were very strongly correlated ($R^2 = 0.99$, data not shown). This is primarily because the annual range of T_a (CV = 0.54) is greater than that of RH (CV = 0.21) (Figure 2). However, it is possible that more complete on-farm RH measurements would have yielded better results for THI [23].

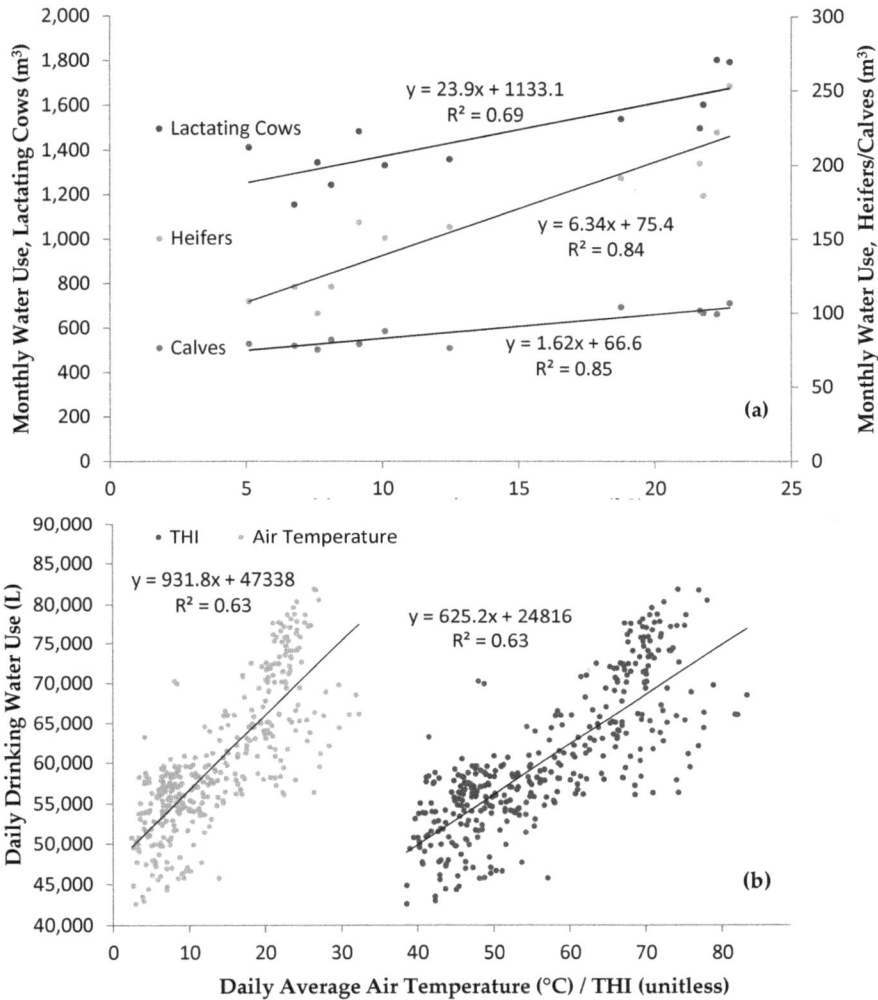

Figure 5. (a) Average monthly air temperature (°C) plotted against average monthly water consumption (m^3) for lactating cows, heifers and calves. (b) Total daily drinking water use (L) plotted against THI (unitless) and average daily air temperature (°C).

3.4. Parlour Wash

The average daily use of the parlour wash was 12,160 ± 1741 L, of which, according to the sanitization protocol, 2560 L was used in the daily washing procedure of the milk pipeline and bulk tank. Of the remaining 9600 L, ~4300 L was used by a high-volume hose for parlour floor cleaning. We can express the final 5300 ± 759 L as 4.2 ± 1.8 L for each cow cleaning instance.

3.5. Recycling Milk Pre-Cooling Water (Plate Cooler)

The plater cooler flow rate was 0.5 L s^{-1} (during milking periods) and corresponded to a daily water use of ~2× the daily milk production, which is in the range of the recommended water:milk plate cooler ratio [24]. Plate cooler flow discharged into the primary reservoir. However, while in use, the plate cooler flow exceeded drinking water consumption and exceeded the reservoir capacity. As a result, 12,702 ± 7900 L overflowed from the primary reservoir into wastewater daily, on average (i.e., overflowed and entered the manure storage). This study observed the effect that plumbing design can have on water conservation. Due to a plumbing change, the daily plate cooler waste increased from 3801 ± 3403 L to 15,604 ± 6685 L. Prior to the change, most of the water destined to the main barn water troughs was drawn through meter 4, from the primary reservoir (into which the plate cooler water was returned). After the change, most of the water was drawn from another line through meter 5, reducing demand on the primary reservoir. As a result, the capacity to reuse plate cooler return water as drinking water was severely reduced, leading to the observed ~11,800 L increase in daily plate cooler waste. This illustrates that plumbing changes in a dynamic farm environment can have unintended effects on seemingly unrelated water components.

Effective plumbing design for plate cooler water recycling should account for water supply and demand dynamics. The plate cooler operates during periods when drinking water demand was lower due to cow movement from the free stall areas into the milk parlour or adjacent holding area (Figure 6). While in use, hourly flow for the plate-cooler into the primary reservoir was ~1719 L h^{-1}, whereas the draw from this reservoir was <500 L h^{-1} at times. Therefore, plate-cooler reservoirs must be designed to handle the intra-day water supply and demand, which are not apparent from typical "guidelines" for water use like Table 3. In other words, the average daily flow is not equally distributed throughout the day, but rather concentrated in short periods of very high flow.

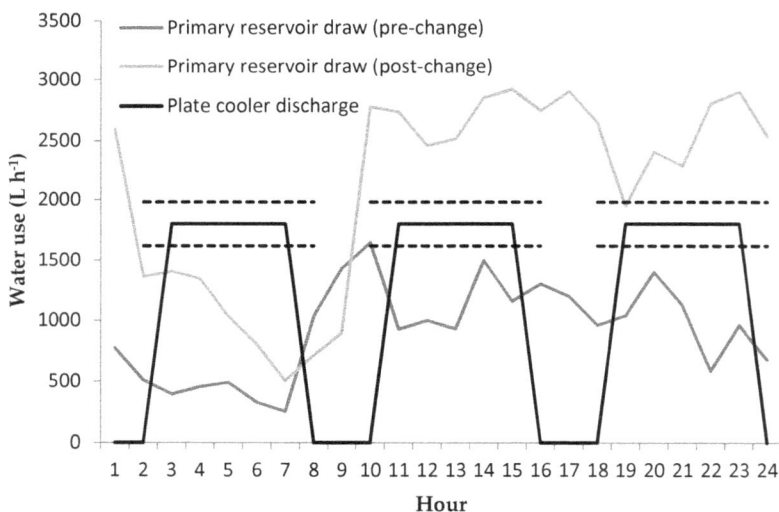

Figure 6. Typical day showing hourly water draw from primary reservoir (L) (*pre-* and *post-* change in plumbing design) and milk pre-cooling water use based on the average flow rate (L) ± 1 SD (dashed lines) at times of operation (0300 h, 1100 h, and 1900 h milking times). Water is wasted as overflow when the plate cooler discharge exceeds the primary reservoir draw.

3.6. Milk Dynamics

The average days in milk (DIM) for the monitoring period was 178 day and the monthly average DIM was slightly greater in the fall and winter months compared to the summer months (Figure 2b). The total milk produced over the year was 5366 t, which converts to 5150 t FPCM. Milk per cow and FPCM per cow were highest in March and April. The lowest per cow production months were October and December for milk and August and September for FPCM (data not shown). Despite these temporal trends, no obvious link between average monthly milk/FPCM production per cow (kg) and average monthly temperature (°C) were observed. However, the total milk fat and protein percentage was negatively correlated with average monthly air temperature (milk fat + protein = $-0.0227 \times T_a + 7.27$, $R^2 = 0.67$, data not shown). This finding is consistent with a previous study of milk fat and protein dynamics in Ontario [25].

The WF scaled by milk production was 6.19 L kg^{-1} milk (6.41 L kg^{-1} FPCM), including contributions from all animal groups and 5.34 L kg^{-1} milk (5.54 L kg^{-1} FPCM) when excluding the water consumption of replacement animals and dry cows. This is higher than the figures determined by Drastig et al. [10] and Capper et al. [9] in their modelling studies.

3.7. Water Conservation Scenarios

In this section a series of water conservation exercises are explored to estimate potential savings. The predicted effect on water consumption of decreased average barn air temperature was modelled based on the relationship between total monthly drinking water use to temperature:

$$W = 33.85 \times T_{a,m} + 1372.1, \tag{5}$$

where W is the predicted total monthly water use (m^3) and $T_{a,m}$ is the average monthly air temperature (°C) (RMSE = 121, $R^2 = 0.77$, $p < 0.001$).

In months where $T_{a,m}$ exceeded 18 °C, the measured total monthly water use was replaced with the predicted total if the average monthly temperature was decreased by 2 °C. This analysis showed that if the average barn air temperature were to be maintained at 2 °C lower without the aid of additional water, 351 m^3 of water could be saved annually. Cows regulate their water consumption along with feed intake [15], which affects milk production [26]. When heat stress is a factor, cows may decrease their feed intake and milk production while at the same time increasing their water intake, amplifying the effect on the milk water footprint (i.e., non-productive increase in water consumption). Maintaining cooler temperature may therefore have beneficial effects on milk production, which we did not account for here. Strategies such as better ventilation [27] or lower stocking density [28] can be used to lower ambient air temperatures without the use of additional water. Both of these strategies may increase the cost of operation, however, increased cow comfort can have a positive effect on milk production, which may balance out these additional costs.

If the plate cooler water and other water losses were fully recycled instead of wasted to manure storage, an additional 4882 m^3 in water savings could be achieved. Some researchers have noted that water reuse is currently the most common water saving strategy employed by the farms they surveyed. As the most impactful strategy, considering 55% of surveyed farms did not employ water reuse strategies, there is still a large capacity for water savings industry-wide [18]. The costs associated with proper recycling may include whole farm plumbing survey and design by qualified professionals with or without additional one-time costs such as increasing the holding capacity of the water delivery systems. It is worth noting that after this study, farmers increased the primary reservoir capacity to increase reuse of plate-cooler water.

As was reported in an earlier section, 5300 L day^{-1} of water was used for cow preparation, which represents 4.2 ± 1.8 L for each cow cleaning instance. According to the literature, moist towel cow preparation can be conducted with only 1.9 L per cow preparation [29], therefore, the water use for cow preparation can theoretically be reduced to ~2400 L day^{-1} if the moist towel cow preparation method was optimized for water efficiency, thereby potentially saving 1061 m^3 annually. Here again, optimizing the cow milking procedure may increase the operational cost by increasing the time requirement per milking event.

Combining all of these strategies could lead to a total potential saving of 6229 m^3 annually, a 19% reduction of the annual water use, and reduce the milk production water footprint to 4.18 L kg^{-1} milk (excluding replacement animals) (Table 4).

Table 4. Theoretical water conservation scenarios and their expected effect on milk production water footprint (WF).

No.	Water Saving Strategy	Annual Farm Water Consumption m^3 year^{-1}	Reduction %	WF Including Replacements L kg^{-1} Milk (FPCM [†])	WF Excluding Replacements L kg^{-1} Milk (FPCM [†])
1	Current water use	33,032	-	6.19 (6.41)	5.34 (5.54)
2	2 °C decrease in air temperature	32,682	1.1	6.12 (6.35)	5.28 (5.47)
3	Reduce cow preparation water requirement	31,971	3.2	5.99 (6.21)	5.14 (5.33)
4	Recovery of water losses	28,208	14.6	5.29 (5.48)	4.44 (4.60)
5	Combination of strategies 2–4	26,796	18.9	5.02 (5.20)	4.17 (4.33)

Note: [†] L kg^{-1} fat-and-protein corrected milk (FPCM) is given in brackets.

In scenario 5 (Table 4), drinking water represents 82% of the total water use, which closely resembles values reported by Drastig et al. [10]. By accurately measuring and partitioning water use our results help to validate the water modelling methods used by previous studies. However, our results also highlight the reality of on-farm blue water waste, which would not be considered by existing theoretical models.

Feed dense in energy and protein are necessary for high milk yields [14] and DMI intake is positively correlated to drinking [15]. Therefore, there is limited potential to alter feed intake for the sake of water conservation without negatively affecting milk production. Reducing mild heat stress and minimizing the size of the replacement herd offer some limited potential for conserving drinking water to meet water conservation goals on dairy farms. These scenarios demonstrate that the non-drinking components of dairy farm water use can be optimized. This was also demonstrated in a case study by Brugger and Dorsey [30], who audited and optimized the water usage on a ~1000 cow dairy. By correcting several sources of waste (leaks, plate cooler flow rate, and cleaning protocol) they were able to conserve ~30,000 m^3 annually.

4. Conclusions

Dairy farm operations withdraw appreciable quantities of sub-surface blue water. Some water savings can be achieved through reducing cow drinking by optimizing cow comfort (i.e., reducing barn temperature). The largest potential for water savings observed in this study was related to improving plumbing design to collect, store and re-use cooling water. The dairy industry is unique in that a greater portion of processing takes place at the farm level. Process optimization to reduce water use practiced in other industrial settings is not well established within the dairy industry framework and this research illustrates that there is potential benefit from such optimization. A measure of the proportion of total water used as drinking water could be used as an indicator of milk production efficiency. For instance, farms where drinking water contributes <80% of the total water use may be operating at a sub-optimal level, from a water efficiency point of view. We know that many dairy farmers are already taking steps to implement water saving strategies on their farms [18]. An industry or government sponsored water use assessment program could identify potential water savings and help selecting water-saving strategies from a cost–benefit point of view.

Acknowledgments: Funding is acknowledged from the Agriculture and Agri-Food Canada Abase project #1236, and funding contributions from Dairy Farmers of Canada, the Canadian Dairy Network and the Canadian Dairy Commission under the Agri-Science Clusters Initiative. As per the research agreement, aside from providing financial support, the funders have no role in the design and conduct of the studies, data collection and analysis or interpretation of the data. Researchers maintain independence in conducting their studies, own their data, and report the outcomes regardless of the results. The decision to publish the findings rests solely with the researchers.

Author Contributions: Andrew C. VanderZaag (A.C.V.) and Robert Gordon (R.G.) conceived the study, A.C.V., R.G., and David R. Lapen (D.R.L.) obtained research funding, Stephen Burtt (S.B.) and A.C.V. designed the experimental apparatus; S.B. obtained samples and data; Etienne L. Le Riche (E.L.L.R.), S.B., and A.C.V. analysed the data; A.C.V., D.R.L., and R.G. contributed sensors/materials/analysis tools; E.L.L.R. prepared figures and wrote the paper, A.C.V., S.B., D.R.L., and R.G. provided comments and edits on drafts of the paper.

Conflicts of Interest: The authors declare no conflict of interest.

References

1. Babkin, V.I. The earth and its physical features. In *World Water Resources at the Beginning of the Twenty-First Century*; Shiklomanov, I.A., Rodda, J.C., Eds.; Cambridge University Press: Cambridge, UK, 2003; pp. 1–17.

2. Hoekstra, A.Y. A critique on the water-scarcity weighted water footprint in LCA. *Ecol. Indic.* **2016**, *66*, 564–573. [CrossRef]

3. Pfister, S. Understanding the LCA and ISO water footprint: A response to Hoekstra (2016) "A critique on the water-scarcity weighted water footprint in LCA". *Ecol. Indic.* **2017**, *72*, 352–359. [CrossRef]

4. Brown, A.; Matlock, M.D. *A Review of Water Scarcity Indices and Methodologies*; White Paper #106; The Sustainability Consortium: Scottsdale, AR, USA; University of Arkansas: Fayetteville, AR, USA, 2011.

5. Kulshreshtha, S.N.; Grant, C. An estimation of Canadian agricultural water use. *Can. Water Res. J.* **2007**, *32*, 137–148. [CrossRef]

6. Beaulieu, M.S.; Fric, C.; Soulard, F. *Estimation of Water Use in Canadian Agriculture in 2001*; Catalogue No. 21-601-MIE—No. 087; Statistics Canada: Ottawa, ON, Canada, 2001.

7. Brugger, M. *Fact Sheet: Water Use on Ohio Dairy Farms*; Ohio State University: Columbus, OH, USA, 2007.

8. Murphy, E.; de Boer, L.J.M.; van Middelaar, C.E.; Holden, N.M.; Shalloo, L.; Curran, T.P.; Upton, J. Water footprinting of dairy farming in Ireland. *J. Clean. Prod.* **2017**, *140*, 547–555. [CrossRef]

9. Capper, J.L.; Cady, R.A.; Bauman, D.E. The environmental impact of dairy production: 1944 compared to 2007. *J. Anim. Sci.* **2009**, *87*, 2160–2167. [CrossRef] [PubMed]

10. Drastig, K.; Prochnow, A.; Kraatz, S.; Klauss, H.; Plöchl, M. Water footprint analysis for the assessment of milk production in Brandenburg (Germany). *Adv. Geosci.* **2010**, *27*, 65–70. [CrossRef]

11. National Research Council. *Nutrient Requirements of Dairy Cattle*, 7th ed.; National Academies Press: Washington, DC, USA, 2001; pp. 178–182.

12. Winchester, C.F.; Morris, M.J. Water intake rates of cattle. *J. Anim. Sci.* **1956**, *15*, 722–740. [CrossRef]

13. Ward, D.; McKague, K. *Water Requirements of Livestock*; Agdex #716/400; Ontario Ministry of Agriculture, Food and Rural Affairs: Guelph, ON, Canada, 2015.

14. Kume, S.; Nonaka, K.; Oshita, T.; Kozakai, T. Evaluation of drinking water intake, feed water intake and total water intake in dry and lactating cows fed silages. *Livest. Sci.* **2010**, *128*, 46–51. [CrossRef]

15. Cardot, V.; Le Roux, Y.; Jurjanz, S. Drinking Behavior of Lactating Dairy Cows and Prediction of Their Water Intake. *J. Dairy Sci.* **2008**, *91*, 2257–2264. [CrossRef] [PubMed]

16. Smith, D.L.; Smith, T.; Rude, B.J.; Ward, S.H. Short communication: Comparison of the effects of heat stress on milk and component yields and somatic cell score in Holstein and Jersey cows. *J. Dairy Sci.* **2013**, *96*, 3028–3033. [CrossRef] [PubMed]

17. Gorniak, T.; Meyer, U.; Südekum, K.-H.; Dänicke, S. Impact of mild heat stress on dry matter intake, milk yield and milk composition in mid-lactation Holstein dairy cows in a temperate climate. *Arch. Anim. Nutr.* **2014**, *68*, 353–369. [CrossRef] [PubMed]

18. Robinson, A.D.; Gordon, R.J.; VanderZaag, A.C.; Rennie, T.J.; Osborne, V.R. Usage and attitudes of water conservation on Ontario dairy farms. *PAS* **2016**, *32*, 236–242. [CrossRef]

19. Legrand, A.; Schültz, K.E.; Tucker, C.B. Using water to cool cattle: Behavioral and physiological changes associated with voluntary use of cow showers. *J. Dairy Sci.* **2011**, *94*, 3376–3386. [CrossRef] [PubMed]

20. Lin, J.C.; Moss, B.R.; Koon, J.L.; Flood, C.A.; Smith, R.C., III; Cummins, K.A.; Coleman, D.A. Comparison of various fan, sprinkler, and mister systems in reducing heat stress in dairy cows. *Appl. Eng. Agric.* **1998**, *14*, 177–182. [CrossRef]

21. Gerber, P.; Vellinga, T.; Opio, C.; Henderson, B.; Steinfeld, H. *Greenhouse Gas Emissions from the Dairy Sector: A Life Cycle Assessment*; FAO: Rome, Italy, 2010.

22. Wright, T. *Water Quality for Dairy Cattle*; Agdex # 410; Ontario Ministry of Agriculture, Food and Rural Affairs: Guelph, ON, Canada, 2012.

23. Schüller, K.L.; Burfeind, O.; Heuwieser, W. Short communication: Comparison of ambient temperature, relative humidity, and temperature index between on-farm measurements and official meteorological data. *J. Dairy Sci.* **2013**, *96*, 7731–7738. [CrossRef] [PubMed]

24. Milk Development Council (MDC). *Effective Use of Water on Dairy Farms*; Milk Development Council: Cirencester, UK, 2007.

25. Ueda, A. Relationship among Milk Density, Composition, and Temperature. Master Thesis, University of Guelph, Guelph, ON, Canada, 1999.

26. Clark, J.H.; Davis, C.L. Some aspects of feeding high producing dairy cows. *J. Dairy Sci.* **1980**, *63*, 873–885. [CrossRef]

27. House, H.K. *Dairy Housing—Ventilation Options for Free Stall Barns*; Agdex #410/721; Ontario Ministry of Agriculture, Food and Rural Affairs: Guelph, ON, Canada, 2016.

28. Cooper, K.; Parsons, D.J.; Demmers, T. A thermal balance model for livestock buildings for use in climate change studies. *J. Agric. Eng. Res.* **1998**, *69*, 43–52. [CrossRef]

29. Holmes, B.J.; Struss, S. *Milking Center Wastewater Guidelines—A Companion Document to Wisconsin NRCS Standard 629*; University of Wisconsin—Extension: Madison, WI, USA, 2009.

30. Brugger, M.; Dorsey, B. Water use and savings on a dairy farm. In Proceedings of the ASABE Annual International Meeting 2006, Portland, OR, USA, 9–12 July 2006; ASABE: St. Joseph, MI, USA, 2006; Paper No. 064035.

water

MDPI

Article

Decomposition Analysis of Water Treatment Technology Patents

Hidemichi Fujii [1],*and Shunsuke Managi [2]

[1] Graduate School of Fisheries and Environmental Sciences, Nagasaki University, Nagasaki 852-8521, Japan
[2] Urban Institute & Department of Urban and Environmental Engineering, Kyushu University, 744 Motooka, Nishi-ku, Fukuoka 819-0395, Japan; managi.s@gmail.com
* Correspondence: hidemichifujii@gmail.com; Tel.: +81-95-819-2756

Received: 25 August 2017; Accepted: 3 November 2017; Published: 6 November 2017

Abstract: Water treatment technology development supports a steady, safe water supply. This study examines trends in water treatment technology innovations, using 227,365 patent granted data published from 1993 to 2016 as an indicator of changing research and development (R&D) priorities. To clarify changes in R&D priorities, we used a decomposition analysis framework that classified water treatment technologies into five types: conventional treatment (117,974 patents, 51.9%), biological treatment (40,300 patents, 17.7%), multistage treatment (45,732 patents, 20.1%), sludge treatment (15,237 patents, 6.7%), and other treatments (8122 patents, 3.6%). The results showed that the number of water treatment technology patents granted increased more than 700% from 1993 to 2016; in particular, the number of multistage water treatment patents granted rapidly grew. The main driver of this growth was expansion in the R&D activity scale and an increase in the priority of multistage water treatment technology in China. Additionally, the trends and priority changes in water treatment technology inventions varied by country and technology groups, which implied that an international policy framework for water treatment technology development should recognize that R&D priorities need to reflect the diverse characteristics of countries and technologies.

Keywords: decomposition analysis; global patent data; research and development strategy; water treatment technology

1. Introduction

Water treatment technology creates steady and safe water resources [1,2]. The global importance of water treatment technology has been increasing, especially in developing countries [3]. According to World Health Organization (WHO) and United Nations Children's Fund (UNICEF) [4], in 2015, 844 million people still lacked basic drinking water services, and 892 million people still practiced open defecation. These low-quality water treatment activities increase the risk of disease through the use of polluted surface water for household activities [5]. To improve drinking water quality and sanitation services, the development and diffusion of efficient and affordable water treatment technologies have attracted attention.

Because of water resource problems, the water management issue was individually established as the goal 6, i.e., "Ensure availability and sustainable management of water and sanitation for all", in the sustainable development goals (SDGs) adopted by the United Nations [6]. To achieve this goal, the development of water treatment technology is a key factor in accelerating improvements in water quality [2]. Additionally, the Chinese government released a water pollution prevention and control action plan (the Water Ten Plan) in 2015. In this plan, the Chinese government vowed to improve nationwide water quality by 2030, also pledging to spend billions of dollars [7].

Against the backdrop of the acceleration in water treatment technology development, the number of patents granted has rapidly increased. Figure 1 shows the number of water treatment patents granted by the patent office (Figure 1a) and technology type (Figure 1b). Figure 1 shows that the number of water treatment patents has increased more than threefold, i.e., from 8843 in 2009 to 28,181 in 2016. In particular, water treatment patents granted in China (SIPO) rapidly increased during this period (Figure 1a).

As shown in Figure 1b, the patent share of each water treatment technology type changed from 2009 to 2016. In 2009, conventional water treatment technology had the largest share of the patented water treatment technologies. However, from 2009 to 2016, the number of patents granted for multistage water treatment technology rapidly increased.

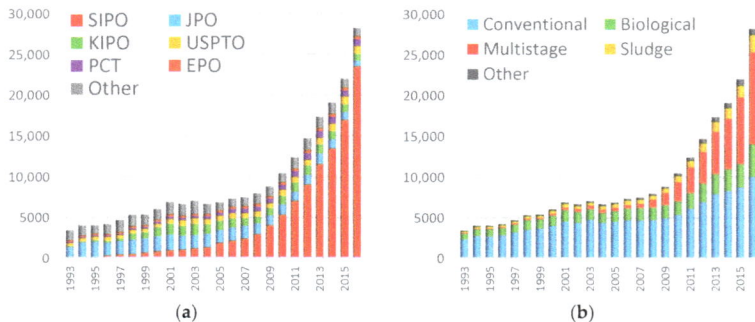

Figure 1. Trends in water treatment technology patents granted from 1993 to 2016 (number of patents). Source: Authors' estimate using the IPC code in Table S1 and the PATENTSCOPE database; Note: SIPO: State Intellectual Property Office of The People's Republic of China; JPO: Japan Patent Office; KIPO: Korean Intellectual Property Office; USPTO: United States Patent and Trademark Office; PCT: Patent Cooperation Treaty; EPO: European Patent Office. (**a**) Water treatment patents granted by country; (**b**) Water treatment patents granted by technology.

Additionally, water treatment technology demands are different in different regions because water is linked to the local lifestyles and weather conditions. According to UN-Water [8], the subjects that are the most challenging for coordination and agreements are the work areas related to integrated water resources management (IWRM), transboundary waters, capacity development, water and sanitation, and climate change. Furthermore, the appropriate water treatment technology differs based on the type of water pollution because contaminants and pollutant substances are diverse.

Thus, the incentives for water treatment technology inventions clearly vary among the regions and types of technology. Clarifying the characteristics of each water treatment technology type is important for formulating an effective policy that encourages water treatment technology research and development. Based on this background, the objective of this study is to clarify the strategy changes in the water treatment technology development using patent data that is categorized by country and technology type.

To consider the differences in the water treatment technology types, we classified the water treatment technology patents based on the World Intellectual Property Organization (WIPO) [9] classification using the International Patent Classification (IPC) code (see Table S1 in Supplementary Materials). In this study, we defined water treatment as the "treatment of water, wastewater, sewage, and sludge", which is the IPC=C02F definition that was introduced by the WIPO [9]. Additionally, we divided the patent data into the following five water treatment technology groups: (1) conventional water treatment (Conventional), (2) biological water treatment (Biological), (3) multistage water treatment (Multistage), (4) sludge treatment (Sludge), and (5) other water treatment technology (Other) (see Table 1).

Table 1. Description of water treatment technology patents.

Technology Code	Technology Group (IPC Code)	Description of Technology Group
Conventional	Conventional treatment (IPC=C02F1)	Conventional water treatment technology includes heating (C02F1/02), degassing (C02F1/20), freezing (C02F1/22), flotation (C02F1/24), ion-exchange (C02F1/42), and oxidation (C02F1/72).
Biological	Biological treatment (IPC=C02F3)	Biological water treatment technology includes aerobic processes (C02F3/02), activated sludge processes (C02F3/12), and anaerobic digestion processes (C02F3/28).
Multistage	Multistage treatment (IPC=C02F9)	Multistage water treatment technology covers combined treating operations. This technology group includes electrochemical treatment (C02F9/06), thermal treatment (C02F9/10), and irradiation or treatment with electric or magnetic fields (C02F9/12).
Sludge	Sludge treatment (IPC=C02F11)	This technology group includes sludge treatment by pyrolysis (C02F11/10), de-watering (C02F11/12), and thermal conditioning (C02F11/18).
Other	Other water treatment technology (IPC=C02F5, C02F7, C02F101, C02F103)	Other water treatment technology includes softening water (C02F5), aeration of stretches (C02F7), nature of the contaminant (C02F101), and nature of the wastewater (C02F103).

Source: Author revised the definitions introduced by the World Intellectual Property Organization (WIPO) [9];
Note: The detail technology grouping is described in Table S1.

Previous literature has mostly focused on the development of water treatment technologies. Most literature is based on natural sciences, especially chemical and engineering research fields. Rodriguez-Narvaez et al. [10] surveyed approximately 200 reports on water treatment technology for emerging contaminants. They indicated that recent research tended to use phase-changing processes, including adsorption onto different solid matrices and in membrane processes, followed by advanced oxidation processes and biological treatment for water treatment. Subramani and Jacangelo [11] published a critical review on emerging desalination technologies for water treatment and focused on thermal-based, membrane-based, and alternative technologies.

Some literature has focused on specific water treatment technologies. Palma et al. [12] investigated the efficiency of membrane technology for water treatment processes. They used nanofiltration membranes and reverse osmosis membranes for three types of water, i.e., irrigation water, municipal supply water, and wastewater. Alzahrani and Mohammad [13] focused on membrane technology implementation for water treatment in the petroleum industry. In addition to these membrane studies, Temesgen et al. [14] reported the trends in micro- and nano-bubble technology for water treatment, which included more than 150 reports.

Limited literature reports are available on water treatment technologies using social science approaches. Fujii and Managi [15] evaluated wastewater treatment efficiencies using a production function approach, and set the water pollution data as the undesirable output factor. Another social science approach is patent data analysis. Hara et al. [16] analyzed the historical development of wastewater and sewage sludge treatment technologies in Japan using patent data. Another patent data analysis was introduced by Fujii and Managi [17], and the analysis clarified the main driver of environmentally related technology in Japan using a decomposition analysis.

While literature about water treatments exists, most studies focus on the efficiencies of the technologies, and studies on the priority changes in technology development are limited. Based on this background, we propose a research framework to investigate the priority changes in water treatment technology using patent data. This research is the first to use patent data that is related to water treatment technologies to clarify priority changes in research and development using a decomposition analysis framework.

Patent data analyses are widely applied to evaluate research and development activities in the fields of engineering, economics, and corporate management [18]. Popp [19] analyzed the effect of energy prices on research and development activities using patent data. He considered the share of energy-related patents granted to the total patents granted as the proxy variable of research and development priority for energy technology. Fujii [20] used this idea to develop the patent decomposition analysis framework.

According to Haščič and Migotto [21], there are several advantages and limitations of using patent data. The advantages of patent data are that the data are widely available from public databases and can be used for quantitative analyses. Additionally, patent data can be disaggregated into specific technological fields, such as water treatments, in this study.

The limitations of patent data include the following. First, "not all innovations are patentable", and "not all patentable inventions are patented". Therefore, patent data does not account for all of the innovations. According to Smith [22], many water treatment innovations have been produced in slum areas (e.g., the SONO water filter and Safe Agua Water System). These frugal technologies are community or need-based, and technology diffusion and adoption by many people is the priority target. The patent system is not useful for these technologies because patent protection affords exclusive rights to the patent holder to exploit the invention. Additionally, in a patent data analysis, identifying the type of water being treated is difficult because water treatment technologies are applied to many types of water, including wastewater, drinking water, and agricultural water. Patent data can distinguish the water treatment method but not the type of water that was treated. Therefore, this study analyzes water treatment technology development by focusing on the water treatment method.

Finally, the true value of patents and their perception in different countries is not the same. This is because guidelines and examination standards are not the same among different countries [23]. Therefore, a comparative analysis among countries should carefully consider this point.

2. Methods

This study uses a decomposition analysis framework to clarify the changing factors that are involved in granting water treatment technology patents. We use the following three indicators to decompose the water treatment technology patents granted: the priority of a specific water treatment technology (PRIORITY), the importance of the water treatment technology among all of the patents granted (WTT), and the research and development (R&D) activity scale (SCALE).

We define the PRIORITY indicator as the number of specific water treatment patents granted, divided by the total number of water treatment patents granted to provide the share of the specific water treatment patents granted among the total water treatment patents. As explained in Table 1, we set five specific water treatment technologies, i.e., conventional treatment, biological treatment, multistage treatment, sludge treatment, and other treatment. The PRIORITY indicator increases if the number of specific water treatment patents granted increases more quickly than the total number of water treatment patents granted, and indicates that inventors are concentrating research resources on specific types of water treatment technology inventions. Inventors are prioritizing specific water treatment technology types over other types when PRIORITY increases.

Similarly, the WTT indicator is defined as the total number of water treatment patents granted, divided by the total number of patents granted, which indicates the share of the total water treatment patents of the total patents. This indicator increases if the number of total water treatment patents granted increases more quickly than the number of total patents granted, indicating that inventors are concentrating research resources on water treatment technology inventions. Inventors are prioritizing the invention of water treatment technology over other types of technology when WTT increases.

The SCALE indicator is defined as the total number of patents granted and represents the scale of the R&D activities. Generally, active R&D efforts promote the invention of new technologies. Thus, the total number of patents granted reflects the active R&D effort level. Additionally, R&D activities in companies depend on corporate financial circumstances because the number of patents granted is associated with the cost of researcher salaries, experimental materials, and applying for and registering patents. SCALE increases as the total number of patents granted increases. If the SCALE score increases, then the number of patents granted for water treatment technology increases with the increase in the overall R&D activities.

Here, we introduce a decomposition approach using the conventional treatment technology patent group as a specific type of water treatment patent granted (Table 1). The number of conventional treatment technology patents granted (CONVENTIONAL) is decomposed using the total water treatment patents granted (ALLWATER) and total patents granted (TOTAL), as in Equation (1).

$$\text{CONVENTIONAL} = \frac{\text{CONVENTIONAL}}{\text{ALLWATER}} \times \frac{\text{ALLWATER}}{\text{TOTAL}} \times \text{TOTAL} = \text{PRIORITY} \times \text{WTT} \times \text{SCALE} \qquad (1)$$

We consider the change in conventional treatment patents granted from year $t - 1$ ($\text{CONVENTIONAL}^{t-1}$) to year t (CONVENTIONAL^t). Using Equation (1), the growth ratio of the conventional treatment patents granted can be represented as follows:

$$\frac{\text{CONVENTIONAL}^t}{\text{CONVENTIONAL}^{t-1}} = \frac{\text{PRIORITY}^t}{\text{PRIORITY}^{t-1}} \times \frac{\text{WTT}^t}{\text{WTT}^{t-1}} \times \frac{\text{SCALE}^t}{\text{SCALE}^{t-1}} \qquad (2)$$

We transform Equation (2) into a natural logarithmic function to obtain Equation (3). Notably, zero values in the dataset cause problems in the decomposition formulation due to the properties of logarithmic functions. To solve this problem, Ang and Liu [24] suggested replacing zero values with a small positive number.

$$\text{lnCONVENTIONAL}^t - \text{lnCONVENTIONAL}^{t-1} = \ln\left(\frac{\text{PRIORITY}^t}{\text{PRIORITY}^{t-1}}\right) + \ln\left(\frac{\text{WTT}^t}{\text{WTT}^{t-1}}\right) + \ln\left(\frac{\text{SCALE}^t}{\text{SCALE}^{t-1}}\right) \qquad (3)$$

Multiplying both sides of Equation (3) by $\omega_i^t = \left(\text{CONVENTIONAL}^t - \text{CONVENTIONAL}^{t-1}\right) / \left(\text{lnCONVENTIONAL}^t - \text{lnCONVENTIONAL}^{t-1}\right)$ yields Equation (4), as follows.

$$\text{CONVENTIONAL}^t - \text{CONVENTIONAL}^{t-1} = \Delta\text{CONVENTIONAL}^{t,t-1} =$$
$$\omega_i^t \ln\left(\frac{\text{PRIORITY}^t}{\text{PRIORITY}^{t-1}}\right) + \omega_i^t \ln\left(\frac{\text{WTT}^t}{\text{WTT}^{t-1}}\right) + \omega_i^t \ln\left(\frac{\text{SCALE}^t}{\text{SCALE}^{t-1}}\right). \qquad (4)$$

Therefore, changes in the number of patents granted for conventional treatment technologies ($\Delta\text{CONVENTIONAL}$) are decomposed by changes in the PRIORITY (first term), WTT (second term), and SCALE (third term). The term ω_i^t operates as an additive weight for the estimated number of patents granted for conventional treatment technologies.

3. Data and Results

3.1. Data

We use the patents granted data from PATENTSCOPE (http://www.wipo.int/patentscope/en/), which is provided by the World Intellectual Property Organization (WIPO). The PATENTSCOPE database covers more than 56 million patents granted from 1978 to 2016. The data coverage by country and period are shown in Table S2 in the Supplementary Materials.

Because the PATENTSCOPE data coverage for Japan, which is a major water treatment technology innovator, began after 1993, we use the patent dataset from 1993 to 2016 (see Table S2). Following Fujii [20], we only use the primary IPC code to categorize the technology group to avoid double counting patent data in each technology group.

The PATENTSCOPE database and search strategy with IPC in Table S1 determined that 227,365 water treatment technology patents were filed from 1993 to 2016. The composition of each technology group is as follows: conventional treatment (117,974 patents, 51.9%), biological treatment

(40,300 patents, 17.7%), multistage treatment (45,732 patents, 20.1%), sludge treatment (15,237 patents, 6.7%), and other treatments (8122 patents, 3.6%).

3.2. Trends in Water Treatment Technology Patent Inventions

Table 2 shows the changes in the water treatment technology patents granted by type of technology for each patent office. The composition of the patents granted shares differs among the countries.

Table 2 shows that conventional water treatment technology represents more than half of the total number of water treatment technology patents granted in most countries, whereas multistage water treatment technology is the major technology type granted by the SIPO. The share of the multistage water treatment technology is only 0.4% for the JPO, which is extremely low when compared with that for the other patent offices.

Table 2. Data description of the water treatment technology patents granted (number of patents).

Patent Office	Technology Type	1993–2016	Share	1993–1998	1999–2004	2005–2010	2011–2016
SIPO	Conventional	39,116	36.6%	952	3699	8730	25,735
	Biological	15,744	14.7%	157	865	3349	11,373
	Multistage	41,055	38.4%	34	641	4743	35,637
	Sludge	6950	6.5%	25	202	969	5754
	Other	4084	3.8%	70	284	757	2973
JPO	Conventional	23,461	67.3%	6607	6955	5494	4405
	Biological	7809	22.4%	2634	2514	1619	1042
	Multistage	144	0.4%	44	17	27	56
	Sludge	2725	7.8%	396	649	1138	542
	Other	706	2.0%	158	236	178	134
KIPO	Conventional	13,263	60.0%	1280	4153	3713	4117
	Biological	4683	21.2%	402	1793	1451	1037
	Multistage	974	4.4%	81	410	236	247
	Sludge	2689	12.2%	245	822	779	843
	Other	485	2.2%	43	149	101	192
USPTO	Conventional	9870	68.2%	1630	2308	2509	3423
	Biological	3013	20.8%	506	752	973	782
	Multistage	727	5.0%	73	119	147	388
	Sludge	311	2.1%	73	70	55	113
	Other	557	3.8%	195	128	113	121
PCT	Conventional	7265	69.3%	593	1254	2180	3238
	Biological	1833	17.5%	240	304	558	731
	Multistage	500	4.8%	10	6	140	344
	Sludge	508	4.8%	29	51	168	260
	Other	376	3.6%	59	71	100	146
EPO	Conventional	4620	65.0%	803	971	1244	1602
	Biological	1431	20.1%	314	361	390	366
	Multistage	307	4.3%	44	53	75	135
	Sludge	385	5.4%	84	80	112	109
	Other	365	5.1%	106	123	73	63
Other patent office	Conventional	20,379	64.9%	4862	5804	4561	5152
	Biological	5787	18.4%	1726	1652	1162	1247
	Multistage	2025	6.4%	303	445	566	711
	Sludge	1669	5.3%	518	442	355	354
	Other	1549	4.9%	420	405	320	404

Source: Authors' estimate using the IPC code in Table S1 and the PATENTSCOPE database; Note: SIPO: State Intellectual Property Office of The People's Republic of China; JPO: Japan Patent Office; KIPO: Korean Intellectual Property Office; USPTO: United States Patent and Trademark Office; PCT: Patent Cooperation Treaty; EPO: European Patent Office.

Next, we consider the numerical changes in the water treatment technology patents granted. As shown in Table 2, all of the patent offices, except for the JPO, had increased water treatment technology patent publications from the period of 1993–1998 to 2011–2016. However, the number of patents granted by the JPO was the largest from 1993 to 1998 for conventional, biological, and sludge treatment technologies. One interpretation of this result is that the water treatment technology demand increased in Japan after the basic environmental law was enforced in 1993 [16].

Notably, the number of patents granted by the SIPO increased more than four times, i.e., from 18,548 during 2005–2010 to 81,472 during 2011–2016. This patent publication growth was observed for all five water technology types in China. One major driver promoting water treatment technology development in China is "a water pollution prevention and control action pan (Water Ten Plan)", which was released by the Chinese government in 2015 [25]. The Chinese government expects the Water Ten Plan to create 1.9 trillion RMB in new investments for water treatment [26]. According to Fujii and Managi [17], technology innovation is induced by future business market expansion. Therefore, innovators have a strong incentive for water treatment technology development because of future business opportunities supported by the Water Ten Plan.

3.3. Results of the Patent Decomposition Analysis

Figure 2 shows the results of a patent decomposition analysis for the five water treatment technologies at all of the patent offices listed in Table S2. The plotted point in red indicates the change in the number of specific patents granted, and the bar chart shows the effects of each decomposed factor on the number of patents granted for specific water treatment technologies. The sum of the bars is equivalent to the value of the plotted point. The figure shows the differences in the driving factors for the patents granted based on the water treatment technology type.

Figure 2 shows that the number of patents granted for multistage and conventional water treatment technologies increased from 1993 to 2016. However, the specific water treatment technology priority differently affects these two technology types. As shown in Figure 2, the relative priority of the conventional water treatment technology was negative, whereas that of the multistage water treatment technology was positive. This result implies that the water treatment technology patent invention priority shifted from conventional water treatment to multistage water treatment. The relative priority of the other three technology types did not significantly change during this research period. The results suggest that the patents granted for those three technologies showed a similar trend to that of the total water treatment technology patents granted.

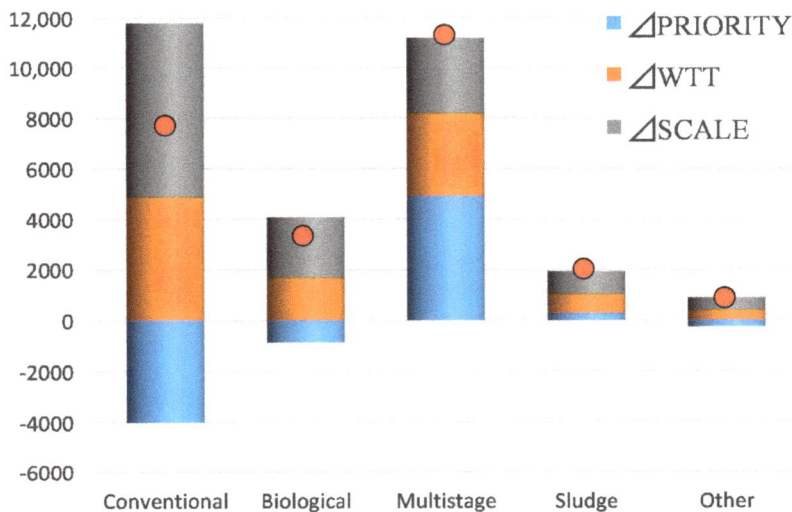

Figure 2. Results of the patent decomposition analysis (number of patents). Note: The vertical axis is standardized by setting the number of changes in the patents granted in 1993 to zero. The sum of the bars is equivalent to the value of the plotted point.

Table 3 shows the patent decomposition analysis results for each patent office. The table shows that the main contributor to the increase in patents granted is China. The scale change in the R&D activity at the SIPO contributes to all five of the water treatment technology types. One interpretation of this result is that the Chinese patent application law revisions in 2001 and 2009 simplified patent applications for domestic companies that use a subsidy program [27]. Hu et al. [28] noted a rapid patent application increase at the SIPO that was caused by external factors, such as the patent law revision and a new subsidy system. Thus, the Chinese patent application system revision contributed to expanded R&D activities (e.g., patent applications) at the SIPO, which increased the number of patents for water treatment technology.

Table 3. Results of the patent decomposition analysis for water treatment technology from 1993 to 2016.

Specific Technology	DECOMPOSED Factor	SIPO	JPO	KIPO	USPTO	PCT	EPO
Conventional	$\triangle$Patent	6816	−386	460	414	505	99
	$\triangle$PRIORITY	−1800	9	48	28	13	19
	$\triangle$WTT	2236	15	−25	−69	29	−99
	$\triangle$SCALE	6380	−410	437	456	463	179
Biological	$\triangle$Patent	3260	−224	132	63	86	7
	$\triangle$PRIORITY	−182	−159	−44	−52	−40	−18
	$\triangle$WTT	975	23	43	−42	15	−22
	$\triangle$SCALE	2467	−88	133	157	111	47
Multistage	$\triangle$Patent	10,905	−3	37	84	58	21
	$\triangle$PRIORITY	1978	1	−4	55	34	17
	$\triangle$WTT	2908	1	17	15	3	−6
	$\triangle$SCALE	6019	−5	23	14	21	10
Sludge	$\triangle$Patent	1779	34	93	13	49	2
	$\triangle$PRIORITY	289	133	−11	0	14	8
	$\triangle$WTT	456	−27	28	−0	4	−11
	$\triangle$SCALE	1034	−73	75	13	30	5
Other	$\triangle$Patent	649	−2	23	−7	10	−13
	$\triangle$PRIORITY	−262	15	6	−19	−18	−14
	$\triangle$WTT	250	−8	7	−13	−0	−10
	$\triangle$SCALE	661	−10	11	25	28	12

Note 1: SIPO: State Intellectual Property Office of the People's Republic of China; JPO: Japan Patent Office; KIPO: Korean Intellectual Property Office; USPTO: United States Patent and Trademark Office; PCT: Patent Cooperation Treaty; EPO: European Patent Office; Note 2: $\triangle$Patent = $\triangle$Priority + $\triangle$WTT + $\triangle$Scale.

4. Conclusions

This study examined the trend and priority changes in water treatment technologies using patents granted data from 1993 to 2016. We focused on the following five technologies: (1) conventional water treatment technology, (2) biological water treatment technology, (3) multistage water treatment technology, (4) sludge treatment technology, and (5) other water treatment technologies. We clarified the priority shifts that were reflected in the patents covering innovations in these five technologies by applying the decomposition analysis. We obtained the following results.

First, the number of water treatment technology patents granted increased from 1993 to 2016. In particular, rapid growth was observed in multistage water treatment technology. The main driver of this growth was the expansion in the R&D activity scale and an increase in the priority of multistage water treatment technology in China. The patent application law revision and subsidy system in China are noted as external factors that promoted R&D activity among Chinese innovators.

Second, the priority placed on multistage water treatment technology innovations decreased in Japan from 1993 to 2016. This result indicated that the R&D strategy for water treatment technology in Japan clearly differs from that in other countries and patent offices. This information indicates that the Japanese government should recognize the necessity of promoting aquaculture technology development in Japan.

Finally, we observed that the priority changes in water treatment technology innovations were diverse across countries and technology groups. The differences in water treatment technology characteristics are useful for clarifying technological advantages and high priority technology types in each country.

Supplementary Materials: The following materials are available online at www.mdpi.com/2073-4441/9/11/860/s1: Table S1. International patent clarification (IPC) related to water treatment technologies. Table S2. Patent data collection periods in the PATENTSCOPE database by country.

Acknowledgments: This research was funded by the Grant-in-Aid for Specially Promoted Research [26000001B] and Grant-in-Aid for Young Scientists (B) [17K12858] from the Ministry of Education, Culture, Sports, Science and Technology (MEXT), Japan. The results and conclusions of this article do not necessarily represent the views of the funding agencies.

Author Contributions: Hidemichi Fujii contributed to the construction of the dataset, the development of the methodology, and the drafting of the article. Shunsuke Managi assisted with the literature revision and conclusions.

Conflicts of Interest: The authors declare no conflicts of interest.

References

1. Howe, K.J.; Hand, D.W.; Crittenden, J.C.; Trussell, R.R.; Tchobanoglous, G. *Principles of Water Treatment*; John Wiley & Sons: Hoboken, NJ, USA, 2012; p. 672.
2. United Nations World Water Assessment Programme (WWAP). *The United Nations World Water Development Report 2017, Wastewater: The Untapped Resource*; United Nations World Water Assessment Programme: Paris, France, 2017; p. 198.
3. Shannon, M.A.; Bohn, P.W.; Elimelech, M.; Georgiadis, J.G.; Mariñas, B.J.; Mayes, A.M. Science and technology for water purification in the coming decades. *Nature* **2008**, *452*, 301–310. [CrossRef] [PubMed]
4. World Health Organization (WHO); United Nations Children's Fund (UNICEF). *Progress on Drinking Water, Sanitation and Hygiene: 2017 Update and SDG Baselines*; WHO and UNICEF: Geneva, Switzerland, 2017; p. 110.
5. United Nations Water (UN-Water). *Towards a Worldwide Assessment of Freshwater Quality*; A UN-Water Analytical Brief; UN-Water: Geneva, Switzerland, 2016.
6. United Nations. Transforming Our World: The 2030 Agenda for Sustainable Development. Outcome Document for the United Nations Summit to Adopt the Post-2015 Development Agenda. Available online: https://sustainabledevelopment.un.org/post2015/transformingourworld (accessed on 25 September 2015).
7. Stanway, D. *In China's Murky Waters, Global Sewage Firms Seek Rewards*; Reuters: London, UK, 2017; Available online: https://www.reuters.com/article/us-China-sewage/in-Chinas-murky-waters-global-sewage-firms-seek-rewards-idUSKBN19U00Z (accessed on 17 October 2017).
8. United Nations Water (UN-Water). *Regional Coordination Mechanisms for Water: A Report of the UN-Water Task Force on Regional-Level Coordination*; UN-Water: Geneva, Switzerland, 2014.
9. World Intellectual Property Organization (WIPO). IPC Green Inventory. Available online: http://www.wipo.int/classifications/ipc/en/est/ (accessed on 18 August 2017).
10. Rodriguez-Narvaez, O.M.; Peralta-Hernandez, J.M.; Goonetilleke, A.; Bandala, E.R. Treatment technologies for emerging contaminants in water: A review. *Chem. Eng. J.* **2017**, *323*, 361–380. [CrossRef]
11. Subramani, A.; Jacangelo, J.G. Emerging desalination technologies for water treatment: A critical review. *Water Res.* **2015**, *75*, 164–187. [CrossRef] [PubMed]
12. Palma, P.; Fialho, S.; Alvarenga, P.; Santos, C.; Brás, T.; Palma, G.; Cavaco, C.; Gomes, R.; Neves, L.A. Membranes technology used in water treatment: Chemical, microbiological and ecotoxicological analysis. *Sci. Total Environ.* **2016**, *568*, 998–1009. [CrossRef] [PubMed]
13. Alzahrani, S.; Mohammad, A.W. Challenges and trends in membrane technology implementation for produced water treatment: A review. *J. Water Process Eng.* **2014**, *4*, 107–133. [CrossRef]
14. Temesgen, T.; Bui, T.T.; Han, M.; Kim, T.I.; Park, H. Micro and nanobubble technologies as a new horizon for water-treatment techniques: A review. *Adv. Colloid Interface Sci.* **2017**, *246*, 40–51. [CrossRef] [PubMed]

15. Fujii, H.; Managi, S. Wastewater management efficiency and determinant factors in the Chinese industrial sector from 2004 to 2014. *Water* **2017**, *9*, 586. [CrossRef]
16. Hara, K.; Kuroda, M.; Yabar, H.; Kimura, M.; Uwasu, M. Historical development of wastewater and sewage sludge treatment technologies in Japan—An analysis of patent data from the past 50 years. *Environ. Dev.* **2016**, *19*, 59–69. [CrossRef]
17. Fujii, H.; Managi, S. Research and development strategy for environmental technology in Japan: A comparative study of the private and public sectors. *Technol. Forecast. Soc. Chang.* **2016**, *112*, 293–302. [CrossRef]
18. Candelin-Palmqvist, H.; Sandberg, B.; Mylly, U.M. Intellectual property rights in innovation management research: A review. *Technovation* **2012**, *32*, 502–512. [CrossRef]
19. Popp, D. Inducing innovation and energy prices. *Am. Econ. Rev.* **2002**, *92*, 160–180. [CrossRef]
20. Fujii, H. Decomposition analysis of green chemical technology inventions from 1971 to 2010 in Japan. *J. Clean. Prod.* **2016**, *112*, 4835–4843. [CrossRef]
21. Haščič, I.; Migotto, M. *Measuring Environmental Innovation Using Patent Data*; OECD Environment Working Papers, No. 89; OECD Publishing: Paris, France, 2015.
22. Smith, C.E. *Design with the Other 90%: Cities*; Cooper-Hewitt, National Design Museum, Smithsonian Institute: Washington, DC, USA, 2011.
23. Japan Patent Office. Comparative Study on Examination Practices among JPO, KIPO and SIPO. 2017. Available online: https://www.jpo.go.jp/torikumi_e/kokusai_e/comparative_study.htm (accessed on 17 October 2017).
24. Ang, B.W.; Liu, N. Handling zero values in the logarithmic mean Divisia index decomposition approach. *Energy Policy* **2007**, *35*, 238–246. [CrossRef]
25. Han, D.; Currell, M.J.; Cao, G. Deep challenges for China's war on water pollution. *Environ. Pollut.* **2016**, *218*, 1222–1233. [CrossRef] [PubMed]
26. China Water Risk. New 'Water Ten Plan' to Safeguard China's Waters. Available online: http://Chinawaterrisk.org/notices/new-water-ten-plan-to-safeguard-Chinas-waters/ (accessed on 18 August 2017).
27. Dang, J.; Motohashi, K. Patent statistics: A good indicator for innovation in China? Patent subsidy program impacts on patent quality. *China Econ. Rev.* **2015**, *35*, 137–155. [CrossRef]
28. Hu, A.G.Z.; Zhang, P.; Zhao, L. China as number one? Evidence from China's most recent patenting surge. *J. Dev. Econ.* **2017**, *124*, 107–119. [CrossRef]

water

MDPI

Case Report

Assessing Risks at a Former Chemical Facility, Nanjing City, China: An Early Test of the New Remediation Guidelines for Waste Sites in China

Yanhong Zhang [1], Shujun Ye [1,*], Jichun Wu [1] and Ralph G. Stahl Jr. [2]

[1] Key Laboratory of Surficial Geochemistry, Ministry of Education, School of Earth Sciences and Engineering, Nanjing University, Nanjing 210023, China; yhzhang618@gmail.com (Y.Z.); jcwu@nju.edu.cn (J.W.)
[2] DuPont Corporate Remediation Group, DuPont Company, Wilmington, DE 19805, USA; Ralph.G.Stahl-Jr@dupont.com
*   Correspondence: sjye@nju.edu.cn; Tel.: +86-25-8968-4150

Received: 30 June 2017; Accepted: 28 August 2017; Published: 1 September 2017

Abstract: China has recognized the need to investigate and remediate former manufacturing facilities and return the land they occupy to a new, productive use. As a result, national guidelines entitled "Technical guidelines for Risk assessment of contaminated sites" were issued in 2014 to guide site investigations, risk assessments, and remedial actions to reduce or mitigate potential exposures of people and ecological receptors to contaminants. This study was pursued to gain experience with the new guidelines at a small, former chemical manufacturing facility in Nanjing City, China. A series of investigations were undertaken to determine the locations and levels of contaminants in soils and groundwater, develop a conceptual site model, and prepare an initial estimate of risks to humans and ecological receptors. Groundwater results revealed several contaminants that were greater than the Dutch Intervention Levels, yet, surprisingly, few, if any, contaminants were found in multiple samplings of soil. Despite the limited investigations of soil and groundwater, data were sufficient to prepare initial risk evaluations for humans, both for systemic toxins and potentially carcinogenic chemicals. The site and nearby area contain industrial facilities and residential neighborhoods; hence, there were too few ecological receptors to warrant an ecological risk assessment. The new guidelines for site investigations and risk assessments proved sufficient for the purposes of this small site; however, more complex sites may require much greater levels of effort and more detailed guidelines for investigations, risk assessments, and remedial actions.

Keywords: guidelines; contaminated sites; risk assessment; China

1. Introduction

Like many nations, China has begun to recognize the need to investigate former manufacturing sites and remediate them so they can be returned to productive use [1]. In China's case, this is particularly important because many former manufacturing sites are quickly becoming isolated within large, newly developed or developing residential areas [2]. These former manufacturing facilities occupy highly desirable land that could be used for residential housing, new manufacturing, or for recreation. Without guidelines, backed by regulatory frameworks and trained staff, it is not possible to conduct the consistent, protective remediation of these former facilities, nor insure that remedial actions are overseen and tracked by trained professionals. Of additional importance to undertaking remedial actions at these facilities is the need for a risk-based decisional process. Such a process would include written guidelines for conducting qualitative and quantitative analyses of potential risks to humans and ecological receptors, as well as details on how the risk assessment should be applied in deciding what remedial action(s) are appropriate for the site under study.

The following details the investigations undertaken at this site, the lessons learned in applying the new guidelines, and suggestions on how the guidelines might be improved. We followed these steps in a risk-based decisional process to undertake, describe, and discuss our approach to work at the Luhe site: (1) Problem formulation; (2) Exposure and Hazard Assessment; and (3) Risk Characterization. Throughout the following text, we identify areas of uncertainty that we encountered while applying the new guidelines to this site. Finally, we have also provided a relatively large Supplemental Materials section that includes many of the guideline details translated from the original Chinese versions.

2. Methods

A major objective of the study was to test the newly issued Chinese national guidelines for addressing contaminated waste sites. Hence for the Luhe site, we generally followed the methods detailed in three of these new guidelines: (1) Site investigations [3]; (2) Risk assessments [4]; and (3) Remedial actions [5]. As the study began in 2011, these three sets of guidelines were still in draft form; however, after some revisions by the Ministry of Environmental Protection (MEP), they became final in 2014 and were ultimately used for this study. Later, we describe one of the major changes when the guidelines were finalized and how that impacted our risk assessment.

2.1. Problem Formulation

The location of the former manufacturing site is shown in Figure 1. The future land use plan for the site is for the eastern part to become a part of the adjacent highway, while the western part will become a municipal landscape area. These details are important for planning data collection and undertaking the risk assessment.

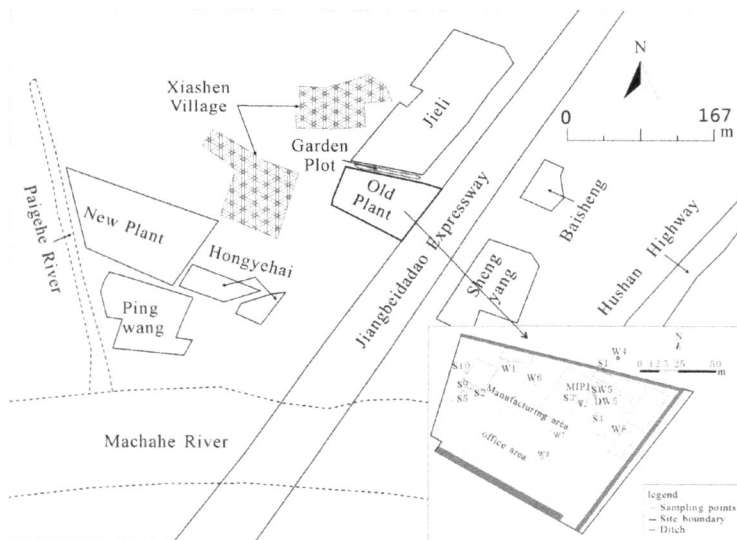

Figure 1. The schematic Location of Luhe site, Nanjing, China.

During our work, we were aware that some of the more detailed guidance and policies applicable for site investigations, risk assessments, and remedial decisions were still being developed in China. That work began before we started our project, and continues today. As a result, our initial screening of soil and groundwater data required us to augment Chinese guidelines with those from other countries, including the United States (US) and The Netherlands.

2.2. Site Setting, Operational History, and Initial Investigations

The Luhe site is approximately 1200 m^2 in size, and included several old buildings and surrounding villages (Figure 1). There are two rivers that border the site: the Paigehe River to the west, and the Machahe River, a branch of the Yangtze River, located about 300 m to the south. These two rivers are used for crop irrigation in this region of China, as well as the transportation of goods into and out of the Yangtze Delta area. Small residential villages are located just north and adjacent to the site, and residents were observed to use a small paved road to cross the site to gain access to the larger streets and nearby highway. Small vegetable gardens are present along the eastern side of this paved road, and another manufacturing complex is located to the east of those gardens. We did not attempt to collect information on operations or investigations at this adjacent manufacturing complex, and thus did not include it in this study. This highlights an important point with respect to contaminated site investigations in China—the guidelines do not yet provide recommendations for how to address potentially contaminated sites near the site under investigation.

From 1999 to 2010, the Luhe site manufactured optical brightener PF (polyester film), 2-Amino-4-methylphenol, and 2-Nitro-4-methylphenol. Unfortunately, the manufacturing history for this site does not include written details sufficient to fully understand how these substances were made and what they were used for. Available information on optical brighteners suggests that their composition can vary, largely depending on their intended use. They are sometimes referred to as fluorescent whitening agents (FWA), and can be used to enhance colors in various textiles, consumer products, paints, etc. [6]. The site ceased the production of PF in 2010, shortly before we undertook our study. The site underwent substantive physical changes during the course of our work, which complicated some aspects of our investigations.

2.3. Collection of Soil and Groundwater Samples

Three field investigations were conducted to determine the nature and extent of soil and groundwater contamination at the Luhe site. The soil sample boreholes were advanced via direct-push technology, using a Geoprobe 6620DT system. The sampling was conducted utilizing a 110 mm diameter, 1.219 m length, stainless steel macro-corer with a new, dedicated, 50.8 mm diameter, 1.219 m long, hollow plastic liner. In addition, an auger was used to collect undisturbed shallow soil samples at the target depth.

Hand bailers or mechanical pumps were used to collect groundwater from wells on the site. Groundwater sampling was performed in general accordance with US Environmental Protection Agency (USEPA) protocols [7] to minimize the potential for cross-contamination. Two groundwater samples were collected in parallel at each well location; the first was for volatile organic compounds (VOCs), followed by a second one for semi-volatile organic compounds (SVOCs).

2.4. Analytical Methodology

All soil and groundwater samples were submitted to ALS (ALS Analytical Testing (Shanghai) Co., Ltd., Shanghai, China) for "typical" chemical analysis in accordance with USEPA SW-846 methods, as presented in Supplemental Materials, Part A: Table S1. A UV-Vis spectrophotometer was used to analyze the optical brightener PF, which absorbs at 363 nm (UV 2300, Shanghai *Tianmei* Science and Technology Corporation, China). In our experience, the optical brightener PF has not been a "typical" contaminant found at contaminated sites and therefore required the use of a different analytical method to measure it in soil and groundwater samples.

2.5. Data Screening and Risk Assessment

When the study began in 2011, there were few or no guidelines available in China for conducting contaminated waste site investigations, risk assessments, or remedial evaluations and selections. However, this information gap was filled by the publication of draft guidelines, followed by final versions in 2014. These relatively new guidelines included procedures for site investigations, risk assessments, and selecting remedial actions. English translations of the models, assumptions, and many details applied to calculate various exposure, hazard, and risk parameters are shown in the Supplemental Materials, Part A.

The soil analytical data were evaluated against the Screening Levels for the Soil Environmental Risk Assessment of Sites (SLSRAS, DB11/T 811-2011) [8], Dutch Soil Quality Standards [9], and USEPA Region 9 Preliminary Remediation Goals PRGs, Version Nov. 2012) [10], respectively. The groundwater analytical data were evaluated against the Chinese Quality Standard for Groundwater (CQSG, GB/T 14848-9) [11], the Dutch Groundwater Quality Standards [9], and the USEPA Region 9 PRGs (Version Nov. 2012) [10], respectively. The use of screening criteria and standards from outside China reflected the lack of such risk-based screening values in China for our work at the Luhe site. In most instances, Dutch Target Values (DTV), which represent generally recognized safe soil and groundwater concentrations, are appropriate screening values where such values are not specified by or available in a specific country. As a result, we used the Dutch Standards (Target Values) for screening contaminants detected in soil and groundwater samples at the Luhe site. Even so, there were instances where there were no screening values readily available, including for the optical brightener PF.

3. Results and Discussion

3.1. Soils

From 2011 to 2014, three field investigations were conducted to evaluate the contamination levels at the Luhe site. For the first and second investigations, a total of nine soil samples were collected. None of the concentrations of VOCs or SVOC contaminants exceeded their applicable SLSRAS, DIV-S, or USEPA PRG screening values, where applicable screening values were available. During the third investigation, an additional 30 soil samples were collected for VOC and SVOC analysis. While low levels of ethylbenzene, chlorobenzene, 1,3-dichlorobenzene, and 1,4-dichlorobenzene were detected in soil samples collected from location S4, no contaminant was observed at a concentration above Dutch Intervention Values for Soils (DIV-S) (Table 1). Unfortunately, there are no relevant Dutch Target Values for Soil (DTV-S) for the contaminants found at the site, and SLSRAS for only a limited number of chemicals. In addition, there are not yet any applicable guidelines in China for the de novo derivation of risk-based screening values. Optical brightener PF was detected in soils at 2 mg/g from location SW5, yet this chemical does not appear to be a "typical" soil or groundwater contaminant found at former manufacturing sites that are or have been investigated, either in China or elsewhere. These results clearly demonstrate the need for China to develop risk-based screening levels for soils and groundwater, or adopt and supplement those already developed by the US, The Netherlands, or other countries.

Table 1. Contaminants in soil samples from the Luhe site, China (Unit: mg/kg in dry weight).

Sampling Points and Depth (m)	Benzene	Toluene	Xylenes	Ethyl Benzene	Chloro Benzene	1,4-Dichloro Benzene	1,2,4-Trichloro Benzene
CAS No.	71-43-2	108-88-3	1330-20-7	100-41-4	108-90-7	106-46-7	120-82-1
S4, 0.7	0.025	0.025	0.025	10.4	1.64	2.03	0.025
DW5, 8.5	0.025	0.025	0.12	0.17	0.025	0.025	1.46
DW5, 10.9	0.025	0.025	0.025	0.025	0.025	0.025	0.16
SW8, 1.5	0.025	0.025	0.025	0.025	0.1	0.025	0.025
SLSRAS	1.4	3300	100	860	64	-	-
DIV-S	1.2	320	17	110	15	19	11
DTV-S	-	-	-	-	-	-	-
PRGs	5.1	4700	280	25	130	11	26

3.2. Groundwater

All potential contaminants of concern (pCOCs) were non-detectable (detection limit = 0.5 µg/L) in the groundwater samples collected from locations W4, SW5, W6, and W7 (Table 2). Concentrations in the groundwater of all pCOCs were above Dutch Intervention Values for Groundwater (DIV-G) screening values (shown in bold type in Table 2) in samples collected from S1, S4, etc. The Dutch Target Values for Groundwater (DTV-G) were used for screening groundwater samples since there were no relevant risk-based screening values in China for groundwater contaminants.

Table 2. Contaminants found in groundwater samples at the Luhe site (Unit: µg/L).

Sampling Points and Depth (m)	Benzene	Toluene	Xylenes	Ethyl Benzene	Chloro Benzene	1,4-Dichloro Benzene	1,2,4-Trichloro Benzene
CAS No.	71-43-2	108-88-3	1330-20-7	100-41-4	108-90-7	106-46-7	120-82-1
W2, 3	0.6	0.25	0.25	2.9	2.9	39.3	**753**
W3, 3	0.25	0.25	0.25	0.25	0.25	0.25	**35.9**
S2, 3	0.25	0.25	0.25	0.25	0.25	0.25	5.5
S3, 3	0.7	0.6	0.25	**6**	1	14.4	**425**
S4, 3	22.1	89.1	0.25	**1200**	**638**	**1625**	**7300**
S1, 3	123	3.5	0.25	**505**	**2400**	**4117**	**4800**
W4, 3.5	0.25	0.25	0.25	0.25	6.1	16.1	5.9
SW5, 3	0.25	0.25	0.25	0.25	0.25	0.25	1.8
DW5, 10	1.2	61	128	166	42	41.9	**1353**
W7, 9	0.25	0.25	0.25	0.25	0.25	0.25	0.25
W8, 4	0.25	0.25	0.25	0.25	19	9.9	**57.4**
CQSG	-	-	-	-	-	-	-
DIV-G	30	1000	70	150	180	50	10
DTV-G	0.2	7	0.2	4	0.3	0.2	0.03
PRGs	0.45	110	19	1.5	7.8	0.48	0.4

Note: Concentrations above Dutch Intervention Values for Groundwater (DIV-G) are shown in bold type.

3.3. Conceptual Model

A conceptual model is one of the key elements of site investigations and the remediation process. It serves as both a communication tool and an illustration of contaminant movements through environmental media and the pathways by which those contaminants reach (or not) important receptors such as humans or wildlife [12,13]. It is evergreen, and updated throughout the investigation phase as new data become available. Over the course of this work, the Luhe site underwent substantial physical changes including the demolition of all site buildings and the removal of subsurface utilities. This activity tended to complicate some of our soil and groundwater investigations, because the entire surficial soils were modified by the use of heavy equipment such as excavators and hauling trucks, among others. Nevertheless, in one aspect, it was fortunate that these modifications happened during our study. It demonstrated how such sites in China may be subject to clearing and redevelopment, regardless of where they might be in relation to the investigation and identification of potential remedial actions. To our knowledge, these changes did not result in any observable changes to or impacts on the nearby residential areas.

The details above are important to the risk assessors and decision makers since they are directly related to developing the appropriate exposure scenarios that need to be considered in the human health risk assessment and for future risk communication to the local public. They also illustrate a fundamental difference in the length of time allowed for investigations and remedial actions in China compared to the process in the US, where investigations and remediation can take many years, especially on larger, more complex sites.

With respect to the groundwater, developing a site conceptual model required that the source, type, and magnitude of contamination be determined, as well as collecting data on the physical characteristics of the surficial and subsurface soils (Figure 2a). A clay layer was found at approximately four meters below the ground surface (bgs), which, fortunately, appears to be acting as a confining unit (i.e., aquitard) to the movement of groundwater and contaminants under the site. However, a discontinuous perched water zone was found above this clay layer. The apparent groundwater flow in this perched zone was found to be from northwest to southeast based on the groundwater level measurements collected from the shallow monitoring wells (Figure 2b). A semi-confined aquifer was found beneath the clay, with an apparent groundwater flow from north to south across the site based on the groundwater level measurements collected from the deeper monitoring wells.

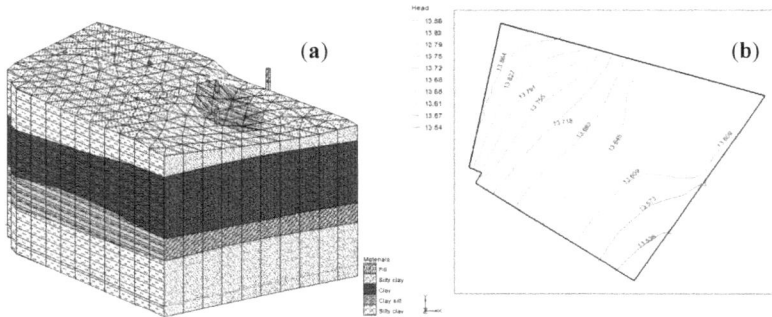

Figure 2. (**a**) The geological model for the LuHe site; (**b**) groundwater contour map of the unconfined aquifer (unit of head: m).

Figure 3 shows the details of the Source-Pathway-Exposure model for soil, based on these various field investigations. The Exposure-Routes-Receptor Model for the site is shown in Table 3. Under Chinese guidelines, contaminated land is divided into two categories: "sensitive", and "non-sensitive". Sensitive land includes land that will be used for non-industrial purposes, such as housing, recreation, etc., whereas non-sensitive land will be used for industrial purposes. In both categories, contaminated land is subject to specific input variables for estimating the potential exposure and risks to human receptors. For the Luhe site, considered non-sensitive land, exposure estimates for adults are characterized by a long exposure duration and high exposure frequency. In addition, exposure estimates are only focused on adults, both for carcinogens and non-carcinogens.

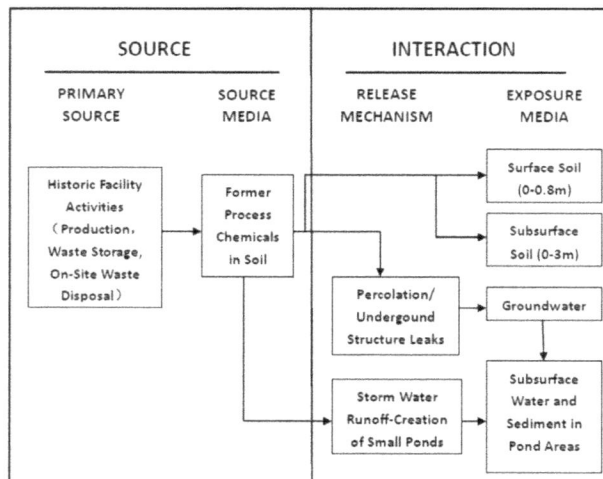

Figure 3. Source-Pathway-Exposure Model for Soil at the Luhe site.

Table 3. Exposure-routes-receptor model for the Luhe site.

Environmental Media	Protect Target	Sources	Receptor	No.	Exposure Routes
				1	Oral Ingestion
				2	Dermal Contact
		Surficial soil	Adults	3	Particle Inhalation
Soil	Human Health			4	Inhalation of Contaminants in Vapor in Outdoor Air
		Subsurface soil	Adults	5	Inhalation of Contaminants in Vapor in Outdoor Air
				6	Inhalation of Contaminants in Vapor in Indoor Air
				7	Consumption
Groundwater	Human Health	Ground-water	Adults	8	Inhalation of Contaminants in Vapor in Outdoor Air
				9	Inhalation of Contaminants in Vapor in Indoor Air

Note: This reflects the Luhe site being categorized as "non-sensitive" land, a designation similar to "industrial" in other national remediation programs.

The completion of exposure and hazard assessments for non-sensitive land presented some challenges and concerns in this study. First, this category of land use only considers potential risks to human health and does not include an evaluation of potential exposures or hazards for ecological receptors. This could be problematic because groundwater can be an important pathway for carrying subsurface contaminants to surficial areas such as ponds, lakes, rivers, and estuaries. Under that scenario, the movement of contaminants in groundwater that is later expressed in surface water bodies could potentially lead to unacceptable exposures to humans and ecological receptors. Because that potential exposure scenario is not required under the current guidelines, potential risks to humans and ecological receptors could go undetected. For our work at the Luhe site, this potential problem will be addressed in our future evaluations given that groundwater is a critical environmental media which should also be considered.

Second, this land-use category only considers human adults, but not children, as receptors that need to be evaluated. During our investigations, we observed children traversing the site using the small paved road, yet their potential exposure to contaminants in soil and groundwater were not evaluated. Hence there is uncertainty about this exposure scenario, as well as the scenario whereby wind-borne dusts that carry contaminants from the site, make their way into homes in the nearby village. This too is an issue that should be evaluated in future studies at the LuHe site.

Third, the model does not consider human exposure through the dietary ingestion route. Our investigations showed that small vegetable gardens have been and may continue to be grown adjacent to the site. In this case, there is some potential for vegetables to be contaminated through the uptake of soil-borne contaminants, or from those contaminants that might be deposited through dust settling on the vegetables. This latter point is particularly relevant in China as the rapid expansion of urban, residential areas may have taken place on former manufacturing lands that have not been properly investigated or remediated. It is also known that urban dust that settles on streets, homes, and perhaps vegetables may contain contaminants posing some level of risk to humans [14,15]. Since the potential for contaminants to reach edible crops was noted, samples of soils from the small garden plots were taken randomly and analyzed by the Supervision and Testing Center of East China Mineral Resources of Ministry of Land and Resources. Collection of the soil samples generally followed the methods outlined previously. The analytical results from sampling those soils are shown in Supplemental Materials, Part A: Table S2.

3.4. Risk Assessment

As a first step in the risk assessment process, we reviewed our results from the soil and groundwater investigations to determine which of the various contaminants would be carried forward into our initial assessment. Typically, the analytical results from sampling environmental media are screened against conservative risk-based values developed by a regulatory agency, or other groups that would be appropriate for the specific country. Those contaminants found to exceed these screening values are then carried into the more detailed risk assessment. It is at this juncture in the risk assessment for the site that we deviated from the typical approach. We elected to select final contaminants of concern (COCs) and carry them into the risk assessment based on two criteria: (1) detecting them in a media at or above the Dutch screening values (DTV); and (2) having sufficient toxicological data on the chemical for conducting the hazard (toxicity) assessment. There were limited toxicity data for both PF and no relevant Dutch screening values for PF. As a result, PF was not carried forward into the risk assessment, and thereby contributed to an uncertainty that may need to be addressed in future efforts. Chemicals that met the criteria above and which became final COCs were benzene, toluene, xylenes, ethylbenzene, chlorobenzene, 1,4-dichlorobenzene, and 1,2,4-trichlorobenzene.

3.4.1. Exposure Assessment

For non-sensitive land, the exposure rate calculations for carcinogenic and non-carcinogenic contaminants in soil and groundwater are shown in Supplemental Materials, Part A: Tables S3 and S4, respectively. The models for calculating exposure rates are found in Supplemental Materials—Part B1, and the values of the parameters used to calculate the exposure rate are listed in Supplemental Materials—Part C. As mentioned above, the values of the parameters applied to calculate the exposure rate were mostly those developed in the US, and as such they may not reflect exactly the characteristic of similar items in China. For example, they do not reflect building type and structure, and the types of soil, etc., found in China. In the future, those parameters will need to be developed and refined based on investigations and basic studies at the various contaminated waste sites in China.

3.4.2. Toxicity Assessment

The potential hazardous effects on human health via different exposure routes were analyzed, including those for carcinogens and non-carcinogens. This analysis also included an evaluation of dose-response associations and mechanisms of toxicity (hazard) for the COCs, as required and/or recommended by the Chinese guidelines. Toxicity parameters for carcinogenic effects include the inhalation unit risk (IUR), inhalation cancer slope factor (SFi), oral ingestion induced cancer slope factor (SFo), and dermal contact-induced cancer slope factor (SFd). The values of the toxicity parameters for carcinogenic effects for the COCs are shown in Supplemental Materials, Part A: Table S5. The SFi is obtained through extrapolation of IUR in Part B2; the SFd is obtained through extrapolation of SFo, also provided in Part B2.

The toxicity parameters for non-carcinogenic effects include the inhalation reference concentration (RfC), inhalation reference dose (RfDi), oral ingestion reference dose (RfDo), and dermal contact reference dose (RfDd). The values of the toxicity parameters for the non-carcinogenic effects of COCs are given in Supplemental Materials, Part A: Table S5. The RfDi and RfDd are also obtained through the Equations provided in Part B2. The toxicity values for contaminants in the new guidelines are limited, particularly for PF and 1,2,4-tricholorobenzene, which are both found at the site. It is expected that as China's national remediation program evolves, much of the data currently needed to refine these input values and support risk assessments will be obtained and incorporated into future regulatory guidelines.

The physical and chemical properties of COCs required for risk assessment calculations include the dimensionless Henry's constant (H'), air diffusion coefficient (Da), water diffusion coefficient (Dw), soil-organic carbon allocation coefficient (Koc), and water solubility (S). The values of the physical and chemical parameters of COCs are given in Supplemental Materials, Part A: Table S5. Other relevant parameters include the digestive tract absorption factor (ABSgi), skin absorption factor (ABSd), and the absorption factor for oral ingestion (ABSo).

3.4.3. Risk Characterization

Characterizing risks generally follows two approaches: the hazard quotient/hazard index, or a probabilistic approach. The Chinese guidelines provide some flexibility for this step in the risk assessment process. The risks are characterized by calculating the hazard quotient (HQ—systemic toxicity) and carcinogenic risk (CR) of COCs from samples (soils, groundwater, etc.) collected at the various sampling points. As shown in Supplemental Materials, Part A: Table S5, there are no recommended toxicity parameters for the potential carcinogenic effects of chlorobenzene. The results shown in Supplemental Materials, Part A: Tables S3 and S4, were then applied to the models needed for calculating the potential risks of the individual COCs. This included applying the results to the six exposure pathways for soils and three pathways for groundwater, as required, and are shown in Part B3.

The initial risk characterization was based on the previous results and are summarized in Table 4, which shows the concentration(s) for each COC that were applied to estimating risks. In this case, only the COCs in soils and groundwater at sampling location S4 were detected at levels above the Dutch DIV screening values. Unfortunately, the data set for this aspect of the risk characterization step was too limited for a more robust evaluation, and in the future and especially for larger and complicated sites, statistical methods will be needed for calculating more representative concentrations or ranges of concentrations of relevant COCs. This would also allow for calculating risk ranges, rather than single point estimates, something that is of interest to risk managers as they determine what action will be needed to protect human health and the environment.

Table 4. Contaminants of concern (COCs) concentrations applied for risk characterization at the Luhe site.

Parameter Symbol	Benzene	Toluene	Xylenes	Ethyl Benzene	Chloro Benzene	1,4-Dichloro Benzene	1,2,4-Trichloro Benzene
C_{sur}	0.025	0.025	0.025	10.4	1.64	2.03	1.46
C_{sub}	0	0	0	0	0	0	0
C_{gw}	0.123	0.0891	0.128	1.2	2.4	4.12	7.3

Notes: C_{sur}: Concentration of contaminants in Surficial soil, Unit: mg/kg; C_{sub}: Concentration of contaminants in Sub surficial soil, Unit: mg/kg; C_{gw}: Concentration of contaminants in Groundwater, Unit: mg/L.

The results of calculating HQ and CR are shown in Table 5. The guidelines define an acceptable risk as an HQ of 1 or less, while the acceptable level of CR (ACR) is 1.00×10^{-6} or less. The bold numbers in Table 5 highlight where the HQ or CR of an individual contaminant was greater than the acceptable risk level. In those instances, the guidelines indicate that the area where the contamination was found (around S4) should be designated as having an unacceptable risk. Among the nine different exposure routes evaluated, the groundwater consumption route was found to contribute most to the potential harm to human health for both non-carcinogenic and carcinogenic effects.

Figure 4 provides a relative contribution from the individual routes, taking 1,4-trichlorobezene as being representative of COCs. The recommended models that were used to develop these estimates are provided in Part B4. Dermal contact with soil (dcs) greatly contributes to the potential.

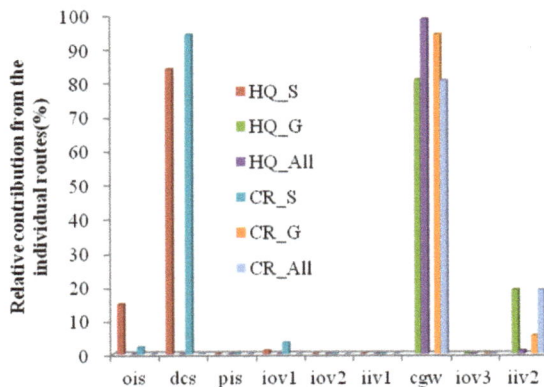

Figure 4. Relative contribution from the individual exposure routes (1,4-trichlorobenzene).

Table 5. Hazard quotient (HQ—systemic toxicity) and carcinogenic risk (CR) of contaminants for different potential human exposure routes at the Luhe site.

Parameter Symbol	Benzene	Toluene	Xylenes	Ethyl Benzene	Chloro Benzene	1,4-Dichloro Benzene	1,2,4-Trichloro Benzene
HQois	7.45×10^{-8}	6.05×10^{-6}	6.05×10^{-6}	6.29×10^{-4}	4.96×10^{-4}	1.75×10^{-4}	3.53×10^{-4}
HQdcs	2.15×10^{-4}	1.08×10^{-5}	4.30×10^{-7}	3.58×10^{-3}	2.82×10^{-3}	9.98×10^{-4}	5.02×10^{-3}
HQpis	2.33×10^{-7}	1.40×10^{-9}	7.00×10^{-8}	2.91×10^{-6}	9.19×10^{-6}	7.11×10^{-7}	2.04×10^{-4}
HQiov1	4.08×10^{-6}	2.45×10^{-8}	1.22×10^{-6}	5.09×10^{-5}	1.61×10^{-4}	1.24×10^{-5}	3.57×10^{-3}
HQiov2	0.00	0.00	0.00	0.00	0.00	0.00	0.00
HQiiv1	0.00	0.00	0.00	0.00	0.00	0.00	0.00
HIn1	2.19×10^{-4}	1.68×10^{-5}	7.77×10^{-6}	4.26×10^{-3}	3.49×10^{-3}	1.19×10^{-3}	9.15×10^{-3}
HQcgw	1.86	6.74×10^{-2}	3.87×10^{-3}	7.26×10^{-1}	7.26	3.56	4.42×10
HQiov3	6.98×10^{-4}	3.28×10^{-6}	2.01×10^{-4}	2.31×10^{-4}	3.84×10^{-3}	2.44×10^{-4}	6.15×10^{2}
HQiiv2	1.03×10^{-1}	4.83×10^{-4}	2.96×10^{-2}	3.41×10^{-2}	5.66×10^{-1}	3.59×10^{-2}	1.06×10
HIn2	1.96	6.79×10^{-2}	3.36×10^{-2}	7.60×10^{-1}	7.83	3.60	6.70×10^{2}
CRois	5.78×10^{-10}	/	/	4.81×10^{-8}	/	4.61×10^{-9}	1.78×10^{-8}
CRdcs	1.82×10^{-9}	/	/	2.43×10^{-7}	/	2.09×10^{-7}	/
CRpis	3.79×10^{-12}	/	/	5.06×10^{-10}	/	4.34×10^{-10}	/
CRiov1	6.63×10^{-11}	/	/	8.84×10^{-9}	/	7.59×10^{-9}	/
CRiov2	0.00	/	/	0.00	/	0.00	/
CRiiv1	0.00	/	/	0.00	/	0.00	/
CRn1	2.47×10^{-9}	/	/	3.01×10^{-7}	/	2.22×10^{-7}	/
CRcgw	2.84×10^{-5}	/	/	5.55×10^{-5}	/	9.35×10^{-5}	8.89×10^{-4}
CRiov3	1.13×10^{-8}	/	/	4.01×10^{-8}	/	1.49×10^{-7}	/
CRiiv2	1.67×10^{-6}	/	/	5.91×10^{-6}	/	2.19×10^{-5}	/
CRn2	3.01×10^{-5}	/	/	6.14×10^{-5}	/	1.16×10^{-4}	/

Notes: /: no calculated value existed; HQois: HQ for route of Oral Ingestion of Soil (ois), Dimensionless; HQdcs: HQ for route of Dermal Contact of Soil (dcs), Dimensionless; HQpis: HQ for route of Inhalation of Soil Particles (pis), Dimensionless; HQiov1: HQ for route of Inhalation of contaminant vapor in outdoor air from surficial soil (iov1), Dimensionless; HQiov2: HQ for route of Inhalation of contaminant vapor inoutdoor air from Sub surficial soil (iov2), Dimensionless; HQiiv1: HQ for route of Inhalation of contaminant vapor in Indoor air from Sub surficial soil (iiv1), Dimensionless; HIn1: The hazard index (HI) for all exposure routes in soil, Dimensionless; HQcgw: HQ for route of groundwater ingestion (cgw), Dimensionless; HQiov3: HQ for route of inhalation of contaminant vapor in Outdoor air from Groundwater (iov3), Dimensionless; HQiiv2: HQ for route of inhalation of contaminant vapor in Indoor air from Groundwater (iiv2), Dimensionless; Hin2: HI for all routes in Groundwater, Dimensionless; CRois: CR for route of ois, Dimensionless; CRdcs: CR for route of dcs, Dimensionless; CRpis: CR for route of pis, Dimensionless; CRiov1: CR for route of iov1, Dimensionless; CRiov2: CR for route of iov2, Dimensionless; CRiiv1: CR for route of iiv1, Dimensionless; CRn1: The total CR for all exposure routes in soil, Dimensionless; CRcgw: CR for route of cgw, Dimensionless; CRiov3: CR for route of iov3, Dimensionless; CRiiv2: CR for route of iiv2, Dimensionless; CRn2: The total CR for all routes in Groundwater, Dimensionless.

The total risk among all soil exposure routes was as high as 84.10% on an HQ basis (HQ_S in Figure 4) and 94.30% on a CR basis (CR_S in Figure 4). For the three routes associated with exposure to groundwater ingestion, cgw contributes the largest portion (80.89%) on an HQ basis (HQ_G in Figure 4), and 94.41% on a carcinogenic risk basis (CR_G in Figure 4). Considering all nine exposure routes, the contribution from groundwater consumption (cgw) is the most significant (98.96 of HQ_All basis and 80.73 of CR_All). These results clearly illustrate that the potential for risk to humans from consuming contaminated groundwater could be problematic, and should be addressed in any proposed risk management actions.

3.4.4. Calculating Risk Control Values

The guidelines provide steps for calculating risk control values, which are those values that will be used to determine the level of remediation that might be required for sites where unacceptable risks have been estimated. Calculating soil and groundwater risk control values, based on non-carcinogenic and carcinogenic effects, follows the same process as noted for estimating risks (either HQ or CR). An acceptable HQ for an individual COC is 1 or less, and for the CR of an individual contaminant, the acceptable level is 10×10^{-6} or less. The recommended models for calculating soil and groundwater risk control values are given in Supplemental Materials—Part B4 and the relevant parameter values are given in Supplemental Materials—Part C.

The calculated risk control values for soil and groundwater based on non-carcinogenic and carcinogenic effects are shown in Table 6. When determining the soil and groundwater remediation target values at the contaminated site, the soil and groundwater risk control values should be calculated as primary reference values based on the risk assessment model. The calculated soil risk control values based on non-carcinogenic and carcinogenic effects, and groundwater risk control values based on non-carcinogenic and carcinogenic effects, are then compared (Table 6). The bold numbers are the risk control values of individual contaminants for soil and groundwater at the Luhe site, respectively. This comparison indicates that the soil remediation target value, as well as the cleanup value for ethylbenzene, is 1.65 mg/kg, and that the groundwater remediation target value is 2.15×10^{-2} mg/L. For benzene, ethylbenzene, and 1,4-dichlorobenzene, the risk control values were calculated based both on non-carcinogenic and carcinogenic effects (Table 6). These results show that the risk control values based on non-carcinogenic effects are almost two orders of magnitude higher than the risk control values based on carcinogenic effects.

Table 6. Risk control values of soil and groundwater based on non-carcinogenic effects and carcinogenic effects at the LuHe site.

Parameter Symbol	Benzene	Toluene	Xylenes	Ethyl Benzene	Chloro Benzene	1,4-Dichloro Benzene	1,2,4-Trichloro Benzene
HCVSois	6.61×10^2	1.32×10^4	3.31×10^5	1.65×10^4	3.31×10^3	1.16×10^4	1.65×10^3
HCVSdcs	1.16×10^2	2.33×10^3	5.81×10^4	2.91×10^3	5.81×10^2	2.03×10^3	2.91×10^2
HCVSpis	2.02×10^8	3.36×10^{10}	6.73×10^8	6.73×10^9	3.36×10^8	5.38×10^9	1.35×10^7
HCVSiov1	1.15×10^7	1.92×10^9	3.85×10^7	3.85×10^8	1.92×10^7	3.08×10^8	7.70×10^5
HCVSiov2	1.13×10^6	2.65×10^8	9.75×10^6	9.23×10^7	5.99×10^6	2.51×10^8	7.45×10^3
HCVSiiv1	3.32×10^5	5.12×10^7	1.20×10^6	9.78×10^6	1.18×10^6	3.19×10^7	2.24×10
HCVSn	9.88×10	1.98×10^3	4.72×10^4	2.47×10^3	4.94×10^2	1.73×10^3	2.04×10
HCVGcgw	2.45×10^3	5.75×10^5	2.12×10^4	2.00×10^5	1.30×10^4	5.45×10^5	1.62×10^5
HCVGiov3	2.25×10^3	3.47×10^5	8.16×10^3	6.64×10^4	8.00×10^3	2.16×10^5	1.30×10^3
HCVGiiv2	6.25×10^{-1}	1.16×10	3.39×10^2	1.38×10	6.65	3.93×10	1.35×10
HCVGn	6.61×10^{-2}	1.32	3.29×10	1.65	3.31×10^{-1}	1.16	1.65×10^{-1}
RCVSois	4.33×10	/	/	2.16×10^2	/	4.41×10^2	8.21×10
RCVSdcs	7.61	/	/	3.81×10	/	7.75×10	1.44×10
RCVSpis	2.58×10^4	/	/	3.16×10^5	/	7.17×10^4	/
RCVSiov1	1.48×10^3	/	/	1.81×10^4	/	4.10×10^3	/
RCVSiov2	1.44×10^2	/	/	4.33×10^3	/	3.34×10^3	/
RCVSiiv1	3.13×10^{-1}	/	/	9.40	/	7.26	/
RCVSn	3.00×10^{-1}	/	/	7.52	/	6.61	1.44×10
RCVGcgw	4.25×10	/	/	4.59×10^2	/	4.24×10^2	/
RCVGiov3	2.88×10^{-1}	/	/	3.11	/	2.88	/
RCVGiiv2	3.05×10^{-2}	/	/	2.16×10^{-2}	/	8.48×10^{-2}	/
RCVGn	4.26×10^{-3}	/	/	2.15×10^{-2}	/	4.34×10^{-2}	8.21×10^{-3}

Notes: /: no calculated value existed; HCVSois: Soil risk control values based on non-carcinogenic effects (HCVS) through routes of ois, Unit: mg/kg; HCVSdcs: HCVS through routes of dcs, Unit: mg/kg; HCVSpis: HCVS through routes of pis, Unit: mg/kg; HCVSiov1: HCVS through routes of iov1, Unit: mg/kg; HCVSiov2: HCVS through routes of iov2, Unit: mg/kg; HCVSiiv1: HCVS through routes of iiv1, Unit: mg/kg; HCVSn: HCVS through all the above six routes in soil, Unit: mg/kg; HCVGcgw: Groundwater risk control values based on non-carcinogenic effects (HCVG) through routes of cgw, Unit: mg/L; HCVGiov3: HCVG through routes of iov3, Unit: mg/L; HCVGiiv2: HCVG through routes of iiv2, Unit: mg/L; HCVGn: HCVG through all the above three routes in groundwater, Unit: mg/L; RCVSois: Soil risk control values based on carcinogenic effects (RCVS) through routes of Oral Soil Ingestion, Unit: mg/kg; RCVSdcs: RCVS through routes of dcs, Unit: mg/kg; RCVSpis: RCVS through routes of pis, Unit: mg/kg; RCVSiov1: RCVS through routes of iov1, Unit: mg/kg; RCVSiov2: RCVS through routes of iov2, Unit: mg/kg; RCVSiiv1: RCVS through routes of iiv1, Unit: mg/kg; RCVSn: RCVS through all the above six routes in soil, Unit: mg/kg; RCVGcgw: Groundwater risk control values based on carcinogenic effects (RCVG) through routes of cgw, Unit: mg/L; RCVGiov3: RCVG through routes of iov3, Unit: mg/L; RCVGiiv2: RCVG through routes of iiv2, Unit: mg/L; RCVGn: RCVG through all the above three routes in groundwater, Unit: mg/L.

The acceptable carcinogenic risk (ACR) of an individual contaminant for humans is 1.00×10^{-6} in the new guidelines, and perhaps deserves some reflection. For example, as China is still developing and working to undertake remediation at multiple sites around the country, one might consider whether an ACR of 1.00×10^{-6} is an appropriate value to apply at this time. Similar to remedial policies, regulations, and guidelines that began and evolved in the US and Europe, the current guidelines in China are likely to be revised over time, and it could be that setting a less stringent ACR at this time might allow more sites to be addressed, and in a shorter time period, than might be the case with a more stringent ACR. An ACR range of 1.00×10^{-4} to 1.00×10^{-5} has been applied generally for industrial sites in other countries, thus providing a less stringent but more financially practical approach. In addition, at this time, the guidelines only provide a single point estimate of potential exposure, rather than ranges or statistically-based estimates. As the experience base in remediating sites in China grows, a risk-range could be considered in determining what will be the appropriate risk control values, as well as the cleanup number, in the future.

4. Summary and Conclusions

Our work over the past three years at the Luhe site, China, has provided a useful and insightful platform for piloting contaminated site investigations, evaluating potential remedial options, and more importantly, an initial test of the new guidelines for addressing contaminated waste sites.

The Luhe site belongs to "non-sensitive land", and our application of the guidelines in this effort gave us an opportunity to determine how they perform at this relatively small site. Topic areas where the guidelines may benefit from additional work include: Future Land Use and Exposure Scenarios, co-located Contaminated Sites, Receptors (children and adults), country-specific Exposure and Screening Values, and Risk-Control Values.

4.1. Future Land Use and Exposure Scenarios

For non-sensitive land, exposures are only focused on adults, resulting in uncertainty with respect to potential harm that might arise from children that might or could frequent the site and thereby be exposed to contaminants. This is especially important given the situation at the site. Despite the site being categorized as "non-sensitive" land, we observed children using a road that went through the former site, and while not observed by us directly, could also indicate that children might also play on the former site. In addition, we observed small vegetable gardens being grown along the site. In this case, there is no provision to consider exposures through vegetables, whether for children or adults. Future revisions to the guidelines would benefit from including the flexibility to modify exposure scenarios based on direct observations at a site, especially if some portions of the "non-sensitive" land will continue to be used by children and for growing vegetables to supplement the diets of nearby residents. Because children can be more sensitive receptors, and are not included in the current guidelines, the guidelines are not sufficiently protective, something which should be corrected in the future.

4.2. Co-Located Contaminated Sites

Undertaking investigations and remediation at a contaminated site that is adjacent to another potentially contaminated site is an issue that is not unique to the site, or China for that matter. At some sites, it is not unusual to find contaminants in soil, groundwater, etc., that cannot be explained on the basis of the past operating history. The question then arises as to the source of the contamination, and who is responsible for dealing with it. Provisions in the US Superfund legislation and regulations [16] allow for investigating the potential for contaminant migration off of the site in question. Where contamination is found off-site, and clearly originates from the site under investigation, responsible parties are then obligated to continue the investigation until the contaminant levels drop to acceptable levels, or to some other pre-determined endpoint. In this case, information on potential impacts at nearby sites can be collected indirectly and evaluated. Where contaminants are found to originate from a nearby site, the owners of the nearby site then become potentially liable for addressing that contamination in connection with the work ongoing at the site under study. In our case, we observed another manufacturing site adjacent to the Luhe site, which may or may not be contributing to some of the contamination we found in soils and groundwater. The potential for this situation to occur, and how to address it, will likely require evaluation by the Chinese government.

4.3. Receptors

Our investigation provided information sufficient to develop an initial site conceptual model for the Luhe site, which helped to illustrate the various exposure pathways and receptors potentially at risk. It is noteworthy that there are no provisions in the current guidelines to include ecological receptors in the conceptual model, and as yet, none either for estimating risks to ecological receptors, regardless of the current or future use of the site. We also observed that there were no provisions to account for potential risks to groundwater itself, except for the potential for humans to be exposed to contaminants through the ingestion of contaminated groundwater. Where groundwater might connect to and be expressed in surficial water bodies, there is no provision for addressing the risks associated with this situation.

4.4. Country-Specific Exposure and Screening Values

Having technically-sound, risk-based, and country-specific screening values for soils, sediments, groundwater, etc., is another area that could require further work. This is also an area not unique to the Luhe site, or to China. It is understandable that for the time being, risk assessors will need to rely on toxicological and risk-based information already developed in other geographical regions. For the Luhe site, we were able to utilize information from The Netherlands, the US (IRIS database), and other countries, that allowed the risk assessment to proceed, albeit with the caveat that some aspects of this approach could benefit from China-specific information. It is reasonable to expect that China-specific screening, exposure, toxicological, and risk control values will be developed in the future, and similar to those in the US and Europe, undergo periodic revision as experience is gained over time.

4.5. Risk-Control Values

Perhaps one of the most vexing policy issues in national environmental or remediation programs is the level of protection that has been chosen for humans and ecological receptors. Despite over 30 years of remedial activities within the US, the debate on what level of protection should be afforded to humans and ecological receptors continues. This has been true for the 1×10^{-6} (one in a million) cancer risk level set for humans, which some believe is too conservative and others might see as too lenient [17]. In this paper, we called attention to the level of protection (1×10^{-6}) selected for humans working on or using non-sensitive sites where carcinogens have been detected. Given the current status of the national remediation program in China, we posed the question as to whether the 1×10^{-6} cancer risk level is an appropriate one, or is something that merits further discussion among policy makers, medical professionals, or other interested groups. Similar to the summary points we have mentioned already, the answer to this question rests with the Chinese government as it is a policy decision and not a technical one.

In conclusion, the work at the Luhe site has allowed us to highlight some of the more prominent issues that may require further evaluation and discussion within the Chinese government and regulatory communities. Taken collectively, the disadvantages in the new guidelines do not suggest that they need to be completely revised, but in their current form, they provide a good starting point for developing a qualitative and quantitative risk assessment system for contaminated waste sites in the nation. As more investigations and risk assessments are completed, and more remedial action experience is gained in China, the current guidelines can be revised and updated to reflect that new knowledge and experience. For that reason, these guidelines, like similar ones in the US, The Netherlands, the United Kingdom, and elsewhere, should be subject to revision on a periodic basis. It is also possible that as technical staff in the Chinese regulatory community become more experienced in contaminated site investigations, risk assessments, and evaluating remedial options, they will have the flexibility to consider each contaminated waste site on a case-by-case basis, and thus accommodate the variability in the sites across the nation.

Supplementary Materials: The following are available online at www.mdpi.com/2073-4441/9/9/657/s1, Part A:Tables, Table S1: Lab analytical method for investigations at the Luhe site, China., Table S2: Heavy metals of soil samples in garden plot at the Luhe site, Nanjing, China (Unit: mg/kg in dry weight), Table S3: Calculation of Exposure rate based on non-carcinogenic effects of COCs at the Site, Table S4: Calculation of Exposure rate based on carcinogenic effects of COCs at the Site, Table S5: Toxicity parameters of COCs at the Site., Table S6: Physical and chemical properties, and other relevant parameters for COCs at the Site; Part B: Models, Part B1: Models for calculating exposure rate based on non-carcinogenic effects and carcinogenic effects, Part B2: Models for toxicity assessment, Part B3:Models for calculation of hazard quotient and carcinogenic risk, Part B4: Models for calculating risk control values based on non-carcinogenic effects and carcinogenic effects, Part B5: Models for analysis on risk exposure contribution; Part C: Risk-based Parameters and values; Part D: List of other Risk-based Parameters.

Acknowledgments: The work was financially supported by DuPont Grant and NSFC No. 41472212. The results and conclusions contained herein are those of the authors and do not represent any official or un-official positions of their affiliated organizations.

Author Contributions: Shujun Ye and Jichun Wu proposed the idea, and designed the framework of the study. Yanhong Zhang conducted the site risk assessment and wrote the draft. The other co-authors revised the draft, especially Ralph G. Stahl Jr., who helped with the English writing.

Conflicts of Interest: The authors declare no conflict of interest.

References

1. He, G.; Zhang, L.; Mol, A.P.; Wang, T.; Lu, Y. Why small and medium chemical companies continue to pose severe environmental risks in China. *Environ. Pollut.* **2014**, *185*, 158–167. [CrossRef] [PubMed]

2. Li, D.; Zhang, C.; Pizzol, L.; Critto, A.; Zhang, H.; Lv, S.; Marcomini, A. Regional risk assessment approaches to land planning for industrial polluted areas in China: The Hulubeier region case study. *Environ. Int.* **2014**, *65*, 16–32. [CrossRef] [PubMed]

3. Chinese Ministry of Environmental Protection (MEP). Technical Guidelines for Environmental Site Investigation (HJ25.1-2014). 2014. Available online: http://kjs.mep.gov.cn/hjbhbz/bzwb/trhj/trjcgfffbz/201402/t20140226_268358.htm (accessed on 28 February 2016).

4. Chinese Ministry of Environmental Protection (MEP). Technical Guidelines for Risk Assessment of Contaminated Sites (HJ25.2-2014). 2014. Available online: http://kjs.mep.gov.cn/hjbhbz/bzwb/trhj/trjcgfffbz/201402/t20140226_268358.htm (accessed on 28 February 2016).

5. Chinese Ministry of Environmental Protection (MEP). Technical Guidelines for Site Soil Remediation (HJ25.4-2014). 2014. Available online: http://kjs.mep.gov.cn/hjbhbz/bzwb/trhj/trjcgfffbz/201402/t20140226_268360.htm (accessed on 28 February 2016).

6. Ciba Specialty Chemicals (CSC). Optical Brighteners: Fluorescent Whitening Agents for Plastics, Paints, Imaging and Fibers. Available online: www.cibasc.com (accessed on 28 February 2016).

7. U.S. Geological Survey (USGS). National Field Manual for the Collection of Water-Quality Data. 2015. Available online: http://water.usgs.gov/owq/FieldManual/ (accessed on 28 February 2016).

8. Beijing Municipal Administration of Quality and Technology Supervision (BMAQTS). Screening Levels for Soil Environmental Risk Assessment of Sites (DB11/T 811-2011), Beijing, 2011. Available online: http://www.doc88.com/p-1738086361850.html (accessed on 28 February 2016).

9. Dutch Ministry of Housing, Spatial Planning and the Environment (VROM). Dutch Standards (Version 2009). 2009. Available online: http://www.esdat.net/Environmental%20Standards/Dutch/annexS_I2000Dutch%20Environmental%20Standards.pdf (accessed on 28 February 2016).

10. U.S. Environmental Protection Agency (USEPA). USEPA Region 9 Preliminary Remediation Goals (USEPA PRGs, Version Nov. 2012). 2012. Available online: http://www.esdat.net/Environmental_Standards.aspx (accessed on 28 February 2016).

11. The Administration of Technical Supervision (ATS). Chinese Quality Standard for Groundwater (GB/T 14848-9). Chinese Standard Press: Beijing, 1993. Available online: http://down.foodmate.net/standard/sort/3/6710.html (accessed on 28 February 2016).

12. Somaratne, N.; Zulfic, H.; Ashman, G; Vial, H.; Swaffer, B.; Frizenschaf, J. Groundwater risk assessment model (GRAM): Groundwater risk assessment models for wellfield protection. *Water* **2013**, *5*, 1419–1439. [CrossRef]

13. Suter, G.W. Developing conceptual models for complex ecological risk assessments. *Hum. Ecol. Risk Assess.* **1999**, *5*, 375–396.

14. Hu, X.; Zhang, Y.; Luo, J.; Wang, T.; Lian, H.; Ding, Z. Bioaccessibility and health risk of arsenic, mercury and other metals in urban street dusts from a mega-city, Nanjing, China. *Environ. Pollut.* **2011**, *159*, 1215–1221. [CrossRef] [PubMed]

15. Bakhshyash, B.E.; Delkash, M.; Scholz, M. Response of vegetables to cadmium-enriched soil. *Water* **2014**, *6*, 1246–1256. [CrossRef]

16. Congress of the United States. *The Comprehensive Environmental Response, Compensation and Liability Act (CERCLA)*; Public Law 96-510, United States Code 42, 9601 et seq.; Congress of the United States: Washington, DC, USA, 1980.

17. Forslund, J.; Samakovlis, E.; Johansson, M.V.; Barregard, L. Does remediation save lives? On the cost of cleaning up arsenic-contaminated sites in Sweden. *Sci. Total Environ.* **2010**, *408*, 3085–3091. [CrossRef] [PubMed]

water

MDPI

Review

The Zeolite-Anammox Treatment Process for Nitrogen Removal from Wastewater—A Review

Mark E. Grismer * and Robert S. Collison

Biological & Agricultural Engineering, University of California, Davis, CA 95616, USA; rscollison@ucdavis.edu
* Correspondence: megrismer@ucdavis.edu

Received: 15 October 2017; Accepted: 13 November 2017; Published: 20 November 2017

Abstract: Water quality in San Francisco Bay has been adversely affected by nitrogen loading from wastewater treatment plants (WWTPs) discharging around the periphery of the Bay. While there is documented use of zeolites and anammox bacteria in removing ammonia and possibly nitrate during wastewater treatment, there is little information available about the combined process. Though relatively large, zeolite beds have a finite ammonium adsorption potential and require periodic re-generation depending on the wastewater nitrogen loading. Use of anammox bacteria reactors for wastewater treatment have shown that ammonium (and to some degree, nitrate) can be successfully removed from the wastewater, but the reactors require careful attention to loading rates and internal redox conditions. Generally, their application has been limited to treatment of high-ammonia strength wastewater at relatively warm temperatures. Moreover, few studies are available describing commercial or full-scale application of these reactors. We briefly review the literature considering use of zeolites or anammox bacteria in wastewater treatment to set the stage for description of an integrated zeolite-anammox process used to remove both ammonium and nitrate without substrate regeneration from mainstream WWTP effluent or anaerobic digester filtrate at ambient temperatures.

Keywords: anammox bacteria; wastewater treatment; nitrification; denitrification; zeolite

1. Introduction

As with many estuaries associated with population centers around the world, San Francisco Bay (SFB) water quality is adversely affected by nitrogen and phosphorous inputs from multiple anthropogenic sources, the greatest being nitrogen loads from wastewater treatment plant (WWTP) discharges on the Bay periphery. Nitrogenous waste (consisting primarily of ammonia and/or nitrate) is of particular concern in SFB, especially in the more shallow reaches subject to tidal flooding/draining processes. Ammonia is directly toxic to fish and marine life, while nitrate stimulates algal growth that depletes dissolved oxygen (DO) levels at night resulting in suffocation of oxygen-breathing organisms. While SFB has shown some resistance to the classic symptoms of nutrient over-enrichment, recent observations suggest that SFB's resistance to nutrient enrichment is weakening. It appears that SFB may be trending toward, or already experiencing, adverse impacts due to high nutrient loads, thereby requiring greater regulation of WWTP nitrogen loading to the Bay [1]. Thus, discharge permitting at WWTPs may require greater removal of both reduced and oxidized nitrogen species. This review considers the development of zeolite and anammox domestic wastewater treatment methods during the past two decades to set the stage for possible commercial development of the integrated zeolite-anammox treatment process capable of transforming WWTP effluent nitrogen loads to nitrogen gas prior to effluent disposal.

"Traditional" nitrogen removal in WWTPs relies on a two-step treatment process of nitrification and denitrification. The nitrification process employs nitrifying bacteria to oxidize ammonia to nitrate using available dissolved oxygen, while denitrification uses denitrifying bacteria to reduce the nitrate to nitrogen gas. Nitrification occurs only under aerobic conditions at dissolved oxygen (DO) concentrations of >1.0 mg/L where *Nitrosomonas*-type bacteria convert ammonium to nitrite; then *Nitrobacter*-type bacteria convert nitrite to nitrate. Nitrification is sensitive to inhibition by high organic concentrations because of bacterial competition and is typically represented by the equation;

$$NH_4^+ + 2.5O_2 => NO_3^- + 2H_2O, \tag{1}$$

Denitrification is an anaerobic process occurring at DO levels < 0.5 mg/L where facultative heterotrophic bacteria reduce nitrate to nitrogen gas that volatilizes to the atmosphere. It requires a carbon source as an electron donor, uses nitrate as an electron acceptor and is represented by the simplified equation;

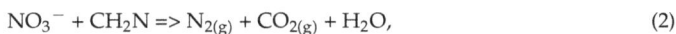

$$NO_3^- + CH_2N => N_{2(g)} + CO_{2(g)} + H_2O, \tag{2}$$

During the past two decades, new approaches to nitrogen treatment methods have developed in the laboratory and some tested in pilot-scale treatment plants; two of the more promising methods include use of zeolite aggregates and anammox bacteria. Zeolites are a relatively commonly found deposit around the world whose aggregates have relatively low density, some internal porosity and unusually large cation-exchange capacity (CEC) for the type of mineral. Some research has explored use of the zeolite aggregates as an ammonium adsorption substrate. Anammox bacteria were discovered in WWTP anaerobic digesters and in several marine environments. They were key towards closing nitrogen balance estimates in WWTP and estuary-marine studies and found to readily convert ammonia ions using nitrite to nitrogen gas. Anammox bacteria prefer anaerobic environments and are relatively slow growing; some ten times slower than nitrifiers for example. Presumably, anammox bacteria congregate at aerobic-anaerobic interfaces where they can combine available nitrite and ammonia to form nitrogen gas with some residual nitrate following the reaction [2]:

$$NH_4^+ + 1.32NO_2^- + 0.066HCO_3^- + 0.13H^+ => 1.02N_{2(g)} + 0.26NO_3^- + 2.03H_2O + 0.066CH_2O_{0.5}N_{0.15}, \tag{3}$$

As anammox bacteria are capable of direct conversion of oxidized and reduced forms of nitrogen in WWTP discharge to nitrogen gas with little sludge production, they provide an interesting opportunity to reduce WWTP nitrogen loads to sensitive receiving waters; however, there are only limited reports of commercial application of this integrated process.

2. Literature Review

This literature review considers the wastewater treatment aspects associated with use of zeolite aggregate as a reactor substrate and cultivation of anammox bacteria for transformation of dissolved aqueous nitrogen species (i.e., nitrate, nitrite and ammonia) found in WWTP discharge to nitrogen gas thereby reducing nitrogen loading to receiving waters. We direct this review towards increasing the development and evaluation of zeolite-anammox treatment systems for commercial-scale applications to improve receiving water quality wherever adversely impacted by WWTP discharges.

2.1. Zeolites and Wastewater Treatment

In the late 1950's, enormous beds of zeolite-rich sediments, formed by the alteration of volcanic ash in lake and marine waters, were discovered in the western United States and elsewhere around the world, notably in Australia, Canada, China, South America and Turkey. Zeolites are characterized by extensive internal porosity, very large surface areas (i.e., both internal and external), and correspondingly high CECs. Zeolites are classified as inclusion compounds of hydrated aluminosilicates having three-dimensional tetrahedral networks of SiO_4 and AlO_4, linked by the shared oxygen atoms. Partial substitution of Al^{3+} for Si^{4+} results in excess negative charge offset by alkali and earth alkaline cations. These cations, along with the water molecules, are located in cavities and channels inside the aluminosilicate macro-anion framework enabling zeolites to function as effective natural ion exchangers. During the past 20 years, there has been a substantial amount of research and application of natural zeolites in environmental remediation schemes that capitalize on their ready availability and ion-exchange properties [3,4].

Several proposed wastewater treatment methods exploit the ammonium adsorption abilities of zeolites across a range of scales, from commercial WWTPs to development of patents for modified septic systems using zeolites [5,6] reviewed studies of natural zeolites from around the world and found varying ion-exchange capacities for ammonium, some anions and organics, and heavy metal ions. Of the 21 zeolites considered, 18 were clinoptilolites with SiO_2 and Al_2O_3 fractions that ranged from 56–71% and 7.5–15.8%, respectively, while CECs ranged from 0.6 to 2.3 meq/mg. Similarly, at temperatures ranging from 20 to 70 °C (when reported), the corresponding ammonium adsorption capacities of the different clinoptilolites ranged from 23 to 3 mg/g with higher values reported using Canadian forms while the USA-derived clinoptilolite value reported was 18.5 mg/g. Widiastuti et al. [7,8] studied use of Australian zeolite for greywater treatment, and, similar to that reported by others, found zeolite ammonium removal capacity increases with increasing initial ammonium concentration [9], presumably as a result of greater aqueous to adsorbed phase concentration gradients. It appears that the ammonium ions can migrate from the external surface to the internal micro-pores of the zeolite within a given contact time. Several studies indicated that the adsorption or ion-exchange process is quite rapid and can be modeled by typical Langmuir and Freundlich isotherms [10–13]. Solution pH affected ammonium removal efficiency by the zeolite as well because the nitrogen dissociation form (NH_3^+ or NH_4^+) depends on pH. For example, ammonium removal efficiency from a 50 mg/L NH_4 solution increased as pH increased from 2 to 5 peaking at about pH 5 and declining thereafter. Similarly, Jorgensen et al. [14] found that zeolite was more selective at pH 5. Conversely, Du et al. [12] reported that an optimal ammonium removal efficiency was achieved at pH 6 while Ji et al. [15] using Ca^{2+}-formed clinoptilolite found a maximum adsorption capacity of 82% at pH 7 and Saltali et al. [16] reported 75% ammonium removal at pH 7 and nearly 79% at pH 8 for Turkish (Yildizeli) zeolite. Together with Karadag et al. [17], Ji et al. [15] and Saltali et al. [16] found the adsorption process to be exothermic and removal efficiency improved with decreasing temperatures. Studies have also considered the influence of other ions or compounds in solution on ammonium uptake by zeolites. Jorgensen and Weatherley [14] found that in most cases studied, the presence of organic compounds enhanced ammonium ion uptake. Similarly, considering adsorption from aqueous solutions having ammonium concentrations of 0–200 mg/L in the presence of Ca, K, Mg and Cl ions, Weatherley and Miladinovic [18] found only minor changes on ammonium uptake by mordenite and clinoptilolite. This was a rather unexpected result since most other work to date had shown clinoptilolite exhibiting a greater affinity for potassium as compared to the ammonium ion. Calcium ions in solution had the greatest effect upon ammonium ion uptake, followed by potassium ions while magnesium ions had the least effect. Most studies considering zeolite ion-exchange properties were conducted using laboratory-scale reactors with controlled environments, though some work has involved larger-scale applications in wastewater treatment.

Misaelides [4] noted in a short review that in addition to the ion-exchange properties of zeolites, zeolite aggregates demonstrated the ability to harbor bacteria that can increase sludge activity in WWTPs. The review by Hedstrom [19] acknowledged the ion-exchange capability of zeolites with respect to wastewater treatment but noted that biological or chemical regeneration methods would be required. The apparent drawback of this use was the slow formation of the bacteria layer on the zeolite surface, which does not become immediately effective, requiring bacterial growth establishment times of 1–2 weeks in the digesters. The modification of zeolites by cation-active polyelectrolytes accelerated the interaction among the bacteria with the zeolite surface further increasing the sludge activity. By 2011, zeolite was recognized for its high CEC and for its ability to preferentially remove ammonium ions from wastewater. Use of zeolite for ammonium removal increased because of its wide availability and low-costs where available, and because ammonium-saturated zeolite can be relatively easily regenerated and re-used. High-strength brine was traditionally the preferred method of regeneration [15], but concerns about high levels of dissolved solids in the spent regenerant liquor led to development of other methods. An electrochemical method of regeneration was also established and used in several applications [20]. One of the more promising methods explored more recently, however, is biological regeneration using microbial action to strip the ammonium from the cation exchange sites.

There are few commercial scale applications of zeolite adsorption reactors to remove ammonium from wastewater. Facing strict regulations associated with treated wastewater disposal to a pristine river, the Truckee Sanitation District deployed a zeolite reactor to remove residual ammonium prior to discharge. Using a relatively short contact time of several hours, the zeolite reactor successfully removed the ammonium from the treated wastewater. However, the zeolite reactor required near daily regeneration using saline water that eventually was disposed with the treated wastewater. Unfortunately, the regenerant addition to the discharge stream increased the salinity beyond acceptable disposal levels to the river and the reactor was decommissioned.

Early discovery of biological regeneration of zeolite by nitrifying bacteria by researchers in Israel [21] suggested a two-stage process where brine removed ammonium from zeolite, followed by brine regeneration using nitrifying bacteria. Later processes exploited the ability of these bacteria to strip the ammonium from the zeolite, thereby simplifying the process [22]. In Norway, "zeolite containing expanded clay aggregate filter media" was used to remove ammonia from domestic wastewater by a combination of nitrification and ion exchange. No chemical regeneration was necessary in addition to the biological regeneration during the four-month experimental period [23]. Zeolites used for stripping ammonium in reactors are typically sand-sized aggregates combining a relatively large exterior surface area with ease of handling. The bacteria presumably could not strip ammonium from exchange sites within the zeolite aggregates since their cells are approximately 1000 times larger than the pores formed by the zeolite lattice structure. Nitrifying biofilm-enhanced zeolite also appears to provide a dampening effect on shocks to digesters associated with peak or variable loads [19,24]. Such early studies considering nitrifying bacteria combined with older knowledge about anammox bacteria found in marine environments led to the possibility of combining these processes with zeolites to enhance nitrogen removal rates from domestic wastewater.

2.2. Anammox and Wastewater Treatment

As nitrogen removal processes and models were refined, WWTP operators and marine environment researchers became aware that nitrogen mass-balance "errors" indicated an unexplained nitrogen loss. Though existence of microorganisms capable of anaerobic ammonium oxidation using nitrite or nitrate as the electron acceptor was predicted in the 1970s [25], they were not discovered until around 1992 in a WWTP in Delft, The Netherlands [26–28], when they were named "anaerobic ammonium oxidation" or "anammox" bacteria. At the same time, the importance of anammox bacteria towards nitrogen cycling in the marine environment was well understood and researchers explored isolation of these bacteria from freshwater and marine environments for other applications. However, it was difficult to isolate this process in the laboratory until Mulder et al. [29] developed laboratory denitrifying fluidized-bed reactors capable of removing nitrogen under anaerobic conditions. As anaerobic autotrophs, it remains difficult to isolate and raise pure cultures of anammox bacteria in the laboratory; DNA-sequencing of the bacteria is largely limited to university and research institute laboratories. However, study of highly enriched cultures obtained from WWTP anaerobic digesters has enabled some understanding of the bacterial cell biology and biochemistry [28]. By 2005, the three genera of anammox bacteria described were quite small (<1 μm) and all shared a similar cellular structure that includes a membrane-bound compartment, known as the anammoxosome, where the anammox process is believed to occur. This membrane is composed of ladderane lipids in part that form a tight proton diffusion barrier, thereby enhancing ATP production within the cell. By 2010, Bae et al. [30] using PCR (polymerase chain reaction) methods identified six anammox genera in activated sludges taken from WWTPs; three freshwater, two marine environment and one mixed species are also generally acknowledged. With discovery of more species and habitats, we anticipate that more versatile species will be identified, but their overall diversity remains relatively unknown [31]. Though surprisingly widespread, anammox bacteria discovered within each ecosystem appear to be dominated by a single anammox genus, indicating specialization for distinct ecological niches [32]. Some have speculated that up to 50% of atmospheric nitrogen is a result of widespread anammox activity [33].

Employment of anammox bacteria can revolutionize domestic wastewater treatment because of their ability to simplify removal of nitrogenous waste at significantly lower costs and with less sludge production than that of conventional WWTP nitrification-denitrification processes [34,35] among others [31] consider the anammox process "as one of the most sustainable alternatives to the conventional costly nitrification-denitrification biological nitrogen removal process" in wastewater treatment, particularly for high nitrogen low BOD wastewater streams. The autotrophic anammox process directly oxidizes ammonium to nitrogen gas utilizing nitrite as the electron acceptor without the need for an organic carbon source as required by heterotrophic denitrification processes [36]. Further, oxygen demand is reduced as the ammonium is only required to be nitrified to nitrite instead of nitrate. As a result, anammox bacterial biomass yield is very low, creating a small amount of excess sludge production and thus lower operational costs [37,38]. Overall, the anammox process can reduce oxygen and exogenous carbon source demand by 64% and 100%, respectively, while reducing sludge production by 80–90% as compared to conventional WWTP nitrogen removal processes [39]. At this point, there are numerous anammox pilot plants currently operating or under construction; however, anammox processes at these plants are limited to treatment of high-ammonium strength wastewater (500 to 3000 mg/L) and operated at relatively warm temperatures (30–40 °C), though marine anammox are known to function at much cooler temperatures (10–15 °C).

Relatively slow growth rates of anammox are seemingly linked to the environments from which they were obtained [28]. For example, anammox exhibit bacterial growth doubling times of about 9–12 days under optimal temperature conditions associated with their origin [40]; that is, about 37 °C for those cultures obtained from wastewater treatment plants while those from cooler anoxic marine environments prefer 12–15 °C. This slow growth rate has limited commercial applications using anammox bacteria at WWTPs [35]. Anammox bacterial growth can be very sensitive to WWTP operational conditions such as dissolved oxygen, temperature, pH and organic matter content thereby requiring considerable direct management or manipulation at the WWTP. While originally thought that nitrate was the oxidant for ammonium by anammox bacteria, nitrogen-isotope labeling experiments confirmed that the bacteria are using the nitrite form where presumably nitrate-reducing bacteria in the environment are converting the nitrate to nitrite prior anammox conversion to N_2 gas. As denitrifying bacteria have much greater growth rates as a competitive advantage over anammox bacteria, the presence of oxygen drastically inhibits the anammox process, though the inhibition process appears to be reversible and the anammox process resumes when anoxic conditions are restored. On the other hand, addition of reduced forms of manganese or iron, as an essential substrate for anammox bacteria, can facilitate growth of anammox bacteria [35], and such additions have been used for culturing anammox sludge [41].

Another important process in possible WWTP applications is linked to anammox ability for dissimilatory nitrate reduction to ammonium (DNRA). This is a microbially mediated pathway transforming nitrate to ammonium and traditionally thought to be involved with fermentation or sulfur oxidation [42] and is a critical process [43] in nitrogen cycling at coastal marine environments. Recently, at least one genus of anammox bacteria appears capable of DNRA, even in the presence of 10 mM ammonium [44,45]. It now appears that, through DNRA, anammox bacteria can also produce nitrogen gas from nitrate, even in the absence of a carbon source (organic or inorganic). Figure 1, taken from Giblin et al. [43], summarizes the key nitrogen transformation processes associated with DRNA as well as the likely associated enzymes.

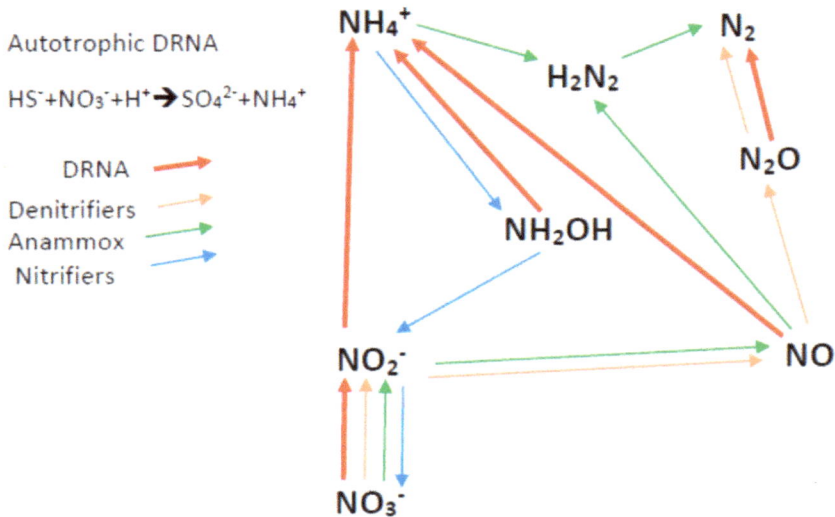

Figure 1. Nitrogen cycle pathways important to the dissimilatory nitrate reduction to ammonium (DNRA) process [43].

2.3. Wastewater Treatment Systems Using Anammox

Although anammox bacteria exist in the nitrification/denitrification "environment" of conventional WWTPs, they seem constrained to micro-sites and are of marginal importance; the slow-growing anammox bacteria are likely out-competed by the faster-growing organo-heterotrophs. The anammox process is primarily anaerobic, though in the absence of DRNA process, enough oxygen must be present to create the nitrite needed to react with NH_4-N to form N_2 gas. Originally considered to be inhibited by organic matter, some anammox species are less inhibited by carbon [46] and some of the most recently discovered species flourish when organic matter is present. Kindaichi [47] postulated that anammox was inhibited by COD; but probably a result of species, pH, temperature, type of carbon, and C:N ratio. Molinuevo's work [48] appeared to indicate that organic matter at high COD concentrations (100 to 250 mg COD/L) negatively affected the anammox process and facilitated heterotrophic denitrification, but at COD concentrations < 100 mg/L, anammox bacteria successfully converted ammonium to nitrogen gas suggesting that anammox removal of nitrogen of already treated wastewater having low COD is quite possible. Dong [49] considered anaerobic digestion of poultry manure and detected active anammox bacteria but determined they were unable to effectively compete with denitrifiers at high CODs (between 2200 and 5400 mg/L COD). Sensitivity to organic matter may be related to the C:N ratio, and wastewater with a BOD_5/N < 1.0 appears to be suitable for anammox treatment. Furukawa [50] successfully treated wastewater having concentrations of 600–800 mg/L BOD, 500–700 mg/L TN, 30–70 mg/L NH_4-N and 4000–4500 mg/L COD. Subsequently, anammox bacteria were found to be much more flexible and capable of competing for organic compounds and nitrate in the environment [51], and may be mixotrophic [51,52] reported that anammox bacteria could use organic acids as electron donors to reduce nitrate and nitrite, and then successfully compete with denitrifiers for use of these compounds. There are also examples of denitrifying bacteria and anammox bacteria existing in dynamic equilibrium to achieve simultaneous nitrogen and COD removal in anaerobic systems [53].

Other research has indicated that anammox bacteria usually find specialized niche environments, though their growth can be inhibited by compounds such as acetylene, phosphate, oxygen, methanol, sulfide at concentrations greater than 1 mM, and organic matter combined with high nitrite concentrations [41,48,52]. There is some research directed at overcoming the relatively slow growth rates of anammox that can delay the full treatment capability of larger-scale systems. Several studies [35,54–57] suggest utilizing external energy fields and/or addition of MnO_2 or ferrous iron to the wastewater stream treated to accelerate anammox growth, though such laboratory-scale augmentations have yet to be validated at the commercial scale. Practically, addition of manganese or iron to the wastewater treatment process, much less large electrical fields, may constitute a substantial cost to the WWTP, especially as uncertainty remains as to the required type of iron or manganese, their related concentration, and the duration supplemental metal additions are needed to maintain desired nitrogen removal.

Much of the anammox process understanding developed from various commercial applications designed to exploit the capability of anammox bacteria [58–61]. Many of these systems involve optimization of a two-step process in which the first reactor or system employs partial nitritation of the available ammonia to nitrite to achieve the 'optimal' 1.2:1 nitrite to ammonia ratio feedstock for the second anammox reactor step converting these to nitrogen gas. Lackner et al. [62] notes the rapid expansion of the partial nitration-anammox process to more than 100 WWTPs worldwide and outlines the operational and process control aspects and concerns described by surveys at 14 installations. The primary commercial systems include the CANON, DEMON and SHARON processes. The CANON process employs natural or engineered wetland systems treating wastewater with high ammonia and low BOD. Under excess ammonium conditions, the cooperation between aerobic (nitrosomonas-like) and anaerobic (planctomycetes) ammonium oxidizing bacteria leave no oxygen or nitrite for aerobic (nitrospira-like) nitrite oxidizing bacteria [27,63]. The DEMON process removes nitrogen from anaerobic co-digestion of urban and industrial sludge liquor using an anammox pathway with aerobic/anaerobic cycling inside a single bioreactor and the DEMON plant in The Netherlands has been operational since 2009. The SHARON process (Single reactor system for High activity Ammonium Removal Over Nitrite) has been developed specifically to treat liquor containing high ammonia concentrations [58]. This is a partial nitrification process where bacteria in the reactor oxidize ammonium to nitrite at temperatures of 30 to 40 °C. An anaerobic ammonium-oxidation process follows this where anammox use the nitrite to oxidize ammonia and produce nitrogen gas. Gonzalez-Martinez et al. [64,65] describe the success of the SHARON process and found a broad range of microbial species completing the nitrogen conversions. In general, such combined partial-nitration anammox reactors have operated successfully and Schmidt et al. [66] and Lackner outline their particular operational advantages or challenges. Overall, the interrelationships between N-removing microbial consortia including nitrifiers, denitrifiers, and anammox have also been documented [67] in wastewater treatment wetlands. Shipin et al. [67] described the role of *Nitrobacter* species in dissimilatory reduction of nitrate to nitrite, providing a major nitrite source for anammox. Clearly interest in applications of anammox bacteria to wastewater treatment continues to grow as Lackner et al. [62] underscored that the number of research publications related to anammox applications in wastewater treatment is also growing rapidly and now to a rate of ~10 articles/year since 2016.

2.4. Wastewater Treatment Using Combined Zeolite-Anammox Systems

Collison [68] reported on bench and pilot-scale linear-channel reactor (wetland flumes) studies investigating several aspects associated with the effects of constructed wetland (CW) substrate and wastewater characteristics on COD and nitrogen removal rates. Collison and Grismer [69] focused more specifically on the role of zeolites in nitrogen removal from these gravity-flow linear reactors. They found that, in the zeolite substrate system, the wastewater NH_4-N was nearly completely removed midway along the first reactor channel prior to an aeration tank leading to the second channel. In the other three aggregate substrate systems, only about a quarter of the NH_4-N was removed prior to an aeration tank with the remaining NH_3-N removed in the aeration tank. That is, the zeolite CW system appeared to remove 98% of the influent nitrogen without using the nitrification-denitrification process. Though zeolite ability to adsorb NH_4-N cations was undoubtedly occurring in the zeolite CW flume, based on the measured zeolite CEC, the calculated mass of NH_4-N ions that could be adsorbed was less than half that added to the system as influent. The failure of ammonium ions to saturate the zeolite adsorption sites indicated that other processes were occurring—most likely biological stripping of the NH_4-N from the aggregate surfaces by anammox bacteria. The ability of anammox to compete effectively in an anaerobic flume with significant organic matter content seemed contentious but promising in terms of developing an efficient long-term nitrogen removal system for domestic wastewater treatment.

As both anammox and nitrifiers bacteria are several orders of magnitude larger (1 to 5 μm) than zeolite pore sizes (0.7 to 1.0 nm), only NH_4 ions can travel to internal CEC sites within the zeolite suggesting that only the NH_4 ions on the aggregate surfaces are available for the bacterial processes. It is also probable that such related bacterial biofilms are very thin, possibly as rudimentary as individual bacteria adhering to the aggregate surface. Quite possibly, influent NH_4 ions can diffuse through the water to the zeolite surface where they were adsorbed at ion-exchange sites and/or ingested by the bacteria. This relatively rapid and efficient process thus only relies on diffusion through water, and neither diffusion through the biofilm or through the aggregate particle is required. Collison and Grismer [69] postulated that the unique performance of the zeolite CW systems in removing nitrogen was a function of the zeolite's ability to rapidly capture NH_4 ions, coupled with the anammox bacteria's ability to strip the NH_4 and regenerate the surface layer of the zeolite substrate. Environmental conditions for the anammox bacteria were further enhanced by the zeolite aggregate ability to soak up water and create an extensive aerobic/anaerobic interface (oxycline), thereby providing conditions where anammox has access to both the nitrite and ammonium ions needed to produce nitrogen gas. We found application of such an approach at the larger scale reported by Pei et al. [70] who created a riparian wetland system that employed a zeolite-anammox treatment process and identified that three primary anammox genera were present and operational when flowrates were such that anaerobic conditions prevailed in the zeolite substrate.

2.5. Commercial Upscaling of the Zeolite-Anammox Wastewater Treatment Process

While considerable laboratory-scale work related to use of zeolite or anammox to remove nitrogen species from various wastewaters has provided insight into the various treatment mechanisms associated with the ion-exchange and autotrophic anammox processes, there has been little work until recently considering the combined processes, especially at the commercial domestic WWTP scale [32]. Building on the proof-of-concept benchtop-scale zeolite-anammox treatment system described by Collison and Grismer [69,71] successfully upscaled this process to remove 25–75 mg/L ammonia-N in secondary WWTP effluent to final discharge ammonia and nitrate concentrations less than 1 and 3 mg/L, respectively. Secondary-treated effluent from east San Francisco Bay region WWTPs was pumped to trailers housing parallel linear-channel reactors assembled from channel sections about 3.7 m long by 0.7 m wide and 0.17 m deep. The channel sections were nearly filled with 20 mm zeolite aggregate and seeded at 3–4% by volume with either anaerobic digester effluent containing annamox bacteria or 'bio-zeolite' (zeolite aggregate having nitrifier/anammox bacteria biofilm) cultured in other reactors. Following a period of several weeks for complete colonization of the reactors, steady flows through the linear channels submerged the lower half of the zeolite substrate maintaining anaerobic conditions, while the upper half was passively aerated through capillary rise, or wicking action by the aggregate. During a roughly one-year period, they found that approximately 22 m of total reactor length was needed to reduce outlet ammonia concentrations to < 1 mg/L; moreover, that these gravity-flow systems required little maintenance and operated across a range of ambient temperatures (10–22 °C). Overall, at inflow rates from about 40 to 110 Lph, the linear-channel reactors removed 21 to 42 g NH_3-N/m^3/day on a bulk-reactor-volume basis (about 1.5 m^3) from the secondary treated wastewater with the greater value associated with the higher nitrogen loading rate. On a total nitrogen mass basis, this removal rate exceeded the zeolite adsorption capacity by more than an order-of-magnitude and could not have occurred by denitrification because there was insufficient carbon in the secondary effluent (i.e., very low BOD/COD) for this process. Determination of the linear channel degradation factors was critical towards development of constructed wetland designs for this tertiary treatment prior to discharge to sensitive waters on the Bay periphery.

In an effort to reduce the zeolite-anammox reactor 'footprint' or total volume and to explore the possibility of using this process to treat much greater ammonia strength wastewater, Collison and Grismer [72,73] investigated use of active aeration methods on nitrogen removal. This effort stemmed in part from needs of the San Francisco Bay area WWTPs and observations from controlled laboratory studies that anammox bacteria based reactors [74] were capable of roughly 1 kg NH_3-N/m^3/day removal when supplied optimal nitrite:ammonia concentration ratio wastewater. In these two studies, Collison and Grismer employed tank reactors using recirculating trickling-filter (RTF) and blown, or forced countercurrent airflow designs to remove ammonia from both secondary-treated effluent and high-strength anaerobic digester (AD) filtrate (~500 mg/L ammonia-N). Nitrogen removal from the AD filtrate can significantly reduce total nitrogen loading in the WWTP facilitating achievement of low effluent discharge targets, however, the AD filtrate treatment posed other problems associated with the very high and variable TSS loading. With the project goal of reducing WW ammonia concentrations to < 100 mg/L, Collison and Grismer [72] first deploy parallel 210 L barrel RTF reactors to assess the feasibility of AD filtrate treatment and investigate effects of aggregate size on ammonia removal. The reactors were operated such that the lower 2/3rds of the reactor depth remained submerged facilitating anammox bacterial growth and function, while the top 1/3rd of the reactor aggregate remained desaturated. The barrel reactors successfully removed about 400 mg/L ammonia from the AD filtrate resulting in discharge concentrations of roughly 70 and 90 NH_3-N mg/L and 100 and 120 NO_3-N mg/L, respectively, for the smaller (10 mm) and larger (20 mm) aggregates. Next, they upscaled the RTF reactor design to a ~68-m^3 (18,000 gal) intermediate-scale 'Baker tank' reactor for treatment of about 10% of the WWTP AD filtrate sidestream. When operated using the two-layer system for an 8-month period, the Baker tank reactor achieved an ~80% removal fraction with a nearly one-day retention time, successfully reducing the average inlet ammonia concentration from about 460 mg/L to about 85 NH_3-N mg/L and 90 NO_3-N mg/L, despite variable inlet ammonia concentrations ranging from 250 to 710 mg/L. Such a removal rate was equivalent to what Mansell [33] achieved with a two-stage partial-nitritation anammox laboratory reactor treating AD filtrate using a 220 day retention time. On a total reactor volume basis, the RTF tank design resulted in an ammonia degradation factor about an order-of-magnitude greater than that in the linear-channel reactors (i.e., 192 to 226 g NH_3-N/m^3/day for the barrel and Baker tank reactors, respectively). The large and highly variable TSS loading associated with the AD filtrate was problematic and contributed to aggregate pore clogging and some flow 'short-circuiting' during testing; not surprisingly, this effect was more apparent in the smaller-aggregate barrel reactors. Efforts to use settling tanks were of limited success and the authors proposed that backflush capabilities be included in the RTF tank reactor designs.

Eventual pore clogging and problems with the recirculation pump in the Baker tank reactor provided the opportunity to operate the tank as a largely anaerobic system for cultivation of biozeolite for other reactors and chance to explore nitrate scavenging potential of the anammox biofilms using DRNA processes. Decreased vertical flows through the top aerated media layer from pore clogging during this stage of the Baker tank reactor experiment, decreased aeration of the lower layer that in turn increased anammox bacterial growth and initially impaired ammonia oxidation in the submerged layer. As described above, had there been an adequate organic food supply, the lower anaerobic layer would have facilitated denitrifying bacterial growth, but the small reactor effluent BOD concentrations (<5 mg/L) indicated that nitrate removal by denitrification was insignificant in this layer. Rather, the absence of nitrate and excess ammonia promoted dissimilatory nitrate reduction to ammonium (DNRA) processes that converted the nitrate back to nitrite. Thus, the anammox bacteria removed about half of the inlet ammonia but practically all influent nitrate such that tank effluent nitrate-N concentrations were averaged ~0.1 mg/L.

Collison and Grismer [73] again explored active aeration methods in the zeolite-annamox process as above, but for treatment of secondary-treated WWTP effluent. Unfortunately, during most of the project period (~13 months), they failed to recognize that the secondary-treated effluent lacked sufficient ferrous iron necessary for anammox bacterial growth because the particular WWTP employed sludge incineration methods that precluded the need to add iron to AD processes to preserve WWTP plumbing infrastructure. As a result, for reactor inlet ammonia and nitrate concentrations of ~30 mg/L and 1 mg/L, reactor discharge ammonia and nitrate concentrations from the RTF and blown-air tank reactors remained disturbingly high at ~3 mg/L and ~25 mg/L, respectively, indicating poor anammox activity and treatment. In the final months of the project, additions of ferric and chelated iron to the secondary effluent had no effect on treatment, though in the very last month, addition of ferrous iron almost immediately resulted in increased anammox activity as reactor discharge nitrate concentrations fell below 4 mg/L. Ultimately, they identified that zeolite aggregate coated with 'black' biofilms was a good indicator that sufficient iron was present in the wastewater to encourage and maintain the anammox bacterial populations in the biofilms necessary for adequate wastewater treatment.

3. Summary and Conclusions

During the past two decades, new approaches to nitrogen treatment methods that include use of available zeolite aggregates as an adsorptive substrate and various strains of newly discovered anammox bacteria capable of converting ammonia to nitrogen gas. Zeolites are a relatively commonly found deposit around the world whose aggregates have relatively low density, internal porosity and unusually large cation-exchange capacity (CEC). Discovered in WWTP anaerobic digesters and in several marine environments, anammox bacteria were key towards closing nitrogen balance estimates in estuary-marine studies. These slow-growing bacteria prefer anaerobic environments and presumably congregate at aerobic-anaerobic interfaces where they can combine available nitrite and ammonia to form nitrogen gas with some residual nitrate, however, in the past few years they appear capable of direct conversion of ammonium to nitrogen gas via H_2N_2 production. As anammox bacteria appear capable of direct conversion of oxidized and reduced forms of nitrogen in WWTP discharge to nitrogen gas, they are an exciting opportunity to reduce WWTP nitrogen loads; however, only limited reports of commercial application zeolites and anammox in domestic wastewater treatment are available. Only recently have reports from Collison and Grismer that build on their previous lab work from 2010 become available describing applications of a zeolite-anammox treatment process in commercial WWTPs of the San Francisco Bay region of California.

Of course, additional laboratory and applied process work remains before the combined capabilities of zeolite substrates and anammox bacteria can be fully exploited at the full-scale domestic WWTP setting. As anammox bacteria are difficult to culture, currently there are no standardized techniques for sampling, preservation and transport of anammox bacterial biofilms from sediment, aggregates or reactor surfaces of practical benefit to facilitate identification of particular strains and DNA sequencing. Bacteria identification and DNA sequencing of what anammox samples are collected are largely limited to university or research institute labs as analytical costs at the very few commercial labs capable of these analyses are prohibitive in practice. No doubt, with such information, several more strains of anammox bacteria may be identified from diverse WWTP and marine environments that could be cultivated for wastewater treatment applications. Lacking such analyses, as a practical measure Collison and Grismer [73] suggest that presence of 'black' biofilms on the aggregate surfaces within WWTP reactors coupled with clear removal of both oxidized and reduced forms of nitrogen from the wastewater is a clear indication of adequate anammox bacteria activity. However, such observation provides little opportunity to identify which anammox strains are present and active.

At the WWTP scale, several operational parameters associated with successful removal of nitrogen species using the zeolite-anammox process remain ambiguous. These operational aspects requiring better definition include bio-zeolite seeding rates in reactors and associated effective start-up times, effective operating temperature ranges, optimal supplemental oxidation rates, and preferred Mn or Fe species supplementation to facilitate anammox growth rates, among others. At the most basic design level, simple gravity-flow zeolite-substrate channel reactors successfully removed nitrogen from secondary treated effluent with little energy or maintenance costs; however, it is not clear that such reactors would function as well at greater flow and nitrogen loading rates. Supplemental aeration through blown-air or recirculating trickling-filter designs appear capable of greater nitrogen removal rates for a particular reactor volume (i.e., greater ammonia degradation factors), but greater operational attention is required to maintain pumps and aerobic-anaerobic layers within the reactors. Nonetheless, preliminary upscaling results thus far are quite promising and additional applied research at the WWTP scale should better refine desirable operational parameters.

As compared to traditional nitrification-denitrification WWTP processes, the primary benefits two-stage partial-nitration anammox or single zeolite-anmmox reactors for wastewater treatment include possibly greater nitrogen removal and far smaller sludge production rates that reduce WWTP operating costs. As compared to the partial-nitration two-stage reactor systems, the single reactor zeolite-anammox systems successfully remove nitrogen across a greater temperature range and wastewater strength variability while also being easier to maintain and operate as they do not require continuous adjustments for wastewater characteristics. On the other hand, as a fixed media bed system, the zeolite-anammox reactors are subject to possible pore clogging and some attention must be directed at either pretreatment removal of recalcitrant solids, or improving back flushing capability within the reactor bed. Finally, from the perspective of WWTP greenhouse-gas generation, anammox bacterial conversions of nitrogen species either directly to nitrogen gas via DRNA processes, or through combination of ammonium and nitrite as outlined in the stoichiometric equations above, bypasses production of CO_2 gas occurring in the traditional nitrification-denitrification treatment process and represents a significant advantage over traditional WWTP processes. However, this aspect also needs further investigation that includes monitoring of the WWTP gases generated by each unit operation across the plant.

Author Contributions: Mark E. Grismer developed and drafted the literature review and Robert S. Collison provided the information from the more recent experiments with anammox-zeolite systems at the WWTP scale.

Conflicts of Interest: The authors declare no conflict of interest.

References

1. SFEI. SF Bay Nutrient Management Strategy Science Plan. March 2016. 68p. Available online: http://sfbaynutrients.sfei.org/sites/default/files/2016_NMSSciencePlan_Report_Sep2016.pdf (accessed on 16 November 2017).
2. Paredes, D.; Kuschk, P.; Stange, F.; Muller, R.A.; Koser, H. Model experiments on improving nitrogen removal in laboratory scale subsurface constructed wetlands by enhancing the anaerobic ammonia oxidation. *Water Sci. Technol.* **2007**, *56*, 145–150. [CrossRef] [PubMed]
3. Jorgensen, S.E.; Libor, O.; Graber, K.L.; Barkacs, K. Ammonia removal by use of clinoptilolite. *Water Res.* **1976**, *10*, 213–224. [CrossRef]
4. Misaelides, P. Application of natural zeolites in environmental remediation: A short review. *Microporous Mesoporous Mater.* **2011**, *144*, 15–18. [CrossRef]
5. Rose, J.A. Zeolite Bed Leach Septic System and Method for Wastewater Treatment. U.S. Patent No. 6,531,063 B1, 11 March 2003.
6. Wang, S.; Peng, Y. Natural zeolites as effective adsorbents in water and wastewater. *Chem. Eng. J.* **2010**, *156*, 11–24.
7. Widiastuti, N.; Wu, H.; Ang, H.M.; Zhang, D. Removal of ammonium from greywater using zeolite. *Desalination* **2011**, *277*, 15–23. [CrossRef]
8. Widiastuti, N.; Wu, H.; Ang, H.M.; Zhang, D. The potential application of natural zeolite for greywater treatment. *Desalination* **2008**, *218*, 271–280. [CrossRef]

9. Sarioglu, M. Removal of ammonium from municipal wastewater using Turkish (Dogantepe) zeolite. *Sep. Purif. Technol.* **2005**, *41*, 1–11. [CrossRef]
10. Rozï Icâ, M.; Cerjan-Stefanovicâ, S.; Kurajica, S.; Ina, V.V.; Icâ, E.H. Ammoniacal Nitrogen Removal from Water by Treatment with Clays and Zeolites. *Water Res.* **2000**, *34*, 3675–3681. [CrossRef]
11. Du, Q.; Liu, S.; Cao, Z.; Wang, Y. Ammonia removal from aqueous solution using natural Chinese clinoptilolite. *Sep. Purif. Technol.* **2005**, *44*, 229–234. [CrossRef]
12. Englert, A.H.; Rubio, J. Characterization and environmental application of a Chilean natural zeolite. *Int. J. Miner. Process.* **2005**, *75*, 21–29. [CrossRef]
13. Motsi, T.; Rowson, N.A.; Simmons, M.J.H. Adsorption of heavy metals from acid mine drainage by natural zeolite. *Int. J. Miner. Process.* **2009**, *92*, 42–48. [CrossRef]
14. Jorgensen, T.C.; Weatherley, L.R. Ammonia removal from wastewater by ion exchange in the presence of organic contaminants. *Water Res.* **2003**, *37*, 1723–1728. [CrossRef]
15. Ji, Z.-Y.; Yuan, J.-S.; Li, X.-G. Removal of ammonium from wastewater using calcium form clinoptilolite. *J. Hazard. Mater.* **2007**, *141*, 483–488. [CrossRef] [PubMed]
16. Saltali, K.; Sari, A.; Aydin, M. Removal of ammonium ion from aqueous solution by natural Turkish (Yıldızeli) zeolite for environmental quality. *J. Hazard. Mater.* **2007**, *141*, 258–263. [CrossRef] [PubMed]
17. Karadag, D.; Koc, Y.; Turan, M.; Armagan, B. Removal of ammonium ion from aqueous solution using natural Turkish clinoptilolite. *J. Hazard. Mater. B* **2006**, *136*, 604–609. [CrossRef] [PubMed]
18. Weatherley, L.R.; Miladinovic, N.D. Comparison of the ion exchange uptake of ammonium ion onto NewZealand clinoptilolite and mordenite. *Water Res.* **2004**, *38*, 4305–4312. [CrossRef] [PubMed]
19. Hedstrom, A. Ion exchange of ammonium in zeolites: A review. *ASCE J. Environ. Eng.* **2001**, *127*, 673–691. [CrossRef]
20. Lei, X.; Li, M.; Zhang, Z.; Feng, C.; Bai, W.; Sugiura, N. Electrochemical regeneration of zeolites and the removal of ammonia. *J. Hazard. Mater.* **2009**, *169*, 746–750. [CrossRef] [PubMed]
21. Güven, D.; Dapena, A.; Kartal, B.; Schmid, M.C.; Maas, B.; van de Pas-Schoonen, K.; Sozen, S.; Mendez, R.; den Camp, H.J.M.O.; Jetten, M.S.M.; et al. Propionate Oxidation by and Methanol Inhibition of Anaerobic Ammonium-Oxidizing Bacteria. *Appl. Environ. Microbiol.* **2005**, *71*, 1066–1071. [CrossRef] [PubMed]
22. Jung, J.-Y.; Chung, Y.-C.; Shin, H.-S.; Son, D.-H. Enhanced ammonia nitrogen removal using consistent biological regeneration and ammonium exchange of zeolite in modified SBR process. *Water Res.* **2004**, *38*, 347–354. [CrossRef] [PubMed]
23. Gisvold, B.; Odegaard, H.; Follesdal, M. Enhancing the removal of ammonia in nitrifying biofilters by the use of a zeolite containing expanded clay aggregate filtermedia. *Water Sci. Technol.* **2000**, *41*, 107–114.
24. Inan, H.; Beler, B.B. Clinoptilolite: A possible support material for nitrifying biofilms for effective control of ammonium effluent quality? *Water Sci. Technol.* **2005**, *51*, 63–70. [PubMed]
25. Jetten, M.S.M.; van Niftrik, L.; Strous, M.; Kartal, B.; Keltjens, J.T.; den Camp, H.J.M.O. Biochemistry and molecular biology of anammox bacteria. *Biochem. Mol. Biol.* **2009**, *44*, 65–84. [CrossRef] [PubMed]
26. Jetten, M.S.M.; Strous, M.; van de Pas-Schoonen, K.T.; Schalk, J.; van Dongen, U.G.J.M.; van de Graaf, A.A.; Logemann, S.; Muyzer, G.; van Loosdrecht, M.C.M.; Kuenen, J.G. The anaerobic oxidation of ammonium. *FEMS Microbiol. Rev.* **1999**, *22*, 421–437. [CrossRef]
27. Sliekers, A.O.; Derwort, N.; Gomez, J.L.C.; Strous, M.; Kuenen, J.G.; Jetten, M.S. Completely autotrophic nitrogen removal over nitrite in one single reactor. *Water Res.* **2002**, *36*, 2475–2482. [CrossRef]
28. Dalsgaard, T.; Thamdrup, B.; Canfield, D.E. Anaerobic ammonium oxidation (anammox) I the marine environment. *Res. Microbiol.* **2005**, *156*, 457–464. [CrossRef] [PubMed]
29. Mulder, A.; van de Graaf, A.A.; Robertson, L.A.; Kuenen, J.G. Anaerobic ammonium oxidation discovered in a denitrifying fluidized bed reactor. *FEMS Microbiol. Ecol.* **1995**, *16*, 177–183. [CrossRef]
30. Bae, H.; Park, K.-S.; Chung, Y.-C.; Jung, J.-Y. Distribution of anammox bacteria in domestic WWTPs and their enrichments evaluated by real-time quantitative PCR. *Process Biochem.* **2010**, *45*, 323–334. [CrossRef]
31. Jetten, M.S.M.; Cirpus, I.; Kartal, B.; van Niftrik, L.; van de Pas-Schoonen, K.T.; Sliekers, O.; Haaijer, S.; van der Star, W.; Schmid, M.; van de Vossenberg, J.; et al. 1994–2004: 10 years of research on the anaerobic oxidation of ammonium. *Biochem. Soc. Trans.* **2005**, *33*, 119–123. [CrossRef] [PubMed]
32. Kassab, G.; Halalsheh, M.; Klapwijk, A.; Fayyad, M.; van Lier, J.B. Sequential anaerobic–aerobic treatment for domestic wastewater—A review. *Bioresour. Technol.* **2010**, *101*, 3299–3310. [CrossRef] [PubMed]

33. Mansell, B.L. Side-Stream Treatment of Anaerobic Digester Filtrate by Anaerobic Ammonia Oxidation. Master's Thesis, Civil & Environmental Engineering at the University of Utah, Salt Lake City, UT, USA, May 2011.

34. Zhang, L.; Zheng, P.; Tang, C.-J.; Jin, R.-C. Anaerobic ammonium oxidation for treatment of ammonium-rich wastewaters. *J. Zhejiang Univ. Sci. B* **2008**, *9*, 416–426. [CrossRef] [PubMed]

35. Liu, Y.; Ni, B.-J. Appropriate Fe (II) Addition Significantly Enhances Anaerobic Ammonium Oxidation (Anammox) Activity through Improving the Bacterial Growth Rate. *Science Rep.* **2015**, *5*, 8204–8208. [CrossRef] [PubMed]

36. Hao, X.; van Loosdrecht, M. Model-based evaluation of COD influence on a partial nitrification-Anammox biofilm (CANON) process. *Water Sci. Technol.* **2004**, *49*, 83–90. [PubMed]

37. Strous, M.; Van Gerven, E.; Zheng, P.; Kuenen, J.G.; Jetten, M.S.M. Ammonium removal from concentrated waste streams with the anaerobic ammonium oxidation (anammox) process in different reactor configurations. *Water Res.* **1997**, *31*, 1955–1962. [CrossRef]

38. Ni, B.-J.; Ruscalleda, M.; Smets, B.F. Evaluation on the microbial interactions of anaerobic ammonium oxidizers and heterotrophs in Anammox biofilm. *Water Res.* **2012**, *46*, 4645–4652. [CrossRef] [PubMed]

39. Bi, Z.; Qiao, S.; Zhou, J.; Tang, X.; Zhang, J. Fast start-up of Anammox process with appropriate ferrous iron concentration. *Bioresour. Technol.* **2014**, *170*, 506–512. [CrossRef] [PubMed]

40. Li, X.-R.; Du, B.; Fu, H.-X.; Wang, R.-F.; Shi, J.-H.; Wang, Y.; Jetten, M.S.M.; Quan, Z.-X. The bacterial diversity in an anaerobic ammonium-oxidizing (anammox) reactor community. *Syst. Appl. Microbiol.* **2009**, *32*, 278–289. [CrossRef] [PubMed]

41. Van de Graaf, A.A.; de Bruijn, P.; Robertson, L.A.; Jetten, M.S.M.; Kuenen, J.G. Autotrophic growth of anaerobic ammonium-oxidizing micro-organisms in a fluidized bed reactor. *Microbiology* **1996**, *142*, 2187–2196. [CrossRef]

42. Burgin, A.; Hamilton, S. Have we overemphasized the role of denitrification in aquatic ecosystems? A review of nitrate removal pathways. *Front. Ecol. Environ.* **2007**, *5*, 89–96. [CrossRef]

43. Giblin, A.E.; Tobias, C.R.; Song, B.; Weston, N.; Banta, G.T.; Rivera-Monroy, V.H. The importance of dissimilatory nitrate reduction to ammonium (DNRA) in the nitrogen cycle of coastal ecosystems. *Oceanography* **2013**, *26*, 124–131. [CrossRef]

44. Kartal, B.; Kuypers, M.M.M.; Lavik, G.; Schalk, J.; den Camp, H.J.M.O.; Jetten, M.S.M.; Strous, M. Anammox bacteria disguised as denitrifiers: Nitrate reduction to dinitrogen gas via nitrite and ammonium. *Environ. Microbiol.* **2007**, *9*, 635–642. [CrossRef] [PubMed]

45. Francis, C.; Beman, J.M.; Kuypers, M.M.M. New processes and players in the nitrogen cycle: The microbial ecology of anaerobic and archaeal ammonia oxidation. *ISME J.* **2007**, *1*, 19–27. [CrossRef] [PubMed]

46. Trimmer, M.; Nicholls, J.C.; Deflandre, B. Anaerobic ammonium oxidation measured in sediments along the Thames estuary, United Kingdom. *Appl. Environ. Microbiol.* **2003**, *69*, 6447–6454. [CrossRef] [PubMed]

47. Kindaichi, T.; Tsushima, I.; Ogasawara, Y.; Shimokawa, M.; Ozaki, N.; Satoh, H.; Okabe, S. In situ activity and spatial organization of anaerobic ammonium-oxidizing (anammox) bacteria in biofilms. *Appl. Environ. Microbiol.* **2007**, *73*, 4931–4939. [CrossRef] [PubMed]

48. Molinuevo, B.; Cruz-Garcia, M.; Karakashev, D.; Angelidaki, I. Anammox for ammonia removal from pig manure effluents: Effect of organic matter content on process performance. *Bioresour. Technol.* **2009**, *100*, 2171–2175. [CrossRef] [PubMed]

49. Dong, Z.; Sun, T. A potential new process for improving nitrogen removal in constructed wetlands—Promoting coexistence of partial-nitrification and ANAMMOX. *Ecol. Eng.* **2007**, *31*, 69–78. [CrossRef]

50. Furukawa, K.; Inatomia, Y.; Qiaoa, S.; Quana, L.; Yamamotoa, T.; Isakab, K.; Suminob, T. Innovative treatment system for digester liquor using anammox process. *Bioresour. Technol.* **2009**, *100*, 5437–5443. [CrossRef] [PubMed]

51. Kartal, B.; Rattray, J.; van Niftrik, L.A.; van de Vossenberg, J.; Schmid, M.C.; Webb, R.I.; Schouten, S.; Fuerst, J.A.; Damste, J.S.; Jetten, M.S.M.; et al. *Candidatus* "Anammoxoglobus propionicus" a new propionate oxidizing species of anaerobic ammonium oxidizing bacteria. *Syst. Appl. Microbiol.* **2007**, *30*, 39–49. [CrossRef] [PubMed]

52. Green, M.; Mels, A.; Lahav, O. Biological-ion exchange process for ammonium removal from secondary effluent. *Water Sci. Technol.* **1996**, *34*, 449–458.

53. Chen, H.; Liu, S.; Yang, F.; Xue, Y.; Wang, T. The development of simultaneous partial nitrification, anammox and denitrification (SNAD) process in a single reactor for nitrogen removal. *Bioresour. Technol.* **2009**, *100*, 1548–1554. [CrossRef] [PubMed]

54. Qiao, S.; Bi, Z.; Zhou, J.; Cheng, Y.; Zhang, J. Long term effects of divalent ferrous ion on the activity of anammox biomass. *Bioresour. Technol.* **2013**, *142*, 490–497. [CrossRef] [PubMed]

55. Qiao, S.; Bi, Z.; Zhou, J.; Cheng, Y.; Zhang, J.; Bhatti, Z. Long term effect of MnO_2 powder addition on nitrogen removal by anammox process. *Bioresour. Technol.* **2012**, *124*, 520–525. [CrossRef] [PubMed]

56. Waki, M.; Yasuda, T.; Fukumoto, Y.; Kuroda, K.; Suzuki, K. Effect of electron donors on anammox coupling with nitrate reduction for removing nitrogen from nitrate and ammonium. *Bioresour. Technol.* **2013**, *130*, 592–598. [CrossRef] [PubMed]

57. Zhang, J.X.; Zhang, Y.B.; Li, Y.; Zhang, L.; Qiao, S.; Yang, F.L.; Quan, X. Enhancement of nitrogen removal in a novel anammox reactor packed with Fe electrode. *Bioresour. Technol.* **2012**, *114*, 102–108. [CrossRef] [PubMed]

58. Van Dongen, U.; Jetten, M.S.M.; Van Loosdrecht, M.C.M. The SHARON-Anammox process for treatment of ammonium rich wastewater. *Water Sci. Technol.* **2001**, *44*, 153–160. [PubMed]

59. Van Loosdrecht, M.C.M.; Hao, X.; Jetten, M.S.M.; Abma, W. Use of anammox in urban wastewater treatment. *Water Sci. Technol. Water Supply* **2004**, *4*, 87–94.

60. Zekker, I.; Rikmann, E.; Tenno, T.; Saluste, A.; Tomingas, M.; Menert, A.; Loorits, L.; Lemmiksoo, V.; Tenno, T. Achieving nitritation and anammox enrichment in a single moving-bed biofilm reactor treating reject water. *Environ. Technol.* **2012**, *33*, 703–710. [CrossRef] [PubMed]

61. Rodriguez-Sanchez, A.; Gonzalez-Martinez, A.; Martinez-Toledo, M.V.; Garcia-Ruiz, M.J.; Osorio, F.; Gonzalez-Lopez, J. The effect of influent characteristics and operational conditions over the performance and microbial community structure of partial nitration reactors. *Water* **2014**, *6*, 1905–1924. [CrossRef]

62. Lackner, S.; Gilbert, E.M.; Vlaeminck, S.E.; Joss, A.; Horn, H.; van Loosdrecht, M.C.M. Full-scale partial nitritation/anammox experiences—An application survey. *Water Res.* **2014**, *55*, 292–303. [CrossRef] [PubMed]

63. Third, K.; Sliekers, A.O.; Kuenen, J.G.; Jetten, M.S.M. The CANON System (completely autotrophic nitrogen-removal over nitrite) under ammonium limitation: Interaction and competition between three groups of bacteria. *Syst. Appl. Microbiol.* **2001**, *24*, 588–596. [CrossRef] [PubMed]

64. Gonzalez-Martinez, A.; Calderon, K.; Albuquerque, A.; Hontorio, E.; Gonzalez-Lopez, J.; Guisado, I.M.; Osorio, F. Biological and technical study of a partial-SHARON reactor at laboratory scale: Effect of hydraulic retention time. *Bioprocess Biosyst. Eng.* **2013**, *36*, 173–184. [CrossRef] [PubMed]

65. Gonzalez-Martinez, A.; Rodriguez-Sanchez, A.; Munoz-Palazon, B.; Garcia-Ruiz, M.J.; Osorio, F.; van Loosdrecht, M.C.M.; Gonzalez-Lopez, J. Microbial community analysis of a full-scale DEMON bioreactor. *Bioprocess Biosyst. Eng.* **2014**. [CrossRef] [PubMed]

66. Schmidt, I.; Sliekers, O.; Schmid, M.; Bock, E.; Fuerst, J.; Kuenen, J.G.; Jetten, M.S.M.; Strous, M. New concepts of microbial treatment processes for the nitrogen removal in wastewater. *FEMS Microbiol. Rev.* **2003**, *27*, 481–492. [CrossRef]

67. Shipin, O.; Koottatep, T.; Khanh, N.T.T.; Polprasert, C. Integrated natural treatment systems for developing communities: Low-tech N-removal through the fluctuating microbial pathways. *Water Sci. Technol.* **2005**, *51*, 299–306. [PubMed]

68. Collison, R.S. Effects of Porous Media and Plants in the Performance of Subsurface Flow Treatment Wetlands. Ph.D. Dissertation, Biological Systems Engineering, Davis, CA, USA, March 2010.

69. Collison, R.S.; Grismer, M.E. Nitrogen and COD Removal from Septic Tank Wastewater in Subsurface Flow Constructed Wetlands: 3. Substrate (CEC) Effects. *Water Environ. Res.* **2014**, *86*, 314–323. [CrossRef] [PubMed]

70. Pei, Y.; Wang, J.; Wang, Z.; Tian, B. Anammox bacteria community and nitrogen removal in a strip-like wetland in the riparian zone. *Water Sci. Technol.* **2013**, *67*, 968–975. [CrossRef] [PubMed]

71. Collison, R.S.; Grismer, M.E. Upscaling the Zeolite-Anammox process: 1. Nitrogen removal from secondary effluent enabling discharge to sensitive receiving waters. *Water Environ. Res.* **2017**, in press.

72. Collison, R.S.; Grismer, M.E. Upscaling the Zeolite-Anammox process: 2. Treatment of high-strength anaerobic digester filtrate. *Water Environ. Res.* **2017**, in press.

73. Collison, R.S.; Grismer, M.E. Upscaling the Zeolite-Anammox process: 3. Effects of Iron Deficiency and Aeration on nitrogen removal from Secondary Effluent. *Water Environ. Res.* **2017**, in press.

74. Kotay, S.M.; Mansell, B.L.; Hogsett, M.; Pei, H.; Goel, R. Anaerobic ammonia oxidation (ANAMMOX) for side-stream treatment of anaerobic digester filtrate process performance and microbiology. *Biotechnol. Bioeng.* **2013**, *110*, 1180–1192. [CrossRef] [PubMed]

water

MDPI

Article

Impact of Combined Sewer Overflow on Wastewater Treatment and Microbiological Quality of Rivers for Recreation

Franz Mascher, Wolfgang Mascher, Franz Pichler-Semmelrock, Franz F. Reinthaler, Gernot E. Zarfel and Clemens Kittinger *

Institute for Hygiene, Microbiology and Environmental Medicine, Medical University of Graz, 8020 Graz, Austria; franz.mascher@medunigraz.at (F.M.); wolfgang.mascher@medunigraz.at (W.M.); franz.pichler-semmelrock@stmk.gv.at (F.P.-S.); franz.reinthaler@medunigraz.at (F.F.R.); gernot.zarfel@medunigraz.at (G.E.Z.)
* Correspondence: clemens.kittinger@medunigraz.at; Tel.: +43-316-3857-3600

Received: 20 October 2017; Accepted: 16 November 2017; Published: 22 November 2017

Abstract: Within the framework of a one-year study the treatment capacity of a municipal wastewater treatment plant (WWTP) was evaluated, with regard to fecal indicator bacteria (FIB) and to their influence on the recipient. The logarithmic reduction rates for fecal coliforms (FC), *Escherichia coli* (EC) and intestinal enterococci (IE) were 2.84, 2.90 and 2.93. In the investigated period of time, the tested treatment plant released 4.3% of the total annual load flow volume as combined sewer overflow (CSO), that is, when the influent into the combined sewer exceeds the capacity of the treatment plant and coarsely cleaned wastewater arrives at the recipient. This CSO discharge increased the number of FIB significantly by 1.2×10^2 MPN/100 mL for EC, and by 1.8×10^1 MPN/100 mL for IE. For the Styrian part of the Mur River (1.6 million inhabitants), a calculation of FIB of all sewage treatment plants estimating the same ratio of CSO (4.3%) and a given mean flow rate (QM) results in a significant increase of the FIB load in the recipient: 3.8×10^3 MPN/100 mL for EC and 5.8×10^2 MPN/100 mL for IE. On the basis of these values the standards of water quality for recreational purposes cannot be met.

Keywords: combined sewer; wastewater treatment; microbiological quality; surface water; river

1. Introduction

Continual efforts in improving the water quality over the last decades and the national implementation of European Directives have been successful. The quality of the Mur River has improved and risen from IV to II according to the saprobic system [1,2], most effectively through the implementation of wastewater treatment plants in the catchment area. Today, 85% of the population within the catchment area are connected to central sewage treatment plants, and this was the main reason for the improvement of the river water quality. Nevertheless, the level of fecal indicator bacteria (FIB) could not be reduced to the quality levels required by the bathing water regulation [3]. A previous study carried out by Kittinger et al. investigated the Mur River concerning its burden of fecal bacteria-like fecal coliforms (FC), *Escherichia coli* (EC), intestinal enterococci (IE) and *Salmonella* spp. [4]. In this study the water quality was investigated monthly over a period of a year at 21 sampling sites on the Mur River. The microbiological data showed a massive burden of FIB, with an increasing load of bacteria in the flow direction and seasonal fluctuation. Interpreting these values on the basis of the European bathing water regulation [5], the Mur River is not suitable for recreation or water sports. Studies on other European rivers show comparable results [6–8]. Outbreak sof zoonosis among sporting events in rivers and the evidence of multiresistant bacteria underline the improper quality of river water for recreational purposes [9–11].

This discrepancy between the good status of the Mur River according to the saprobic system and the high values of FIB provided the impulse for the present study. The state of a specific technology wastewater treatment plant (WWTP) was investigated according to its purification capacity and the influence of the treated wastewater on the recipient. On the basis of validated, mathematical methods, the acquired data thus made it possible to quantify the treatment capacity of a municipal WWTP concerning FIB. Furthermore, the results of this study pinpointed the FIB load that is delivered by the influent of communal WWTPs to the recipient (Mur River).

2. Material and Methods

2.1. Description of the Wastewater Treatment Plant (WWTP)

The investigated WWTP is of the mechanical-biological type with anaerobic sludge stabilization, providing municipal and industrial wastewater services for 50,000 population equivalents at N 46°43′3″/E 15°37′31″. It uses a single stage with denitrification, biological and chemical phosphor elimination, activated sludge separation and finally postfiltration for micro-flocculation retention. The sewage system is a partly mixed but mostly separated draining system; detailed information is given in Table 1.

Table 1. Specific parameters of the investigated wastewater treatment plant (WWTP).

Parameters	Detail Information
Connected Inhabitants	50,000
Hydraulic load Q_{DW}	8000 m^3/day
Hydraulic load Q_{RW}	16,000 m^3/day
Max. Q_{DW}	140 L/s
Max.Q_{RW} (total input activation reservoir)	280 L/s
BOD$_5$ load	3000 kg/day
COD load	6000 kg/day
N$_{tot.}$ load	550 kg/day
P$_{tot.}$ load	75 kg/day

Notes: Max. Q_{DW}: Maximum dry weather influent; Max. Q_{RW}: Maximum rainy weather influent; BOD5 load: biological oxygen demand (5 days); COD load: chemical oxygen demand; N$_{tot.}$ load: total nitrogen; P$_{tot.}$ load: total phosphor.

Characteristic parts of the water treatment line:

- Coarse gravel, combined sewage basin (3000 m^3)
- Influent lifting: Archimedean screws (three aggregates with 140 L/s max. capacity each)
- Rake system: Two lines (3 mm gap size) 280 L/s each
- Sand washing plant, sewage take-up, fecal, foreign sludge/mud and grease take-up
- Sand and grease catchment; two lines
- Two primary clarifiers with 126 m^3 each, and a distribution building
- Aeration basins (two lines) for single-step activated sludge processing, with combined pre-installed and simultaneous denitrification, as well as biological phosphor elimination; $V_{tot.}$ = 5.100 m^3
- Two rectangular, secondary clarification basins with vertical flow-through: $V_{tot.}$ = 4.032 m^3 each
- Precipitation station: iron-aluminum combination
- Filter system: disc filter system for solid retention, three units of 60 m^2 each and with a max. load of 1000 m^3/h.

Characteristic parts of the sludge-line:

- Pre-thickener of the sewage sludge

- Sludge mixing container for transfer to the digestion tower (100 m^3)
- Digestion tower: V = 1500 m^3
- High-performance centrifuge for sludge dewatering, and a belt thickener for surplus sludge 30 m^3/h, sludge storage
- Sewer gas unit consisting of a digesting tower (anaerobic treatment), condensation dryer, active-carbon plant, measuring unit, gas desulphurization, gas tank, gas flare, heating boiler and a gas power station.

2.2. Sample Collection and Investigated Parameters

First, 250 mL of wastewater and 100 g of sludge were collected weekly for one year (October 2012–September 2013, N = 54). The samples were both taken from the untreated and treated wastewaters and from the sludge before and after anaerobic stabilization. Transport to the laboratory was cooled and microbiological parameters were analyzed the same day.

Fecal indicator bacteria (FIB) were determined out of 100 mL (after preparing appropriate dilutions) using the Colilert 18 System for fecal coliforms (FC) and *Escherichia coli* (EC) (ISO 9308-2, [12]), and the Enterolert 18 System for intestinal enterococci (IE) (ISO 7899-2, [13]) according to the manufacturer's instructions (IDEXX, Ludwigsburg, Germany). *Salmonella* were detected out of 100 mL wastewater and 10 g of sludge by an enrichment procedure according to (ISO 19250 [14]).

Statistical significance for all data, as well as median and standard deviation of the median, were calculated with GraphPadPrism™ 6.0 for Windows, GraphPad Software, San Diego, CA, USA, www.graphpad.com.

3. Results

3.1. Bacterial Load and Reduction Rate of the Wastewater

The annual median of the influent concentration of FC (5.79 × 10^6 MPN/100 mL) and EC (4.10 × 10^6 MPN/100 mL) showed the same order of magnitude. The median for IE was one log below (7.26 × 10^5 MPN/100 mL). Although the samples were taken weekly for one year with different meteorological conditions, the values were within a narrow range (0.61 MPN/100 mL to 0.82 MPN/100 mL log for the inter percentile range, and 0.26 MPN/100 mL to 0.38 MPN/100 mL log for inter quartile). The influent of FIB to the WWTP was constant over the investigation period (Figure 1, Table 2).

The median values at the discharge site were also quite similar, with 8.36 × 10^3 MPN/100 mL for FC and 5.21 × 10^3 MPN/100 mL for EC. The median for IE was one log below at 7.50 × 10^2 MPN/100 mL. The reduction rates for the investigated FIB were 2.84 log for FC, 2.90 log for EC and 2.93 log for IE (calculated on base of median values of influent and discharge). Although IE are present on a lower level in the influent, their reduction rate does not differ (significantly) from the reduction rates of FC and EC. Compared to the inflow, values for the discharge site are more disperse: 1.05 log MPN/100 mL to 1.39 log MPN/100 mL for percentiles, and 0.62 log MPN/100 mL to 0.75 log MPN/100 mL for the quartile (Figure 1, Table 2). This difference was probably caused by weather influence or plant-specific conditions.

Qualitative detection of *Salmonella* spp. in 100 mL wastewater led to over 50% of influent samples testing positive. After the biological treatment stage at the discharge side, 30% of the samples could test positive for *Salmonella* spp.

Table 2. Range (Logarithmic steps) between 10% and 90% Perzentile and Quartiles (25% and 75%) for fecal coliforms (FC), *Escherichia coli* (EC) and intestinal enterococci (IE) in the waste water untreated and waste water treated WWu and WWt, and linear and logarithmic reduction rates (RR) (MPN/100 mL).

Investigated Parametere	WWu (10–90/25–75)	WWt (10–90/25–75)	RR (log)
FC	0.82/0.38	1.05/0.62	2.84
EC	0.76/0.32	1.07/0.58	2.90
IE	061/0.26	1.39/0.75	2.93

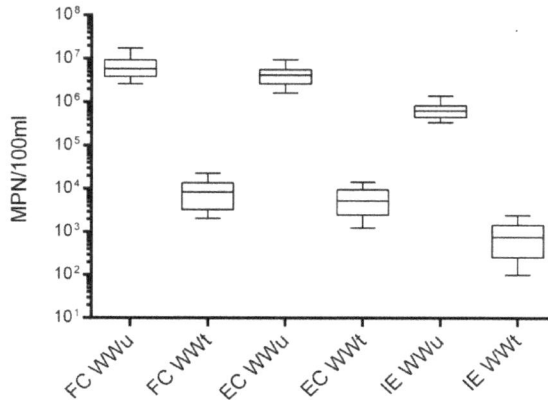

Figure 1. Number (MPN/100 mL) of FC, EC und IE in the untreated wastewater (WWu) and in the treated wastewater (WWt) (annual average, 90th percentile, $p = 0.0001$). Values are presented in a box (25th, 50th and 75th percentile) and whiskers (10th and 90th percentile) chart, (N = 54).

3.2. Bacterial Load and Reduction Rate of the Sewage Sludge

Over a one-year investigation period, the median values of FIB for raw sewage sludge (SSr) were 7.94×10^5 MPN/g for FC, 4.41×10^5 MPN/g for EC and 1.60×10^5 MPN/g for IE, all in a comparable magnitude.

Median values for stabilized sewage sludge (SSs) were 3.84×10^2 MPN/g for FC, 2.20×10^2 MPN/g for EC and 1.01×10^3 MPN/g for IE. The number of IE was high in the SSs, which was the reverse of the treated wastewater. The reduction rate was 3.22 log MPN/g for FC, and EC. IE were reduced only by 2.2 log MPN/g. The high reduction rates had also high standard deviations of up to two log orders. This may be due to changing conditions in the course of sludge stabilization (Figure 2 and Table 3).

SSr sludge was tested positive for *Salmonella* in 66% of all samples, and sludge stabilization reduced this ratio to 26%.

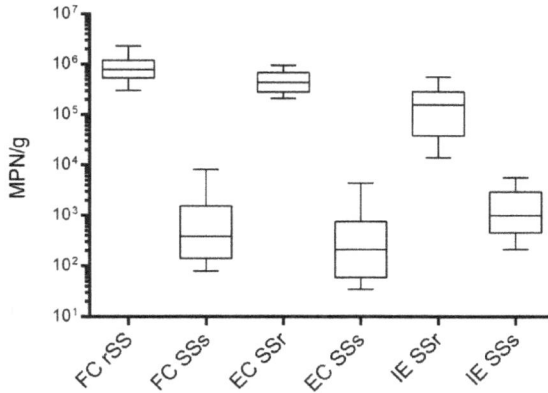

Figure 2. Concentrations of FC, EC and IE in the raw sewage sludge (SSr) and stabilized sludge (SSs), (N = 54).

Table 3. 10th and 90th percentile and quartile (25% and 75%) values for FC, EC and IE in the SSr and SSs, and reduction rates for FC, EC and IE (median, logarithmic vs. linear) in SSs in comparison to SSr (MPN/g).

Investigated Parameter	SSr (10–90/25–75)	SSs (10–90/25–75)	RR (log)
FC	0.88/0.35	2.01/1.04	3.32
EC	0.66/0.39	2.11/1.11	3.32
IE	1.61/0.88	1.43/0.82	2.20

3.3. Calculation of FIB Increase in the Recipient on the Basis of WWt and Combined Sewer Overlfow (CSO)

The results of this study form the basis for developing a formula for the calculation of FIB impact on rivers. If the boundary conditions are known, this formula can be used to predict every single FIB impact on surface waters, no matter if the impact was caused by one single WWTP site or by the total discharge of WWTPs along a river. Different scenarios, from totally WWu to partially WWt and different contamination sources, can be simulated.

The results can then be used to simulate the efficiency of wastewater treatment along a river.

$$C_R = \frac{Cww_t \times WW_v \times X + Cww_u \times WW_v \times (1-X)}{R_v} \tag{1}$$

C_R (MPN/100 mL): Increase of FIB in the recipient as a result of influent of treated and untreated WW

C_{WWt} (MPN/100 mL): Concentration of FIB in treated waste water WW

C_{WWu} (MPN/100 mL): Concentration of FIB in untreated WW

WW_V (m^3/t): total volume of WW per time in the WWTP

R_V (m^3/t): discharge of the recipient per time

X: Clearance ratio regarding treated and untreated WW; value between 0 and 1

3.4. WWTP Purification Efficiency and Its Influence on the Recipient

The mean reduction rate of FIB of the investigated WWTP is high and consistent compared to results from other countries [7,15]. In these studies, the reduction rate was between 0.7 and 3.5 log orders, depending on the size and technical features of the WWTPs. The good reduction efficacy is due to multi-step treatment and retention time within the WWTP. The total one-year flow volume of the WWTP was 3113.092 m^3, which resulted in a daily flow-through of 8.529 m^3 = 0.1 m^3/s, and this volume is directly discharged into the recipient (Mur River). On the basis of the mean flow rate (QM) of the Mur River of 144.6 m^3/s, and the flow rate for low water of 63.82 m^3 ($Q_{95\%}$) [16], a dilution factor (DF) for the discharged treated wastewater for QM and $Q_{95\%}$ of 1446 and 638 can be calculated. Based on this dilution factor, a negligible elevation of FC, EC and IE, on the basis of median values, can be observed for mean flow rates, as well as for $Q_{95\%}$ (Table 4).

Table 4. Increase of FIB-number in the recipient (MPN/100 mL) by the WWTP discharge (median and 90%-Percentile for FC, EC and IE) calculated on basis of the mean flow rate (QM) or $Q_{95\%}$.

Flow Rate	FC (Median/90th Percentile)	EC (Median/90th Percentile)	IE (Median/90th Percentile)
QM	5.8/15.5	3.6/9.8	0.5/1.7
$Q_{95\%}$	13.1/35.2	8.2/22.1	1.2/3.8

3.5. Combined Sewer Overflow (CSO) and Its Influence on the Recipient

Based on the calculation of treated wastewater, the same calculation was carried out for CSO, which leads to a dramatic increase of the FIB in the recipient (Table 4). Such conditions can occur when heavy rainfall causes too much influent which exceeds the volume capacity of the WWTP. Thus, untreated or coarsely treated wastewater flows directly into the recipient. In the investigated WWTP, CSO with a volume of 4.3% of the annual flow occurred. Converted into the annual load this leads to 30- to 37-fold increase of the recipient Mur River (Tables 5 and 6). The calculated bacterial load is probably slightly lower than in practice because the coarse mechanical pre-purification reduces the number of microorganisms by sedimentation. When we investigated FIB of CSO samples in our study (on a random basis), the FIB number did not differ from untreated wastewater. This might be due to the limited retention volume of the WWTP, which leads to a dilution factor that is just two- to threefold and, therefore, it is not that strongly reflected in the FIB number. A study carried out by Kistemann et al. 2008, also calculated a 20-fold increase of FIB in the recipient for CSO (15).

Table 5. Increase of the FIB number (MPN/100 mL) caused by the direct discharge of untreated wastewater (median and 90th Percentile of FC, EC, IE) at QM and $Q_{95\%}$.

Flow Rate	FC (Median/90th Percentile)	EC (Median/90th Percentile)	IE (Median/90th Percentile)
QM	4004/11,964	2835/6420	436/947
$Q_{95\%}$	9075/27,116	6426/14,551	989/2147

Table 6. Annual load of the Mur River for FC, EC and IE (median) for WWt, combined sewer overflow (CSO), and impact factor WWt/CSO (MPN/100 mL).

Annual Load	FC	EC	IE
WWt	2.6×10^{14}	1.6×10^{14}	2.3×10^{13}
CSO	7.8×10^{15}	5.5×10^{15}	8.5×10^{14}
Factor (CSO/WWt)	30	34	37

3.6. Anaerobic Sludge-Stabilization

Sludge stabilization reduced the investigated fecal indicator bacteria by 3.32 log MPN/g (FC, EC, on the basis of the median-values), but only 2.20 log MPN/g for IE, which is due to the higher environmental resistance of IE. The high reduction rates for FC and EC are accompanied by a high dispersion of the values (two log orders, Table 2), which indicates inhomogeneous and changing conditions during sludge stabilization. This hypothesis is also supported by the lower reduction rate of IE, which was not seen in the wastewater. This must therefore be related to the ecological conditions in the sewage sludge. The short residence time (about one day) during wastewater treatment does not inactivate bacteria, but only separates the bacterial load in sewage sludge and wastewater. The annual load for IE for the wastewater intake was 9.0×10^{15} MPN/100 mL, and the annual load for raw sludge was 2.1×10^{15} MPN/g. A substantial part of IE of the untreated wastewater could be identified in the raw sludge.

The Styrian sewage sludge regulation [17] defines the epidemiological harmlessness of sewage sludge with an absence of *Salmonella* spp. and a maximum of 100 EC per g. In our study, on the other hand, *Salmonellae* could be detected quantitatively in 26.4% of the stabilized sludge samples, and the median for *E. coli* was 210/g. Therefore, despite the reduction of FC and EC of three log orders the final product is not sanitized and harmless.

4. Discussion and Conclusions

The discharge of treated wastewater into the recipient with a CSO percentage of 4.3% of the total discharge increased the number of FIB significantly by 1.2×10^2 MPN/100 mL for EC and by 1.8×10^1 MPN/100 mL for IE. This need not automatically lead to exceeding the bathing water limits of the recipient regarding low basal levels of FIB in the recipient and a fairly even distribution of CSO over the whole year. This result, however, is only valid for the investigated WWTP and the corresponding stretch of the river.

For the evaluation of the whole course of the river all other WWTPs and CSO events have to be taken into account. The Styrian part of the Mur River (around 300 km) harbors wastewater treatment plants for 1.6 m people, regardless of the tributaries. A calculation of FIB of all sewage treatment plants estimating the same ratio of CSO (4.3%) and a mean flow rate (QM) near the Slovenian border of 149 m^3/s (long-term annual average) results in a significant increase of the FIB load in the recipient (Table 6). A CSO proportion of 4.3% of the total water flow reduces the purification efficacy from 2.90 to 1.37 log MPN/100 mL. Considering CSO, the microbiological standards required for bathing waters as required by the European Union [5] cannot be achieved for the Mur River. Similar results were obtained by studies on other European rivers as well [7,18–21].

The sole discharge of treated wastewater without CSO would increase the water quality of the Mur River substantially; even if all WWTP are taken into account, bathing water quality would be maintained. The influent of only untreated WW, on the other hand, would decrease the water quality dramatically (Table 7).

Table 7. Impact of 4.3% CSO, 100% treated wastewater and 100% untreated wastewater on the total course of the Mur River. Increase of EC and IE (MPN/100 mL) was calculated out of the median values of the mean discharge for the investigated WWTP and the sum of all WWTP ($\sum$).

Wastewater Treatment	EC (MPN/100 mL)		IE (MPN/100 mL)	
	WWTP	$\sum$ WWTP	WWTP	$\sum$ WWTP
Treated WW (95.7%) CSO (4.3%)	1.2×10^2	3.8×10^3	1.8×10^1	5.8×10^2
CSO (0%) Treated WW (100%)	3.5×10^0	1.1×10^2	0.5×10^0	1.2×10^1
CSO (100%) Treated WW (0%)	2.8×10^3	8.7×10^4	4.3×10^2	1.3×10^4

This calculation does not consider other factors that also lead to a decrease of FIB, like mortality rates of bacteria in natural aquatic systems, biotic factors like grazing by protozoan-plankton, or abiotic factors like the influence of sunlight on them (summed up under the term of "river self-purification"). Sunlight is the most important factor as it reduces FIB by one log within hours, as long as the turbidity of the water is low [22–25].

Neither does it take factors into account that lead to an increase of FIB, for example, non-point sources like surface runoff and soil leaching. The influence of these sources is due to the usage of the surrounding landscape (woods, or livestock pastures). Moreover, the intensity and amount of rainfall and the resulting soil runoff also severely influence the input of fecal bacteria into the river. Studies on tributaries of the River Seine showed that 50% of the fecal bacterial load were caused by non-point sources, and this influence has to be considered to a much higher extent when we observe freshly distributed liquid manure in combination with heavy rainfall [26–28].

Even without other contamination sources taken into consideration, the current practice to accept CSOs does not meet the microbiological standards required for bathing waters, a finding which is underlined by studies carried out on other European rivers [18,29,30]. Also, the repeated outbreaks of leptospirosis among sporting events underline the improper microbiological quality of river water for recreational purposes [10,31].

Wastewater treatment according to state of the art techniques leads to a massive reduction of organic and inorganic pollutants and is able to improve the quality of rivers and lakes. Additional treatment steps in terms of wastewater disinfection, as used for river restoration of the Isar River [32], have proved ineffective as the increase of the bacterial load of the recipient by the discharge of proper state of the art cleaned wastewater is low anyway. Instead of additional disinfection measures of treated wastewater, substantial improvement regarding the reduction of FIB in the course of WWt can only be achieved by a drastic reduction of CSO, which reduces the peak load as well as the annual bacterial load. Further measures to reduce runoff of fecal bacteria and organic compounds could be the establishment of protected areas along the river side with a more careful agricultural management. Summing up, the reduction or avoidance of CSO in combination with careful riverside management would represent the most effective measures to reduce FIB in the recipient.

Author Contributions: C.K., F.M. wrote the manuscript and designed the study, F.P.-S., F.F.R. and G.E.Z. proofread the manuscript, W.M. laboratory work, experiments.

Conflicts of Interest: The authors declare no conflict of interest.

References

1. The Austrian Federal Ministry of Agriculture, Forestry Environment and Water Management. *National Water Management Plan 2009—NWMP 2009*; Federal Ministry of Agriculture, Forestry, Environment and Water Management: Vienna, Austria, 2009.

2. European Community. *Directive 2000/60/EC of the European Parliament and of the Council of 23 October 2000 Establishing a Framework for Community Action in the Field of Water Policy*; European Community: Copenhagen, Denmark, 2000.

3. Mose, J.R.; Thiel, W. Evidence of Salmonellae in the Mur River; Comparison 1969–1981. *Zentralbl. Bakteriol. Mikrobiol. Hyg. B* **1982**, *176*, 562–570. [PubMed]

4. Kittinger, C.; Marth, E.; Reinthaler, F.F.; Zarfel, G.; Pichler-Semmelrock, F.; Mascher, W.; Mascher, G.; Mascher, F. Water Quality Assessment of a Central European River—Does the Directive 2000/60/EC Cover all the Needs for a Comprehensive Classification? *Sci. Total Environ.* **2013**, *447*, 424–429. [CrossRef] [PubMed]

5. European Community. *Directive 2006/7/EC OF the European Parliament and of the Council of 15 February 2006 Concerning the Management of Bathing Water Quality and Repealing Directive 76/160/EEC*; European Community: Copenhagen, Denmark, 2006.

6. Schindler, P.R. Hygiene of Bathing Waters. *Gesundheitswesen* **2001**, *63* (Suppl. 2), S142–S150. [CrossRef] [PubMed]

7. Servais, P.; Garcia-Armisen, T.; George, I.; Billen, G. Fecal Bacteria in the Rivers of the Seine Drainage Network (France): Sources, Fate and Modelling. *Sci. Total Environ.* **2007**, *375*, 152–167. [CrossRef] [PubMed]

8. Kistemann, T.; Rind, E.; Koch, C.; Classen, T.; Lengen, C.; Exner, M.; Rechenburg, A. Effect of Sewage Treatment Plants and Diffuse Pollution on the Occurrence of Protozoal Parasites in the Course of a Small River. *Int. J. Hyg. Environ. Health* **2012**, *215*, 577–583. [CrossRef] [PubMed]

9. Morgan, J.; Bornstein, S.L.; Karpati, A.M.; Bruce, M.; Bolin, C.A.; Austin, C.C.; Woods, C.W.; Lingappa, J.; Langkop, C.; Davis, B.; et al. Outbreak of Leptospirosis among Triathlon Participants and Community Residents in Springfield, Illinois, 1998. *Clin. Infect. Dis.* **2002**, 34, 1593–1599. [CrossRef] [PubMed]
10. Radl, C.; Muller, M.; Revilla-Fernandez, S.; Karner-Zuser, S.; de Martin, A.; Schauer, U.; Karner, F.; Stanek, G.; Balcke, P.; Hallas, A.; et al. Outbreak of Leptospirosis among Triathlon Participants in Langau, Austria, 2010. *Wien. Klin. Wochenschr.* **2011**, 123, 751–755. [CrossRef] [PubMed]
11. Zarfel, G.; Lipp, M.; Gurtl, E.; Folli, B.; Baumert, R.; Kittinger, C. Troubled Water Under the Bridge: Screening of River Mur Water Reveals Dominance of CTX-M Harboring Escherichia Coli and for the First Time an Environmental VIM-1 Producer in Austria. *Sci. Total Environ.* **2017**, 593, 399–405. [CrossRef] [PubMed]
12. *Water Quality—Enumeration of Escherichia coli and Coliform Bacteria—Part 2: Most Probable Number Method (ISO 9308-2)*; International Organization for Standardization: Geneva, Switzerland, 2012.
13. *Water Quality—Detection and Enumeration of Intestinal Enterococci—Part 2: Membrane Filtration Method (ISO 7899-2)*; International Organization for Standardization: Geneva, Switzerland, 2000.
14. *Water Quality—Detection of Salmonella spp., ISO 19250*; International Organization for Standardization: Geneva, Switzerland, 2010.
15. Kistemann, T.; Rind, E.; Rechenburg, A.; Koch, C.; Classen, T.; Herbst, S.; Wienand, I.; Exner, M. A Comparison of Efficiencies of Microbiological Pollution Removal in Six Sewage Treatment Plants with Different Treatment Systems. *Int. J. Hyg. Environ. Health* **2008**, 211, 534–545. [CrossRef] [PubMed]
16. Ministry for Agriculture, Forestry, Environment and Water Management. *Hydrographical Annual of Austria*; Ministry for Agriculture, Forestry, Environment and Water Management: Vienna, Austria, 2013.
17. Styrian Government. *The Legal Information System of the Republic of Austria 2007*; Styrian Government: Styria, Austria, 2007.
18. Kistemann, T.; Schmidt, A.; Flemming, H.C. Post-Industrial River Water Quality-Fit for Bathing again? *Int. J. Hyg. Environ. Health* **2016**, 219, 629–642. [CrossRef] [PubMed]
19. Rechenburg, A.; Koch, C.; Classen, T.; Kistemann, T. Impact of Sewage Treatment Plants and Combined Sewer Overflow Basins on the Microbiological Quality of Surface Water. *Water Sci. Technol.* **2006**, 54, 95–99. [CrossRef] [PubMed]
20. Rechenburg, A.; Kistemann, T. Sewage Effluent as a Source of Campylobacter Sp. in a Surface Water Catchment. *Int. J. Environ. Health Res.* **2009**, 19, 239–249. [CrossRef] [PubMed]
21. Strathmann, M.; Horstkott, M.; Koch, C.; Gayer, U.; Wingender, J. The River Ruhr—An Urban River under Particular Interest for Recreational use and as a Raw Water Source for Drinking Water: The Collaborative Research Project "Safe Ruhr"—Microbiological Aspects. *Int. J. Hyg. Environ. Health* **2016**, 219, 643–661. [CrossRef] [PubMed]
22. Deller, S.; Mascher, F.; Platzer, S.; Reinthaler, F.F.; Marth, E. Effect of Solar Radiation on Survival of Indicator Bacteria in Bathing Waters. *Cent. Eur. J. Public Health* **2006**, 14, 133–137. [PubMed]
23. Kim, G.; Hur, J. Mortality Rates of Pathogen Indicator Microorganisms Discharged from Point and Non-Point Sources in an Urban Area. *J. Environ. Sci.* **2010**, 22, 929–933. [CrossRef]
24. Chigbu, P.; Gordon, S.; Strange, T. Influence of Inter-Annual Variations in Climatic Factors on Fecal Coliform Levels in Mississippi Sound. *Water Res.* **2004**, 38, 4341–4352. [CrossRef] [PubMed]
25. Menon, P.; Billen, G.; Servais, P. Mortality Rates of Autochthonous and Fecal Bacteria in Natural Aquatic Ecosystems. *Water Res.* **2003**, 37, 4151–4158. [CrossRef]
26. Ibekwe, A.M.; Ma, J.; Murinda, S.E. Bacterial Community Composition and Structure in an Urban River Impacted by Different Pollutant Sources. *Sci. Total Environ.* **2016**, 566, 1176–1185. [CrossRef] [PubMed]
27. Franke, C.; Rechenburg, A.; Baumanns, S.; Willkomm, M.; Christoffels, E.; Exner, M.; Kistemann, T. The Emission Potential of Different Land Use Patterns for the Occurrence of Coliphages in Surface Water. *Int. J. Hyg. Environ. Health* **2009**, 212, 338–345. [CrossRef] [PubMed]
28. George, I.; Anzil, A.; Servais, P. Quantification of Fecal Coliform Inputs to Aquatic Systems through Soil Leaching. *Water Res.* **2004**, 38, 611–618. [CrossRef] [PubMed]
29. Touron, A.; Berthe, T.; Gargala, G.; Fournier, M.; Ratajczak, M.; Servais, P.; Petit, F. Assessment of Faecal Contamination and the Relationship between Pathogens and Faecal Bacterial Indicators in an Estuarine Environment (Seine, France). *Mar. Pollut. Bull.* **2007**, 54, 1441–1450. [CrossRef] [PubMed]

30. Kirschner, A.K.; Kavka, G.G.; Velimirov, B.; Mach, R.L.; Sommer, R.; Farnleitner, A.H. Microbiological Water Quality Along the Danube River: Integrating Data from Two Whole-River Surveys and a Transnational Monitoring Network. *Water Res.* **2009**, *43*, 3673–3684. [CrossRef] [PubMed]

31. Brockmann, S.; Piechotowski, I.; Bock-Hensley, O.; Winter, C.; Oehme, R.; Zimmermann, S.; Hartelt, K.; Luge, E.; Nockler, K.; Schneider, T.; et al. Outbreak of Leptospirosis among Triathlon Participants in Germany, 2006. *BMC Infect. Dis.* **2010**, *10*, 91. [CrossRef] [PubMed]

32. Huber, S.; Popp, W. *Reduction Rates of Pathogenes and Faecal Indicators after UV Disinfection at Bad Tölz Waste Water Treatment Plant*; IUVA news: Bethesda, MD, USA, 2003; Volume 3.

Review

Carbamazepine as a Possible Anthropogenic Marker in Water: Occurrences, Toxicological Effects, Regulations and Removal by Wastewater Treatment Technologies

Faisal I. Hai [1],*, Shufan Yang [1], Muhammad B. Asif [1], Vitor Sencadas [2], Samia Shawkat [3], Martina Sanderson-Smith [3], Jody Gorman [3], Zhi-Qiang Xu [4] and Kazuo Yamamoto [5]

[1] Strategic Water Infrastructure Lab, School of Civil, Mining and Environmental Engineering, University of Wollongong, Wollongong, NSW 2522, Australia; sy527@uowmail.edu.au (S.Y.); mba409@uowmail.edu.au (M.B.A.)

[2] School of Mechanical, Materials and Mechatronics Engineering, University of Wollongong, Wollongong, NSW 2522, Australia; victors@uow.edu.au

[3] Illawarra Health and Medical Research Institute, University of Wollongong, Wollongong, NSW 2522, Australia; samias@uow.edu.au (S.S.); martina@uow.edu.au (M.S.-S.); jody_gorman@uow.edu.au (J.G.)

[4] Centre for Medical and Molecular Bioscience, University of Wollongong, Wollongong, NSW 2522, Australia; zhiqiang@uow.edu.au

[5] Environmental Science Centre, Department of Urban Engineering, University of Tokyo, Tokyo 113-0033, Japan; yamamoto@esc.u-tokyo.ac.jp

* Correspondence: faisal@uow.edu.au; Tel.: +61-2-42213054

Received: 8 December 2017; Accepted: 21 January 2018; Published: 26 January 2018

Abstract: Carbamazepine (CBZ), a pharmaceutical compound, has been proposed as an anthropogenic marker to assess water quality due to its persistence in conventional treatment plants and widespread presence in water bodies. This paper presents a comprehensive literature review on sources and occurrences of CBZ in water bodies, as well as toxicological effects and regulations of the drug. Given the documented side effects of CBZ on the human body when taken medicinally, its careful monitoring in water is recommended. CBZ residues in drinking water may provide a pathway to embryos and infants via intrauterine exposure or breast-feeding, which may cause congenital malformations and/or neurodevelopmental problems over long term exposure. An in-depth technical assessment of the conventional and advanced treatment technologies revealed the inadequacy of the standalone technologies. Compared to conventional activated sludge and membrane bioreactor processes, effective removal of CBZ can be achieved by nanofiltration and reverse osmosis membranes. However, recent studies have revealed that harsh chemical cleaning, as required to mitigate membrane fouling, can often reduce the long-term removal efficiency. Furthermore, despite the efficient performance of activated carbon adsorption and advanced oxidation processes, a few challenges such as cost of chemicals and regeneration of activated carbon need to be carefully considered. The limitations of the individual technologies point to the advantages of combined and hybrid systems, namely, membrane bioreactor coupled with nanofiltration, adsorption or advanced oxidation process.

Keywords: advanced oxidation processes (AOPs); activated carbon adsorption; carbamazepine toxicity; conventional treatment processes; membrane technology; occurrence

1. Introduction

The occurrence of pharmaceutically active compounds (PhACs) in environmental systems such as freshwater bodies has become a topic of growing concern over the last decade due to their potential detrimental impacts on aquatic life and human health [1,2]. Because a large proportion of PhACs ends up in sewage via bodily excretion and indiscriminate disposal of unwanted/expired pharmaceuticals, disposal of untreated or ineffectively treated wastewater is considered a major source of their occurrence in environmental systems [3–6]. The need for effective removal of PhACs has resulted in the emergence of various advanced wastewater treatment technologies such as membrane technology and advanced oxidation processes [4].

The widespread occurrence of PhACs in wastewater and wastewater-impacted freshwater has triggered the establishment of water quality standards for their regular monitoring [5,7]. In this context, carbamazepine (CBZ) has been proposed as an anthropogenic marker of sewage contamination in freshwater bodies [8,9]. CBZ is an anticonvulsant and mood stabilizing drug used primarily in the treatment of epilepsy and bipolar disorder [10]. Table 1 summarizes the salient physicochemical properties of CBZ. CBZ is one of the most frequently detected pharmaceutical compounds in environmental systems [4,11]. It is ubiquitously present in raw wastewater in the high ng/L to low μg/L range and is only poorly removed by the conventional wastewater treatment plants (WWTP) [12,13]. Almost all the known advanced technologies have been tested for CBZ removal, however, none has appeared as a universal solution [14,15].

Table 1. Physicochemical and pharmacological properties of carbamazepine (modified after [16,17]).

Structure	
Formula	$C_{15}H_{12}N_2O$
CAS No.	298-46-4
Molecular Weight	$236.2686 \text{ g·mol}^{-1}$
Usage	Anticonvulsant/mood stabilizing drug
Water solubility	17.7 mg/L (25 °C)
Log P (octanol-water partition coefficient)	2.45
Log D at pH = 7 [a]	1.32
Henry's Law Constant	$1.09 \times 10^{-5} \text{ Pa·m}^3\text{·mol}^{-1}$ (25 °C)
Half-life ($t_{1/2}$)	25–65 h
Excretion	72% absorbed and metabolized in liver, 28% excreted in feces
Metabolites in urine	CBZ, CBZ-epoxide, CBZ-diol, CBZ-acridan, 2-OH-CBZ, 3-OH-CBZ
Dosage	800–1200 mg/day
Other information	Autoinduction i.e., induces its own metabolism during continued intake

Note: [a] Log D is the logarithm of the distribution coefficient, which is the ratio of the sum of concentrations of all forms of the compound (ionised and unionised) in octanol and water at a given pH [3].

A number of interesting review papers have been published on the occurrences of micropollutants in environmental systems such as wastewater [18], surface water [4] and groundwater [5], as well as the performance of conventional treatment technologies for the removal of micropollutants [19–21]. The persistence of CBZ in conventional treatment processes leads to its widespread occurrence in water bodies. Thus, CBZ has been proposed as an anthropogenic marker to assess water treatment quality. CBZ removal by conventional WWTPs has been reviewed previously by Zhang et al. [16]. However, the efficacy of advanced treatment technologies has not been critically analyzed to date. In addition, toxicity of CBZ to aquatic species and human and relevant regulations have not been comprehensively reviewed.

In this paper, the occurrence of CBZ in wastewater and freshwater bodies (e.g., surface and groundwater) along with the associated influencing factors are systematically analyzed. In addition, the toxicological effects of CBZ on the aquatic ecosystem are critically discussed. Importantly, the factors governing the resistance of CBZ to the available wastewater treatment processes are elucidated, and the efficacy of advanced/emerging treatment processes is comprehensively discussed.

2. Occurrence in Aquatic Systems

Following consumption, up to 10% of CBZ is excreted from human body [4,22]. Recent studies have reported a few tens to several thousands of ng/L CBZ in municipal wastewater [22–24]. CBZ is poorly removed (typically less than 10%) by the conventional WWTPs [25–27]. Hence, treatment plant effluents are an important gateway for CBZ to enter surface and groundwater. Table 2 depicts the reported levels of occurrence of CBZ in WWTP effluent, surface water and groundwater. Generally, CBZ concentration has been reported to be higher in WWTP effluents as compared to surface water (Table 2) because dilution and natural attenuation can significantly reduce the concentration of pollutants [28]. CBZ is most likely to reach groundwater via bank infiltration of WWTP effluent [5,29]. In addition, seepage of landfill leachate and combined sewer overflows can contaminate groundwater [30]. In this section, the factors influencing the occurrence of PhACs including CBZ in raw wastewater, WWTP effluent and freshwater bodies are critically discussed.

Table 2. Occurrences of carbamazepine (CBZ) in municipal wastewater treatment plant (WWTP) effluent and freshwater.

Country	W WWTP Effluent		Surface Water	Groundwater
	No. of WWTPs *	Concentration (ng/L)	Concentration (ng/L)	Concentration (ng/L)
Canada	7	33–426 [a]	0.7–126 [b]	10–49 [c]
Germany	5	1075–6300 [d]	81–1100 [e]	1–100 [f]
Japan	20	81–86 [g]	0.1–34.7 [h]	1.64–97 [i]
South Korea	11	73–729 [j]	6–61 [k]	NA
Taiwan	4	290–960 [l]	0.5–120 [m]	NA
UK	3	152–4596 [n]	9–327 [o]	425–3600 [p]
USA	16	33–270 [q]	2–172 [r]	1.5–42 [s]

Notes: Data sources: [a] [31–34]; [b] [31,32,35]; [c] [36,37]; [d] [29,38–40]; [e] [38,40,41]; [f] [42,43]; [g] [44–46]; [h] [45, 47]; [i] [46,47]; [j] [48,49]; [k] [48–51]; [l] [52,53]; [m] [52]; [n] [54–56]; [o] [54,55,57,58]; [p] [30,59,60]; [q] [28,61–63]; [r] [61,63–65]; [s] [66,67]. * number of WWTPs surveyed for analyzing the concentration of CBZ; NA: not available.

2.1. Wastewater Treatment Plant Effluent

Since a number of factors can affect CBZ occurrence in wastewater, its concentration in WWTP effluent has been observed to be highly variable (30–6300 ng/L) (Table 2). The variation in WWTP effluent CBZ concentration (Tables 2 and 3) can be attributed to a number of factors such as CBZ production/consumption rate, environmental regulations, effectiveness of the WWTPs and the seasonal factors affecting WWTP performance [4,68,69].

Table 3. Seasonal variations in CBZ concentration along with its annual consumption rate.

Country	WWTP Effluent			Consumption Rate (tons/year)
	No. of WWTPs*	Concentration (ng/L)		
		Winter	Summer	
Australia	3	1480 [70]	NA	10 [16]
Austria	11	952 [71]	1337–1594 [3,71]	6 [16]
	11	1000 [72]	1500 [72]	
Canada	4	426 [32]	300 [33]	28 [16]
Finland	12	500 [6]	NA	4.8 [16]
	3	380–470 [73]	NA	
Germany	3	1900 [38]	2100 [40]	76 [16]
Korea	4	103–195 [48]	5–6 [48]	9.2 [48]
Switzerland	2	1000 [74]	950 [74]	4.1 [75]
	3	400–800 [26]	200–600 [26]	
UK	3 (Winter) 2 (Summer)	637–950 [56]	2499 [54]	40 [16]

Note: * Number of WWTPs surveyed for analyzing the concentration of CBZ in winter and summer; NA: not available.

Some studies have reported good correlation between CBZ concentration in wastewater with its production and consumption rate at the corresponding locations. For instance, Kasprzyk-Hordern et al. [22] and Choi et al. [48] studied the occurrence of PhACs such as carbamazepine, acetaminophen, sulfamethoxazole and codeine in wastewater from selected locations of UK and Korea, respectively. They observed that the concentration of pharmaceuticals in wastewater correlated well with their consumption rates in the respective countries [22,48]. By contrast, CBZ concentration in the WWTP effluent of Canada, Finland and Switzerland (Table 3) did not correlate well with CBZ consumption rates, indicating that there are other influencing factors.

PhAC occurrence in wastewater can be governed by, among other factors, their excretion rate after metabolism within human body. However, it is important to note that low excretion rate from human body does not necessarily lead to their detection in low concentration or frequency in water. For instance, excretion rate of CBZ and a few other PhACs such as ibuprofen, clofibric acid and gemfibrozil is generally low (1–10%), while excretion rates of up to 70% were reported for some PhACs such as atenolol and paracetamol [76,77]. However, as noted above, CBZ is detected in significantly high concentration in raw and treated wastewater because of its low removal by the natural attenuation process and by the WWTPs [4,78].

Seasonal variations that affect the flow pattern of wastewater in combined sewerage system can lead to the change in wastewater composition. In a study by Kasprzyk-Hordern et al. [22], an increase of up to two-folds in the concentration of PhACs including CBZ was observed during dry weather. Moreover, WWTP effluent CBZ concentration in some countries such as Australia [70], Switzerland [26,74] and USA [79] were reported to be consistently higher in winter than in summer (Table 3), which can also be attributed to drier weather conditions in winter [22,26].

2.2. Surface Water

The major source of CBZ in surface water is the disposal of WWTP effluent [80,81]. After its release to freshwater bodies, different natural attenuation processes such as photolysis, aerobic biodegradation, sorption onto sediments and dilution in surface water play an important role in reducing CBZ concentration [81,82]. However, in-stream attenuation rate varies depending on the physicochemical properties of the PhACs and the local environmental conditions. For instance, Kunkel and Radke [83] observed different attenuation rates for 10 pharmaceuticals including CBZ in river water, and this variation was generally attributed to the physicochemical properties of the compounds. Similarly, in a study that investigated the relationship between attenuation rate and physicochemical properties of 225 micropollutants [82], high attenuation rate was obtained for compounds having medium to low volatility ($-4 < \log K_{aw} < -2$) and significant hydrophilicity ($0 < \log K_{ow} < 4.5$). This is because these micropollutants are better exposed to in-stream biotic (e.g., biotransformation) and abiotic (e.g., photolysis) attenuation processes as compared to hydrophobic micropollutants that are adsorbed onto river sediments [84]. Since $\log K_{ow}$ value for CBZ falls between 0-4.5 [17,81], its concentration is expected to be reduced via in-stream attenuation processes [85].

Water dilution can reduce CBZ concentration in surface water. Indeed, higher concentrations of PhACs including CBZ was reported in surface water bodies during dry weather as compared to that observed in wet weather [86]. In a study by Wang et al. [87], CBZ concentration in surface water was lower in samples that were collected during summer than those collected during winter. This is probably because of the enhanced biodegradation rate due to higher temperature in summer [4,81]. Heavy rainfall can often cause increased leaching of PhACs from openly dumped municipal and hospital solid waste. Storm water runoff can lead these compounds to surface water, consequently increasing the concentration of PhACs in surface water [23,88].

Sorption onto river sediments has been reported to reduce the aqueous phase concentration of hydrophobic PhACs along the river segment. However, in this case, concentration of these pollutants is not significantly reduced by the in-stream biotic and abiotic attenuation processes [81,84,89]. In a study by Riml et al. [90], concentration of two PhACs, namely bezafibrate and metoprolol, was observed to be mainly reduced by sorption onto sediments, while the role of biotransformation and photolysis was insignificant. Since it is a moderately hydrophobic compound (log K_{ow} = 2.45), reduction in CBZ concentration has been attributed to photolysis and sorption [85,91].

Residues of CBZ are introduced into sea water via surface runoff and groundwater discharge [92]. Concentrations of CBZ in sea water are very low. Weigel et al. [93] detected CBZ in sea water at a concentration of 2 ng/L by using a method that comprised solid-phase extraction and GC/MS quantification with a detection limit of 0.1 and 0.7 ng/L. Although this value seems insignificant, the fact that CBZ is detectable in sea water indicates that it is an extremely persistent compound. Also interesting to note, is that residues of CBZ can accumulate in soil through seepage of irrigation water, and due to sewage sludge used as fertilizer [94,95]. A study conducted by the US Geological Survey found an average CBZ concentration of 41.6 ng/mg in the sediment of 44 rivers across the US [96].

2.3. Groundwater

Groundwater constitutes approximately 30% of the total freshwater resources in the world. Because 70% of the freshwater resources are frozen, groundwater represents 97% of freshwater available for human use [97]. Groundwater is the major source of freshwater for domestic and industrial use in many countries. PhACs can contaminate groundwater through different pathways such as percolation of landfill leachate, artificial recharge, percolation of storm water runoff and leakages from sewers and septic tanks [5]. Depending on the organic fraction of the soil, high attenuation of some PhACs such as hydrophobic compounds (log D > 3) can occur in soil strata en route to groundwater [98,99]. Nevertheless, CBZ concentration in groundwater has been reported to be in the range of 1–100 ng/L in available studies from Canada, Germany, Japan and USA (Table 2). However, a higher concentration of CBZ (425 to 3600 ng/L) in groundwater was observed at a site in UK [30,59,60]. This is probably because CBZ concentration was also higher in WWTP effluents and surface water in that location (Tables 2 and 3). Indeed, Stuart et al. [30] reported that the occurrence of CBZ in groundwater is most likely to be derived from the bank infiltration of WWTP effluent or through surface water/groundwater interaction. Concentration of CBZ in groundwater may not be as high as in surface water but it is still an issue that should be addressed on a priority basis.

3. Toxicological Effects

CBZ is widely detected in water bodies, hence it is essential to evaluate its effects on the ecosystems. A number of studies have assessed the ecotoxicity of CBZ (Table 4). In an experiment conducted by Ying et al. [25], the respiratory quotient value for CBZ was 4.69, indicating potential risks to aquatic organisms. However, other experimental studies have shown that CBZ may not pose an immediate risk. For example, Ferrari et al. [100] studied the toxicological effects of CBZ on bacteria, algae, microcrustaceans and fish. It was observed to have a relatively limited acute ecotoxicity on the tested organisms.

Table 4. Effect of CBZ exposure on aquatic species under different exposure conditions.

Species	Critical Effects	Exposure Time	LC$_{50}$ (mg/L)	EC$_{50}$ (mg/L)	NOEC (mg/L)	LOEC (mg/L)	References
Water flea							
Daphnia magna	Mortality	2 days	111	-	-	-	[101]
Chironomus dilutus	Survival	10 days	47.3	10.2	-	-	[102]
Hyalella azteca	Growth	10 days	1.5	9.5	-	-	[102]
Brachionus calyciflorus	Reproduction inhibition	2 days	-	-	0.4	0.8	[100]
Ceriodaphnia dubia	Reproduction inhibition	7 days	-	-	0.025	0.1	[100]
Daphnia magna	Mobility inhibition	2 days	-	77.7	-	-	[100]
Daphnia magna	Immobilization	2 days	-	97.8	-	-	[103]
Bacteria							
Aliivibrio fischeri	Bioluminescence	30 min	-	64.2	-	-	[100]
Vibrio fischeri	Bioluminescence	5 min	-	87.4	-	-	[103]
Algae							
Chlorella vulgaris	Growth inhibition	24 h	-	110.9	-	-	[103]
Desmodesmus subspicatus	Inhibition of average growth rate	3 days	-	74	-	-	[104]
Raphidocelis subcapitata	Growth inhibition	4 days	-	89	>100	>100	[100]
Rainbow trout							
Oncorhynchus mykiss	Condition factor	42 days	-	-	-	2	[105]
Oncorhynchus mykiss	Antioxidant responses in muscle	42 days	-	-	-	0.001	[105]
Oncorhynchus mykiss	Changes in RNA-DNA ratio	42 days	-	-	-	2	[105]
Zebrafish							
Danio rerio	Developmental effects	3 days	>245	85.6	30.6	-	[106]
Danio rerio	Embryos and larvae mortality	10 days	-	-	25	50	[100]
Mycorrhizal fungus							
Glomus intraradices	Spore production	28 days	-	0.1	-	-	[107]
Duckweed							
Lemna minor	Inhibition of average growth rate	7 days	-	25.5	-	-	[104]
Cnidarian							
Hydra attenuate	Morphological changes	4 days	29.4	15.5	1	5	[108]

Notes: LC$_{50}$: lethal concentration to kill/inactivate 50% of population; EC$_{50}$: effective concentration that gives half-maximal response; NOEC: no-observed effect concentration; and LOEC: lowest-observed effect concentration. "-": not available.

Possible human health effects of long-term exposure to different pharmaceuticals may include endocrine disruption, induction of antibiotic resistance in human pathogens, genotoxicity, carcinogenicity, allergic reactions, and reproductive and/or developmental effects [109,110]. Limited research has been conducted on the potential human health risks of long-term exposure to CBZ residues in water. Risk assessments conducted to date have generally shown that the trace concentrations of CBZ detected in drinking water does not pose an unacceptable health risk to humans [111]. However, careful monitoring must continue, given the documented side effects of CBZ on the human body when taken medicinally.

CBZ is the main cause of the Stevens–Johnson syndrome and its associated disease toxic epidermal necrolysis in Southeast Asian countries due to its intake medicinally [112]. These are two forms of a life-threatening skin condition with an overall mortality rate of 30%, in which cell death causes the epidermis to separate from the dermis [113]. Recent studies have also revealed that intrauterine exposure of CBZ is associated with spina bifida [114] and neuro developmental problems [115] of human embryo when gravidas were exposed to CBZ monotherapy. Atkinson et al. [116] also reported higher fetal losses and congenital malformation rates among women who were prescribed carbamazepine during pregnancy. Because the residue of CBZ in drinking water may provide a pathway to embryo and infant via intrauterine exposure or breast-feeding, the presence of CBZ in groundwater, and drinking water remains a significant concern warranting further systematic risk assessment studies.

4. Regulations

Strict regulations were introduced in the 1960s in many countries for pharmaceutical production [117]. For example, in Australia, the Therapeutic Goods Administration (TGA) undertakes assessments, similar to those of the US Food and Drug Administration (FDA), to "ensure that prescription and 'over-the-counter' medicines, medical devices, and related products, supplied in or exported from Australia, meet appropriate standards" [118].

Despite the release of a significant amount of PhACs into the environment, there is limited literature available on regulations for the presence of pharmaceuticals in water [117]. The regulatory framework set out for pharmaceuticals governs the quality and safety of use, rather than the health and environmental risks of long-term exposure to drinking water containing trace concentrations of pharmaceuticals. In the U.S., the FDA requires that an environmental assessment report be carried out when the expected concentration of the active ingredient of the pharmaceutical in the aquatic environment is equal to or higher than $1 \mu g/L$. However, some state departments, such as the Minnesota Department of Health regulates that the CBZ concentration in drinking water must not exceed $40 \mu g/L$ [119]. Australian regulations require CBZ concentration in drinking water to be less than $100 \mu g/L$ [120]. In Europe, authorization for pharmaceutical production requires an environmental risk assessment [118]. In many countries, however, health risk-based standards and limit values for the presence of pharmaceuticals in drinking water have either not been set or are insufficient [111].

5. Biological Treatment Technologies for CBZ Removal

5.1. Activated Sludge Based Processes

Removal of PhACs by different biological treatment processes has been studied extensively [3,4]. In conventional activated sludge (CAS) processes, microorganisms generate energy by utilizing bulk organics present in wastewater as a primary source of food (also known as substrate). A part of this energy is used by the microorganisms for their cell growth and remaining energy is used for cell maintenance [121–123]. Since some PhACs such as antibiotics can be toxic to microorganisms and can inhibit their growth, an additional growth substrate (i.e., co-metabolism) is required to maintain microbial growth and diversity for adequate biodegradation [124,125].

The CAS process involves the application of microorganisms for the degradation of pollutants [16]. A membrane bioreactor (MBR) is an integration of the CAS process with an ultrafiltration (UF) or a microfiltration (MF) membrane for effective solid-liquid separation [3,126,127]. Removal of CBZ by CAS and MBR at different operating conditions such as hydraulic retention time (HRT), solids retention time (SRT) and initial concentration is presented in Table 5.

Table 5. CBZ removal by conventional activated sludge (CAS) and membrane bioreactor (MBR) under different operating conditions.

Process Type	Influent Concentration (ng/L)	SRT (days)	HRT (h)	Aerobic/Anoxic	Removal (%)	References
CAS	240	3	12	aerobic	negligible	[128]
	156	10	11.5	both	negligible	[126]
	350–1850	52–114	12.5–13.6	both	negligible	[71]
	350	2–20	1.5–20	both	negligible	[6]
	15–270	3.8–8.4	7.1–9.4	aerobic	negligible–80	[129]
	1000	10	7.3	both	negligible	[74]
	670–704	19	NA	aerobic	0	[130]
	10–20	11–15	9–17	both	negligible–25	[131]
	200–600	15–25	16–24	aerobic	negligible	[132]
MBR	240	infinite	14	aerobic	negligible	[128]
	156	>60	15	aerobic	negligible	[126]
	156	>60	7.2	aerobic	negligible	[126]
	704–1850	10–55	0.5–4	both	negligible	[71]
	1000	16	13	both	25	[74]
	704–1200	22	NA	both	negligible	[130]
	750,000	infinite	24	near-anoxic	68	[133]
	5	88	26	aerobic	40	[27]
	1400	70	24	aerobic	10	[134]

Notes: CAS: conventional activated sludge process; MBR: membrane bioreactor; SRT: solid retention time; HRT: hydraulic retention time; NA: not available.

In the CAS process, a settling tank is used to separate the treated water from the sludge. In MBR, this solid-liquid separation is performed by filtration via MF or UF membranes. Effective retention of the activated sludge by the membrane in MBR decouples SRT from HRT, thereby allowing the operation of the activated sludge based bioreactor at higher mixed liquor suspended solid concentration (MLSS) and longer SRT [135,136]. It has been reported in several studies that MBR provides better aqueous phase removal of moderately biodegradable PhACs as compared to the CAS process [126,137]. For example, removal of the nonsteroidal anti-inflammatory drug diclofenac by MBR was 56%, while its removal was 26% in CAS [138]. Similarly, MBR achieved up to 20% better removal of another nonsteroidal anti-inflammatory drug naproxen [126]. However, CBZ removal by both CAS and MBR has been reported to be poor and unstable (Table 5). Poor removal of CBZ can be attributed to its physicochemical properties such as molecular structure and hydrophilicity [71,134,139,140].

Sorption onto the activated sludge can increase the overall removal of PhACs. CAS and MBR are observed to achieve high removal (>80%) for hydrophobic PhACs (log D > 3) but lower removal (typically < 20%) for hydrophilic PhACs [141,142]. Since CBZ is moderately hydrophobic, its removal via sorption onto activated sludge has been reported to range between 5 to 20% only [16,27]. This suggests that CBZ removal depends on its intrinsic biodegradability, which is governed by its molecular properties. In general, simple structured PhACs, especially without branched/multi chain groups, are readily degradable [134,143]. Moreover, PhACs containing an electron withdrawing functional group (EWG), such as carboxyl, halogen and amide, are resistant to biological treatment [134]. Indeed, CBZ contains an EWG (i.e., amide) that makes it resistant to biodegradation.

It is important to note that operating conditions such as SRT, HRT and MLSS concentration can also influence the removal of some PhACs by activated sludge [3,139]. However, because CBZ is a hardly biodegradable compound, available reports indicate limited influence of these parameters on CBZ removal. Zhang et al. [16] did not observe any CBZ removal by CAS even at an SRT of 100 days.

Similarly, Radjenovic et al. [128], observed no improvement in CBZ removal following the increase in the SRT of both CAS and MBR. By contrast, Wijekoon et al. [27] achieved 40% removal of CBZ in MBR at a SRT of 88 days. Notwithstanding the fact that the experimental conditions may have been different in these studies, the observations here suggest that the removal and fate of CBZ during biological treatment processes depend on multiple factors.

The effect of redox conditions or dissolved oxygen on the removal of PhACs in MBR has been reported in a few studies [144–146]. In a study by Suarez et al. [147], PhACs were classified based on their biodegradation potential under aerobic and anoxic conditions: they observed that readily degradable PhACs such as fluoxetine and ibuprofen were biodegradable under both anoxic and aerobic conditions, while a few such as roxithromycin, naproxen, diclofenac and erythromycin were persistent in anoxic conditions but highly biodegradable under aerobic conditions. However, hydrophilic PhACs including CBZ were resistant to biodegradation in both aerobic and anoxic conditions [147], and a negligible difference in CBZ removal by sequential anoxic—aerobic MBR versus conventional aerobic MBR was noted. Notably, however, Hai et al. [139] reported that near-anoxic conditions (DO = 0.5 mg/L) can be a favorable operating regime for CBZ removal. They explained that 'sequential anoxic-aerobic' and 'continuous near-anoxic (DO = 0.5 mg/L)' operation modes were different. In the former, oxygen transfer from the aerated compartments to the anoxic zone due to the sludge recirculation may influence the removal efficiency [139].

5.2. White-Rot Fungi and Their Extracellular Enzymes

White-rot fungi (WRF) can degrade a variety of recalcitrant pollutants (e.g., poly-aromatic hydrocarbons and PhACs) that are poorly degraded by bacteria-dominated activated sludge [148–152]. In presence of a readily degradable substrate, WRF produce one or more type of extracellular enzymes such as laccase and lignin peroxidases (LiP). These enzymes catalyze the degradation of recalcitrant pollutants over a wide range of pH [153,154]. In addition to the extracellular enzymes, Golan-Rozen et al. [155] observed that the intracellular enzyme viz cytochrome P450 plays a vital role in CBZ degradation by whole-cell WRF. They demonstrated that the degradation of CBZ reduced from 99% to approximately 15% when cytochrome P450 was inhibited [155]. Whole-cell WRF and their extracellular enzymes have been studied extensively for enhanced removal of PhACs as depicted in Table 6.

Table 6. Removal of CBZ by white-rot fungi and their extracellular enzymes.

Bioreactor Type	WRF Species/Enzyme Type	HRT (h)/Incubation Time (days)	Initial Concentration (ng/L)	Removal Efficiency (%)	References
Removal by whole-cell WRF					
Stirred tank (batch)	*Bjerkandera* sp. R1	14	1,000,000	99	[156]
Stirred tank (batch)	*B. adusta* (Laccase, LiP and MnP)	14	1,000,000	99	[156]
Stirred tank (batch)	*T. versicolor* (Laccase, LiP and MnP)	2	10,000	75	[157]
Stirred tank (batch)	*T. versicolor* (Laccase, LiP and MnP)	1	100,000	2	[158]
Stirred tank (batch)	*P. ostreatus* (*Florida N001*) (Laccase, MnP)	32	1000	50	[155]
Stirred tank (batch)	*P. ostreatus* (*Florida F6*) (Laccase, MnP)	32	1000	60	[155]
Stirred tank (batch)	*P. ostreatus* (*PC9*) (Laccase, MnP)	32	1000	99	[155]
Stirred tank [a] (continuous)	*P. chrysosporium* (MnP, LiP)	24	500,000	25–60	[159]
Fluidized bed [a] (continuous)	*T. versicolor* (Laccase, MnP, LiP)	72	200,000	95.6	[160]
Membrane bioreactor (continuous) [a]	*T. versicolor* (Laccase, MnP, LiP)	48	5000	20	[161]

Table 6. *Cont.*

Bioreactor Type	WRF Species/Enzyme Type	HRT (h)/Incubation Time (days)	Initial Concentration (ng/L)	Removal Efficiency (%)	References
Removal by extracellular enzymes					
Stirred tank (batch)	Laccase from *A. oryzae*	1	100,000	<5	[162]
Stirred tank (batch)	Laccase from *T. versicolor*	2	10,000	5	[157]
Stirred tank (batch)	LiP from *P. chrysosporium*	4	NA	10–15	[163]
Membrane bioreactor (continuous)	Laccase from *A. oryzae*	8	5000	<5	[162]
Membrane bioreactor (continuous)	Laccase from *A. oryzae*	8	5000	7	[164]

Notes: [a] Indicates fungal bioreactor operated under non-sterile conditions; NA: not available.

Depending on the fugal species, efficient removal of CBZ has been achieved by whole-cell WRF in sterile batch bioreactors (Table 6). Since WRF species produce different combinations of extracellular enzymes, their performance of CBZ degradation might be different. For instance, Rodarte-Morales et al. [156] observed that *Bjerkandera* sp. R1 and *Bjerkandera adusta* both achieved almost complete removal (99%) of CBZ at an initial concentration of 1 mg/L in a batch bioreactor over an incubation time of 14 days. On the other hand, *Trametes versicolor* was reported to achieve less than 5% removal when CBZ was incubated in a whole-cell batch bioreactor at an initial concentration of 0.1 mg/L and an incubation time of 24 h [158]. Difference in performance was not only observed in case of different WRF species but also in different strains of a WRF species. For instance, Golan-Rozen et al. [155] studied the removal of CBZ by three different strains of *Pleurotus ostreatus* under identical operating conditions. They observed that *P. ostreatus* (PC9) achieved 99% CBZ removal, while a moderate removal (50–60%) was achieved by other two strains, namely *P. ostreatus* (Florida N001) and *P. ostreatus* (Florida F6).

Extracellular enzymes produced by WRF species have been studied for the removal of PhACs including CBZ in both batch and continuous-flow enzymatic bioreactors (Table 6). Degradation of PhACs by extracellular enzymes such as laccase occurs due to the transfer of a single electron from the pollutant to the active sites of the enzyme. Similar to the activated sludge based treatment process, the extent of degradation by an enzyme also depends on the molecular properties of the PhACs. Since CBZ contains a recalcitrant EWG (i.e., amide), its degradation by extracellular enzymes has been reported to range only between 5–15% [157,162]. High removal of CBZ in whole-cell fungal bioreactor as compared to enzymatic membrane bioreactor was explained by Golan-Rozen et al. [155]. They observed that the intracellular enzyme viz cytochrome P450 plays a vital role in CBZ degradation by whole-cell WRF. They demonstrated that the degradation of CBZ reduced from 99% to approximately 15% when cytochrome P450 was inhibited [155].

Performance of WRF for PhAC removal has been predominantly assessed under sterile conditions to avoid bacterial contamination [158,165,166]. This is because bacterial contamination under non-sterile conditions can negatively affect the performance of whole-cell WRF [154,167]. Indeed, poor removal of CBZ has been reported in fungal bioreactors operated under non-sterile conditions as compared to sterile fungal bioreactors [168–170]. For instance, Nguyen et al. [161] reported only 5% CBZ removal in a whole-cell fungal membrane bioreactor. In another study, no CBZ removal was observed in a non-sterile fluidized bed fungal bioreactor during the treatment of hospital wastewater [171]. To avoid bacterial contamination, a number of strategies such as fungal biomass replacement/renovation and pre-treatment of wastewater have been proposed. These strategies have been reviewed by Asif et al. [154].

6. CBZ Removal by Advanced Physicochemical Treatment Technologies

6.1. Performance of Nanofiltration and Reverse Osmosis Membranes

Nanofiltration (NF) and reverse osmosis (RO) are pressure-driven membrane filtration technologies [172,173]. They utilize semi-permeable membranes to primarily target the removal of dissolved contaminants. Both NF and RO have been studied for the removal of PhACs from secondary treated wastewater and freshwater, producing excellent quality effluent [173–175]. Several studies have shown that both NF and RO membranes can effectively retain CBZ, with a typical removal efficiency of greater than 95% [176,177]. Table 7 illustrates representative examples of CBZ removal by NF/RO membranes under a wide range of operating conditions.

Table 7. CBZ removal from various water matrices by nanofiltration (NF) and reverse osmosis (RO) membranes under different operating conditions.

Membrane Type (Pore Size)	Configuration	Water Matrix	Initial CBZ Concentration (ng/L)	Applied Pressure (psi)	Removal (%)	References
NF (0.34 nm)	Flat-sheet	Groundwater	84.5	NA	>98	[176]
NF (0.27 nm)	Flat-sheet (tight)	MBR permeate	150	150	97.3 ± 0.6	[178]
NF (0.42 nm)	Flat-sheet (loose)	MBR permeate	150	75	71.2 ± 3.1	[178]
NF	Flat-sheet	WWTP effluent	500–850	4.35	6	[179]
NF	Flat-sheet	WWTP effluent	500–850	10	8	[179]
NF (0.84 nm)	Flat-sheet	Primary effluent	2000	72	74	[180]
NF	Spiral-wound	Hospital wastewater	1000	98	88	[181]
NF (0.84 nm)	Flat-sheet	Synthetic wastewater	750,000	261	80 [a], 60 [b]	[182]
NF (0.68 nm)	Flat-sheet	Synthetic wastewater	750,000	261	95 [a], 90 [b]	[182]
NF (0.84 nm)	Flat-sheet	Synthetic wastewater	750,000	986	70 [a], 20 [b]	[183]
NF (0.84 nm)	Flat-sheet	Synthetic wastewater	750,000	261	80 [a], 90 [b]	[184]
RO (0.34 nm)	Flat-sheet	Groundwater	84.5	NA	>98	[176]
RO	Flat-sheet	MBR permeate	150	250	91 ± 8.4	[178]
RO	Flat-sheet	MBR permeate	150	150	97.9 ± 1.5	[178]
RO	Flat-sheet	Primary effluent	1000	72	100	[180]
RO	Spiral-wound module	Hospital wastewater	1000	196	99	[181]

Notes: [a] Virgin membrane; [b] fouled membrane; NA: not available.

Conceptually, NF membranes can retain PhACs via following mechanisms: (i) sorption of a solute on the membrane surface; (ii) size exclusion i.e., the sieving property of the membrane; and (iii) charge repulsion. However, electrostatic interaction cannot contribute to CBZ removal by the charged NF membrane, since CBZ remains neutral over a wide range of pH [14]. Hence, the molecular weight cut-off (MWCO) of NF/RO membranes is an important parameter for CBZ removal. In a study by Bellona et al. [177], efficient retention of CBZ by the RO membrane was attributed to size exclusion mechanism because the molecular weight of CBZ (i.e., 236 g/mole) was greater than the MWCO of the RO membrane. In a study by Comerton et al. [178], a loose NF membrane (MWCO = 400 g/mole) and a tight NF (MWCO = 200 g/mole) achieved 7 and 67% CBZ removal, respectively, thus, exemplifying the role of membrane MWCO in CBZ removal. In another study by Nghiem & Hawkes [185], efficient rejection of CBZ was reported for the NF90 membrane (MWCO < 200 g/mole) as compared to the NF270 membrane (MWCO > 300 g/mole).

Membrane fouling can affect the rejection of PhACs by NF membranes due to change in membrane surface properties (Table 7). Notably, CBZ rejection by the NF membranes is governed by the type of foulants. For instance, CBZ rejection by the NF270 membrane was reduced by 5 and 10% due to fouling caused by humic acid and sodium alginate, respectively [186]. This reduction in CBZ removal can be attributed to its diffusion into permeate following its adsorption on the fouling layer formed on the membrane surface. A more dramatic reduction in CBZ rejection (by 50%) was reported when MBR permeate (comprising multiple foulant materials) was fed to the NF-filtration system [186,187]. In general, the combination of membrane fouling and scaling can affect pollutant removal more severely. Chemical cleaning is performed to clean fouled membranes. However, recurrent chemical cleaning can affect membrane properties and, in turn, CBZ rejection. For example, significant reduction in CBZ rejection was observed due to cleaning the NF membrane by using caustic soda [184].

Natural organic matter (NOM) are ubiquitously present in surface water bodies. Since NF/RO processes are widely used for surface water treatment [188], the effectiveness of NF/RO for CBZ rejection in presence of NOM is vital. Comerton et al. [189] observed a statistically significant improvement in CBZ rejection during nanofiltration of pure water spiked with NOM. This is because CBZ, which is a moderately hydrophobic compound, can adsorb on NOM. However, a significant decrease in the rejection of CBZ was observed with concentration of the cations (calcium, magnesium, sodium) doubled. It has been hypothesized that increases in ionic strength and divalent cation concentration can cause conformational changes to NOM macromolecules. This may alter the presentation of sites for compound association leading to a reduction in NOM–compound complexation [189]. Therefore, the decrease in CBZ rejection in the natural waters with increase in cation concentration may be due to reduced association of CBZ with NOM.

6.2. Adsorption of CBZ by Activated Carbon

Activated carbons including granular activated carbon (GAC) and powdered activated carbon (PAC) are widely used as tertiary treatment processes primarily for color and odor removal from drinking water. GAC and PAC have also shown great potential for the removal of PhACs from secondary (i.e., biologically treated) wastewater [4,190,191]. Adsorption/removal of a pollutant by GAC/PAC is governed by the following mechanisms: (i) the electron donor–acceptor complex; (ii) the π–π dispersion interactions; (iii) hydrophobic interactions; and (iv) solvent effects that controls the solubility, reactivity and reaction kinetics [192,193]. The key properties of an adsorbent that can affect the efficacy of adsorption process include but are not limited to surface area, dose, surface chemistry and morphology, while water partitioning coefficient (log K_{ow}), acid dissociation coefficient (pK$_a$), molecular structure and size of the pollutants can influence the extent of adsorption by GAC/PAC [194]. In previous studies, efficient removal of PhACs has been achieved by GAC having larger pore size, because it can effectively adsorb pollutants with different shapes and size. Moreover, it was noted that pore volume has more influence on the adsorption of PhACs than specific area, and larger pore volume can achieve higher removal efficiency [195,196]. Representative examples of CBZ removal by GAC/PAC from a variety of water matrix (e.g., surface water and MBR effluent) is provided in Table 8.

Table 8. Performance of granular activated carbon (GAC) and powdered activated carbon (PAC) for CBZ removal from various water matrices.

AC Type	Water Matrix	Initial CBZ concentration (ng/L)	Contact Time (min)	Removal (%)	References
GAC	Synthetic wastewater	1000	1440	99–80 [a]	[197]
	Ozonation effluent	36	–	88 [b]	[198]
	Groundwater	9	15	>75 [c]	[199]
	Surface water	25	–	99	[200]
	Disinfected surface water	600	1.5–3	79 [d]	[201]
	WWTP effluent	67	–	30 [e]	[202]
	GAC added to an activated sludge based bioreactor	22,000	1440	43 [f]	[203]
	WWTP effluent	4000	100	80 [g]	[204]
	WWTP effluent	30–100	130	>99	[190]
PAC	Surface water	78	300	95 [h]	[205]
	Surface water	78	300	36 [i]	[205]
	MBR permeate	1000	30	99.4 [j]	[206]
	Surface water	50	240	80	[207]
	WWTP effluent	30–100	20–40	95–100	[190]

Notes: [a] The specific throughput of CBZ in a carbon layer of 80 cm was 50 m^3/kg when removal decreased to 80%; [b] the daily production of this drinking water treatment plant was 28,000 m^3, and this removal was observed in winter; [c] the system comprised 20 granular activated filters (volume = 150 m^3 each), experimental duration was 4 months; [d] the daily production of this drinking water treatment plant was 235,000 m^3 (experimental duration = 3 weeks); [e] total GAC volume was 1900 m^3; [f] the GAC concentration in the bioreactor was 1000 ppm, and experimental duration was 33 days; [g] the initial GAC concentration was 20 mg/L; [h] the PAC dose was 35 mg/L; [i] the PAC dose was 5 mg/L; and [j] the PAC dose was 10 g/L, "–": not available.

Since CBZ is neutral at pH ranging from 0–14, its removal by activated carbon is governed by hydrophobic interaction that depends on water partitioning coefficient [4,17]. In a study Yu et al. [17], better adsorption of CBZ (log K_{ow} = 2.45) by activated carbon as compared to naproxen (log K_{ow} = 3.18) and 4-*n*-nonylphenol (log K_{ow} = 5.8) was reported. Better adsorption of CBZ can be attributed to influence of pH on hydrophobicity of ionizable micropollutants [3]. Indeed, Yu et al. [17] demonstrated that activated carbon achieved better CBZ (log K_{ow} = 2.45) removal as compared to naproxen because actual log K_{ow} value for naproxen was 0.89 at the operating pH (i.e., 6.4). On the other hand, 4-*n*-nonylphenol contains both hydrophobic and hydrophilic functional groups in its molecule. The hydrophilic groups of 4-*n*-nonylphenol can affect its adsorption by activated carbon [208,209]. Therefore, better adsorption of other micropollutants including CBZ as compared to 4-*n*-nonylphenol is possible [17].

Effectiveness of GAC/PAC was also investigated at pilot-scale plants treating secondary effluent. For instance, Ternes et al. [197] observed almost complete removal (>99%) of CBZ in a pilot-scale GAC plant. In another study, CBZ removal was 95% in pilot-scale GAC/PAC systems treating secondary effluent at an initial activated carbon dose of 30–100 mg/L [190]. GAC was observed to provide better removal of PhACs including CBZ as compared to two other water treatment processes, namely, ozonation and sand filtration [198].

Increasing activated carbon dose can improve CBZ removal. For instance, CBZ removal improved from 36 to 97% when the PAC dose was increased seven folds [205]. In a study by Nguyen et al. [210], instead of using PAC as a post-treatment, it was added directly to the mixed liquor of an MBR. In that study, CBZ removal improved from 50% to 90% following the increase in PAC dose to MBR from 0.1 to 0.5 g/L [210].

Compared to GAC, PAC has a larger surface area that can conceptually provide faster reaction kinetics and better removal efficiency. However, survey of the available literature suggests that both GAC and PAC are effective for CBZ removal (Table 8). Notably, to date the performance of PAC and GAC has been assessed in short term experiments [204,211]. Since saturation of binding sites reduces the removal of pollutants over time, research is required to investigate PAC/GAC regeneration aspects. Grover et al. [202] monitored the performance of a full-scale post-treatment GAC plant over a period of seven months. They observed that CBZ removal reduced over time, and GAC plant could only achieve 30% removal of CBZ during long term operation. In addition to the saturation of GAC binding sites, impurities such as humic substances can compete for GAC/PAC binding sites that may result in ineffective CBZ removal [4,202].

6.3. CBZ Degradation by Advanced Oxidation Processes

Due to the molecular properties of CBZ, conventional biological processes are not effective for its removal (see Section 5). On the other hand, despite the effective removal of CBZ by NF/RO membrane filtration and activated carbon adsorption (Tables 7 and 8), an additional step is required for the treatment of the produced concentrate. In this context it is noteworthy that advanced oxidation processes (AOP) may achieve effective degradation of CBZ (Table 9). Post treatment of biologically treated wastewater by AOPs may simultaneously achieve disinfection and PhAC removal [212].

Formation of hydroxyl radicals (OH$^\bullet$) are mainly responsible for the degradation of PhACs by AOPs, while formation of ozone radicals (O$_3^\bullet$) in ozonation process can also contribute to the degradation process [213,214]. Some PhACs such as naproxen are degraded by both OH$^\bullet$ and O$_3^\bullet$ radicals, while some are only susceptible to degradation by OH$^\bullet$ radicals. CBZ, which is resistant to biological treatment, is effectively degraded by both OH$^\bullet$ and O$_3^\bullet$ radicals. For instance, Ternes et al. [197] reported almost complete removal (>99%) of CBZ (35-1000 ng/L) during ozonation process at an initial dose of 0.5 mg/L.

Table 9. Performance of various advanced oxidation processes (AOPs) for CBZ removal.

AOP Type	CBZ Initial Concentration (ng/L)	Operating Conditions	Removal (%)	References
Ozonation	1000	Dose = 0.5 mg/L Contact time = 20 min	>99	[197]
	35	Dose = 1-1.5 mg/L Contact time = 10 min	>97	[197]
	9	Dose = 0.2 mg/L Contact time = 15 min	>99	[199]
	8×10^5	Dose = 1 mg/L Contact time = 10 min	>99	[215]
	3.8	Dose = 1.5–2 mg/L Contact time = 20 min	80–99	[216]
	1.18×10^5	Dose = 0.1–2 mg/L Contact time = 10 h	80–99	[217]
	170	Dose = 0.8 mg/L Contact time = 24 min	100	[218]
	4.72×10^5	Dose = 1.5–4 mg/L Contact time = 20 min	>99	[219]
UV alone	5000	UV Wavelength = 254 nm Energy output = 83 W Irradiation time = 60 min	20	[220]
	NA	UV Wavelength = 200–280 nm Energy output = 120 W Irradiation time = NA	7	[221]
	1.5×10^7	UV Wavelength = 254 nm Energy output = 220 W Irradiation time = 2 h	16	[222]
	5×10^6	UV Wavelength = 254 nm Energy output = 400 W Irradiation time = 30 min	<5	[223]
	19–59	UV Wavelength = 254 nm Energy output = 10 W Irradiation time = 3 min	<10	[224]
UV/H_2O_2	4.72×10^6	UV Wavelength = 254 nm Energy output = 83 W H_2O_2 dose = 170 mg/L Irradiation time = 60 min	90	[187]
	210	UV Wavelength = 254 nm Energy output = 30 W H_2O_2 dose = 2–20 mg/L Irradiation time = 20 min	14–74	[225]
	1000	UV Wavelength = 254 nm Energy output = 20 W H_2O_2 dose = 5 mg/L Irradiation time = 15 min	60	[226]
	50,000	UV Wavelength = 254 nm Energy output = 83 W H_2O_2 dose: 10–200 mg/L Irradiation time = 60 min	90–99	[220]

Table 9. *Cont.*

AOP Type	CBZ Initial Concentration (ng/L)	Operating Conditions	Removal (%)	References
	19–59	UV Wavelength = 254 nm Energy output = 10 W H_2O_2 dose = 5 mg/L Irradiation time = 3 min	20	[224]
	2.36×10^5	UV Wavelength = 254 nm Energy output = 1000 W H_2O_2 dose = 5 mg/L Irradiation time = NA	90	[227]
UV/Cl$_2$	1000	UV Wavelength = 254 nm Energy output = 80 W Cl_2 dose = 1 mg/L Irradiation time = 20 min	55	[226]
	19–59	UV Wavelength = 254 nm Energy output = 10 W Cl_2 dose = 5 mg/L Irradiation time = 1.5 min	30–60	[224]
	19–59	UV Wavelength = 254 nm Energy output = 10 W Cl_2 dose = 3 mg/L Irradiation time = 3 min	50	[224]
UV/TiO$_2$	4×10^6	UV Wavelength = 200–296 nm Energy output = 1000 W TiO_2 dose = 100 mg/L Irradiation time = 9 min	99	[228]
	5×10^6	UV Wavelength = 254 nm Energy output = 250 W TiO_2 dose = 20–500 mg/L Irradiation time = 30 min	80–98	[223]
	5×10^6	UV Wavelength = 254 nm Energy output = 400 W TiO_2 dose = 20–500 mg/L Irradiation time = 30 min	90–99	[223]
Photo-Fenton	5×10^7	UV Wavelength = 254 nm Energy output = 400 W Fe^{2+} dose = 5 mg/L Irradiation time = 15 min	>99	[229]
	1×10^5	UV Wavelength = 200–296 nm Energy output = 30 W Fe^{2+} dose = 5 mg/L Irradiation time = 1.5 h	95	[230]

Note: "NA": not available.

Although UV photlysis alone has been observed to be ineffective for CBZ removal (0–20%) in a number of studies [220,231], removal can be significantly improved by adding H_2O_2 or a photocatalyst such as TiO_2 [214,220]. For instance, adding a single dose of H_2O_2 (5–15%) to UV photolysis process resulted in an enhanced CBZ removal of 60–75% [225,226]. In another study, H_2O_2-concentration dependent increase in CBZ removal by UV/H_2O_2 process was reported, and 99% removal was achieved at an initial H_2O_2 concentration of 120 mg/L [220]. CBZ removal can improve with increasing H_2O_2 concentration but it will reach a plateau beyond a threshold H_2O_2 concentration [220].

Fenton's reagent has been reported to efficiently oxidize (>99%) CBZ (Table 9). Notably, Fe^{+2} based Fenton process achieved better CBZ removal than UV alone, UV/TiO_2 and UV/H_2O_2. However, the requirement of acidic conditions for Fenton's process is a considerable drawback for its practical application [232].

AOP are undoubtedly very efficient for the removal of CBZ but their practical applications are constrained by associated high cost of chemicals. Moreover, transformation products formed following the oxidation of PhACs including CBZ can be more toxic than the parent compound [220]. To overcome this issue, the use of biological filters or ACs can be a suitable option [233,234].

6.4. Combined/Hybrid Treatment Systems

A single treatment option may not be universally applicable, so a combined (sequential) or integrated treatment may be more effective for CBZ removal. Combined or hybrid treatment options are also conceptually beneficial, usually leading to improved treatment efficiencies [235–237]. Examples of possible combined/hybrid water treatment processes include membrane filtration followed by activated carbon, MBR followed by activated carbon, activated carbon adsorption followed by UV, integrated MBR-TiO$_2$ photocatalysis, and integrated MBR-PAC adsorption. A summary of CBZ removal by combined/hybrid treatment systems is presented in Table 10.

Table 10. CBZ removal by combined and integrated treatment systems.

Treatment Systems	CBZ Initial Concentration (ng/L)	Operating Conditions	Removal (%)	References
Integrated MBR-UV/TiO$_2$	1×10^7	SRT = 60 d HRT = 50 h UV wavelength = 360 nm TiO$_2$ dose = NA	up to 95	[238]
Integrated MBR-PAC	5000	SRT = infinite HRT = 24 h PAC dose = 0.1 g/L	50	[210]
Integrated MBR-PAC	5000	SRT = infinite HRT = 24 h PAC dose = 0.5 g/L	90	[210]
Integrated Gamma radiation-CAS (batch experiment)	1.7×10^7	Incubation time = 10 d Radiation dose = 800 Gy	>99	[239]
Integrated PAC-UF	3×10^5	PAC dose = 5–10 mg/L Contact time = 1.5 h	40	[240]
Integrated MBR-PAC	7.5×10^5	SRT = infinite HRT = 24 h PAC dose = 0.1 and 1 g/L	34 and 90	[211]
Integrated MBR-PAC	2×10^4	SRT = infinite HRT = 24 h PAC dose = 1 g/L	>99	[241]
Integrated MBR-PAC	390–1800	SRT = 20 d HRT = 24 h PAC dose = 1 g/L	99.4	[206]
Integrated CAS-GAC	2×10^4	SRT = NA HRT = NA GAC dose = 100–1000 mg/L	10–50	[203]
MBR followed by GAC	5000	SRT of MBR = infinite HRT of MBR = 24 h GAC contact time = 7 min	98	[242]
Ozonation followed by biological sand column	2.06	HRT of sand column = 5–6 d Ozone contact time = NA	80	[243]

Note: NA: not available.

Kleywegt et al. [244] investigated the removal of CBZ by GAC adsorption followed by UV photolysis that achieved effective removal of CBZ (93%). Serrano et al. [203] investigated the removal of PhACs including CBZ in a CAS process followed by GAC adsorption. They reported that adsorption onto activated carbon improved the overall removal of CBZ by as much as 40% [203]. Nguyen et al. [210] also reported an overall CBZ removal of 98% by an MBR followed by GAC adsorption. When a sequencing batch reactor coupled to an external microfiltration membrane was investigated by Serrano et al. [241], up to 93% CBZ removal was achieved following the addition of a single dose (1 g/L) of PAC directly into the bioreactor. Likewise, the integrated MBR-PAC system was also reported to to achieve 92% CBZ removal [211]. CBZ removal by MBR treatment followed by GAC filtration was investigated by Nguyen et al. [210]. While MBR alone showed a removal of less than 20%, MBR-GAC achieved an extremely efficient removal of 98% [210]. This result demonstrates that GAC post-treatment could significantly improve the removal of PhACs, which are resistant to degradation by the activated sludge.

Other integrated/hybrid technologies have also shown effective CBZ removal. AOPs cannot mineralize PhACs, however, it is important to note that the metabolites formed following the oxidation of CBZ have been reported to be readily mineralized by the activated sludge [245–247]. For instance, Hübner et al. [243] studied CBZ removal by combining ozonation with a sand column mimicking a soil aquifer treatment (SAT) systems. They observed that the degradation products of CBZ formed after ozonation were significantly mineralized (>80%) in the sand column at an HRT of 5-6 days [243]. In another study [239], a combination of gamma radiation and activated sludge based biological treatment achieved up to 79% CBZ mineralization. They reported that significant CBZ removal (>99%) was mainly achieved by gamma radiation at an initial dose of 800 Gy, while activated sludge was responsible for mineralization of CBZ [239]. In a study by Laerae et al. [238], an integrated MBR-TiO$_2$/UV system achieved up to 95% CBZ removal from pharmaceutical industrial effluent, showing that the integration of biological and chemical oxidation processes can be an effective strategy for enhanced CBZ removal. Despite the efficacy of combined/integrated oxidation and biological processes for mineralization, the cost associated with the application of oxidation processes needs to be considered.

Carbamazepine has been consistently shown to be poorly removed by coagulation despite being a neutral compound [21,197,248–250]. However, a properly designed coagulation/flocculation unit can efficiently remove suspended solids and can thereby enhance the performance of a subsequent activated carbon adsorption unit by reducing competitive adsorption [151]. Coagulation pre-treatment has been found to significantly enhance carbamazepine removal efficiency by adsorption [197]. Because high concentrations of suspended or colloidal solids in the wastewater may impede the advanced oxidation processes, sufficient prior removal of these materials by a physicochemical treatment such as coagulation is required [251,252].

7. Fate and Metabolites of CBZ

CBZ transformation products following its metabolism in human body or following its degradation by different treatment processes have been reported (Figure 1). Figure 1 sheds light on different pathways of CBZ metabolism/degradation by AOPs, human liver, and microorganisms. Huerta-Fontela et al. [199] reported that CBZ degradation by ozonation occurs via a ring opening mechanism due to the attack of ozone on the non-aromatic carbon-carbon double bond of CBZ, forming the metabolite epoxy-carbamazepine. Compared to the CBZ degradation products reported by McDowell et al. [253] in UV photolysis (Figure 1), Vogna et al. [187] observed that the addition of H$_2$O$_2$ to UV photolysis yields different degradation products (Figure 1), indicating a difference in the degradation pathway of CBZ in presence of H$_2$O$_2$. In biological systems (such as human liver, fungus and activated sludge), CBZ degradation products usually contain the azepine structure (Figure 1).

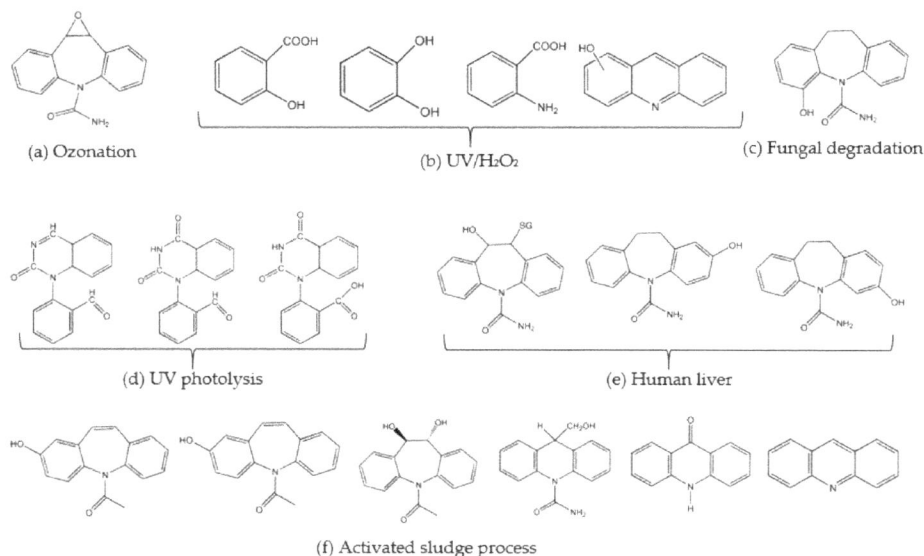

Figure 1. Metabolites/degradation products formed following CBZ degradation by (**a**) ozonation [199]; (**b**) UV/H$_2$O$_2$ [187]; (**c**) fungal degradation [254]; (**d**) UV photolysis [253]; (**e**) human liver [255]; (**f**) activated sludge process [256].

As depicted in Figure 1, several metabolites/degradation products have been reported for CBZ treatment by different processes. Toxicity of CBZ degradation products has been reported in a number of studies as summarized here. Based on the Yeast Estrogen Screen (YES) assay, Mohapatra et al. [257] reported that CBZ and its degradation products showed no estrogenic activity. In another study, Jelic et al. [160] investigated the toxicity of the media in a pulsed fluidized bed bioreactor containing whole-cell *T. versicolor* and CBZ at an initial concentration of 500 µg/L. The acute toxicity test (Microtox) showed the toxicity induced by CBZ was reduced from 95% to 24% after an incubation time of 10 days, suggesting that the degradation products were non-toxic [258]. Similarly, degradation products following the oxidation of CBZ by ozonation, UV and UV/H$_2$O$_2$ processes exhibited no genotoxic, cytotoxic or estrogenic effects [258]. CBZ metabolites including 10,11-dihydro-trans-10,11-dihydroxy-carbamazepine, acidone and acridine formed during CAS process [256] have also been reported to be non-toxic [258,259]. Although degradation products or metabolites of CBZ formed during CAS and AOPs are non-toxic, Bu et al. [255] reported that CBZ-2,3-arene (one oxide intermediate) is believed to cause idiosyncratic effect after CBZ consumption for medicinal purposes [255].

8. Conclusions and Outlook

Although its occurrence in freshwater may not pose an immediate threat to aquatic ecosystems or human health, effective removal of CBZ is still required for safe water reuse applications and drinking water treatment. Biological wastewater treatment processes such as conventional activated sludge and membrane bioreactor are not effective for CBZ removal due to its resistance to biodegradation. However, advanced wastewater treatment processes seem to be effective for efficient CBZ removal. For instance, CBZ removal by the reverse osmosis and nanofiltration membranes is above 90%. Similarly, post-treatment with granular and powdered activated carbon provides efficient CBZ removal ranging from 90–99%. However, membrane fouling in case of membrane technologies and regeneration of activated carbon are obstacles that warrant technical solutions. Depending on the type of fungal species and operating condition, white-rot fungi can achieve almost complete removal of CBZ but bacterial contamination may affect the efficacy of fungal bioreactors during long term operations. Although advanced oxidation processes such as ozonation, UV/H_2O_2, UV/TiO_2 and Fenton processes are effective for CBZ removal, costs associated with the addition of chemicals, and separation of catalysts needs to be carefully considered. Finally, the literature to date suggests that degradation products formed following the degradation of CBZ by biological and chemical oxidation processes may not induce toxic effects on aquatic ecosystems.

Acknowledgments: This research has been conducted with the support of the Australian Government Research Training Program Scholarship. This study was partially funded by the GeoQuEST Research Centre and Faculty of EIS strategic partnership grant, University of Wollongong, Australia.

Author Contributions: F.I.H. conceived and led the project. F.I.H., M.B.A. and S.Y. planned and conducted the literature survey and prepared the manuscript in consultation with the coauthors. The co-authors contributed to specific sections of the manuscript.

Conflicts of Interest: The authors declare no conflict of interest.

References

1. Hughes, S.R.; Kay, P.; Brown, L.E. Global synthesis and critical evaluation of pharmaceutical data sets collected from river systems. *Environ. Sci. Technol.* **2012**, *47*, 661–677. [CrossRef] [PubMed]
2. Gavrilescu, M.; Demnerová, K.; Aamand, J.; Agathos, S.; Fava, F. Emerging pollutants in the environment: Present and future challenges in biomonitoring, ecological risks and bioremediation. *New Biotechnol.* **2015**, *32*, 147–156. [CrossRef] [PubMed]
3. Hai, F.I.; Nghiem, L.D.; Khan, S.J.; Price, W.E.; Yamamoto, K. Wastewater reuse: Removal of emerging trace organic contaminants. In *Membrane biological Reactors: Theory, Modeling, Design, Management and Applications to Wastewater Reuse*; Hai, F.I., Yamamoto, K., Lee, C., Eds.; IWA publishing: London, UK, 2014; pp. 165–205. ISBN 9781780400655.
4. Luo, Y.; Guo, W.; Ngo, H.H.; Nghiem, L.D.; Hai, F.I.; Zhang, J.; Liang, S.; Wang, X.C. A review on the occurrence of micropollutants in the aquatic environment and their fate and removal during wastewater treatment. *Sci. Total Environ.* **2014**, *473*, 619–641. [CrossRef] [PubMed]
5. Lapworth, D.; Baran, N.; Stuart, M.; Ward, R. Emerging organic contaminants in groundwater: A review of sources, fate and occurrence. *Environ. Pollut.* **2012**, *163*, 287–303. [CrossRef] [PubMed]
6. Vieno, N.; Tuhkanen, T.; Kronberg, L. Elimination of pharmaceuticals in sewage treatment plants in finland. *Water Res.* **2007**, *41*, 1001–1012. [CrossRef] [PubMed]
7. Brack, W.; Dulio, V.; Ågerstrand, M.; Allan, I.; Altenburger, R.; Brinkmann, M.; Bunke, D.; Burgess, R.M.; Cousins, I.; Escher, B.I. Towards the review of the european union water framework directive: Recommendations for more efficient assessment and management of chemical contamination in european surface water resources. *Sci. Total Environ.* **2017**, *576*, 720–737. [CrossRef] [PubMed]
8. Kumar, A.; Batley, G.E.; Nidumolu, B.; Hutchinson, T.H. Derivation of water quality guidelines for priority pharmaceuticals. *Environ. Toxicol. Chem.* **2016**, *35*, 1815–1824. [CrossRef] [PubMed]
9. Osorio, M.V.; Reis, S.; Lima, J.L.; Segundo, M.A. Analytical features of diclofenac evaluation in water as a potential marker of anthropogenic pollution. *Curr. Pharm. Anal.* **2017**, *13*, 39–47. [CrossRef]
10. Arye, G.; Dror, I.; Berkowitz, B. Fate and transport of carbamazepine in soil aquifer treatment (sat) infiltration basin soils. *Chemosphere* **2011**, *82*, 244–252. [CrossRef] [PubMed]
11. Ferrer, I.; Thurman, E.M. Analysis of 100 pharmaceuticals and their degradates in water samples by liquid chromatography/quadrupole time-of-flight mass spectrometry. *J. Chromatogr. A* **2012**, *1259*, 148–157. [CrossRef] [PubMed]
12. Wick, A.; Fink, G.; Joss, A.; Siegrist, H.; Ternes, T.A. Fate of beta blockers and psycho-active drugs in conventional wastewater treatment. *Water Res.* **2009**, *43*, 1060–1074. [CrossRef] [PubMed]
13. Alvarino, T.; Suarez, S.; Lema, J.; Omil, F. Understanding the removal mechanisms of ppcps and the influence of main technological parameters in anaerobic uasb and aerobic cas reactors. *J. Hazard. Mater.* **2014**, *278*, 506–513. [CrossRef] [PubMed]

14. Siegrist, H.; Joss, A. Review on the fate of organic micropollutants in wastewater treatment and water reuse with membranes. *Water Sci. Technol.* **2012**, *66*, 1369–1376. [CrossRef] [PubMed]

15. Zhou, S.; Xia, Y.; Li, T.; Yao, T.; Shi, Z.; Zhu, S.; Gao, N. Degradation of carbamazepine by uv/chlorine advanced oxidation process and formation of disinfection by-products. *Environ. Sci. Pollut. Res.* **2016**, *23*, 16448–16455. [CrossRef] [PubMed]

16. Zhang, Y.; Geißen, S.-U.; Gal, C. Carbamazepine and diclofenac: Removal in wastewater treatment plants and occurrence in water bodies. *Chemosphere* **2008**, *73*, 1151–1161. [CrossRef] [PubMed]

17. Yu, Z.; Peldszus, S.; Huck, P.M. Adsorption characteristics of selected pharmaceuticals and an endocrine disrupting compound-naproxen, carbamazepine and nonylphenol-on activated carbon. *Water Res.* **2008**, *42*, 2873–2882. [CrossRef] [PubMed]

18. Deblonde, T.; Cossu-Leguille, C.; Hartemann, P. Emerging pollutants in wastewater: A review of the literature. *Int. J. Hyg. Environ. Health* **2011**, *214*, 442–448. [CrossRef] [PubMed]

19. Bolong, N.; Ismail, A.; Salim, M.R.; Matsuura, T. A review of the effects of emerging contaminants in wastewater and options for their removal. *Desalination* **2009**, *239*, 229–246. [CrossRef]

20. Verlicchi, P.; Al Aukidy, M.; Zambello, E. Occurrence of pharmaceutical compounds in urban wastewater: Removal, mass load and environmental risk after a secondary treatment—A review. *Sci. Total Environ.* **2012**, *429*, 123–155. [CrossRef] [PubMed]

21. Alexander, J.T.; Hai, F.I.; Al-aboud, T.M. Chemical coagulation-based processes for trace organic contaminant removal: Current state and future potential. *J. Environ. Manag.* **2012**, *111*, 195–207. [CrossRef] [PubMed]

22. Kasprzyk-Hordern, B.; Dinsdale, R.M.; Guwy, A.J. Illicit drugs and pharmaceuticals in the environment—Forensic applications of environmental data. Part 1: Estimation of the usage of drugs in local communities. *Environ. Pollut.* **2009**, *157*, 1773–1777. [CrossRef] [PubMed]

23. Singer, H.; Jaus, S.; Hanke, I.; Lück, A.; Hollender, J.; Alder, A.C. Determination of biocides and pesticides by on-line solid phase extraction coupled with mass spectrometry and their behaviour in wastewater and surface water. *Environ. Pollut.* **2010**, *158*, 3054–3064. [CrossRef] [PubMed]

24. Dvory, N.Z.; Kuznetsov, M.; Livshitz, Y.; Gasser, G.; Pankratov, I.; Lev, O.; Adar, E.; Yakirevich, A. Modeling sewage leakage and transport in carbonate aquifer using carbamazepine as an indicator. *Water Res.* **2018**, *128*, 157–170. [CrossRef] [PubMed]

25. Ying, G.-G.; Kookana, R.S.; Kolpin, D.W. Occurrence and removal of pharmaceutically active compounds in sewage treatment plants with different technologies. *J. Environ. Monit.* **2009**, *11*, 1498–1505. [CrossRef] [PubMed]

26. Tixier, C.; Singer, H.P.; Oellers, S.; Müller, S.R. Occurrence and fate of carbamazepine, clofibric acid, diclofenac, ibuprofen, ketoprofen, and naproxen in surface waters. *Environ. Sci. Technol.* **2003**, *37*, 1061–1068. [CrossRef] [PubMed]

27. Wijekoon, K.C.; Hai, F.I.; Kang, J.; Price, W.E.; Guo, W.; Ngo, H.H.; Nghiem, L.D. The fate of pharmaceuticals, steroid hormones, phytoestrogens, uv-filters and pesticides during mbr treatment. *Bioresour. Technol.* **2013**, *144*, 247–254. [CrossRef] [PubMed]

28. Guo, Y.C.; Krasner, S.W. Occurence of primidone, carbamazepine, caffeine and precursors for n-nitrosodimethylamine in drinding water sources impacted by wastewater. *J. Am. Water Resour. Assoc.* **2009**, *45*, 58–67. [CrossRef]

29. Heberer, T. Occurrence, fate, and removal of pharmaceutical residues in the aquatic environment: A review of recent research data. *Toxicol. Lett.* **2002**, *131*, 5–17. [CrossRef]

30. Stuart, M.; Lapworth, D.; Crane, E.; Hart, A. Review of risk from potential emerging contaminants in UK groundwater. *Sci. Total Environ.* **2012**, *416*, 1–21. [CrossRef] [PubMed]

31. Metcalfe, C.D.; Miao, X.-S.; Koenig, B.G.; Struger, J. Distribution of acidic and neutral drugs in surface waters near sewage treatment plants in the lower great lakes, canada. *Environ. Toxicol. Chem.* **2003**, *22*, 2881–2889. [CrossRef] [PubMed]

32. Miao, X.-S.; Metcalfe, C.D. Determination of carbamazepine and its metabolites in aqueous samples using liquid chromatography−electrospray tandem mass spectrometry. *Anal. Chem.* **2003**, *75*, 3731–3738. [CrossRef] [PubMed]

33. Miao, X.-S.; Yang, J.-J.; Metcalfe, C.D. Carbamazepine and its metabolites in wastewater and in biosolids in a municipal wastewater treatment plant. *Environ. Sci. Technol.* **2005**, *39*, 7469–7475. [CrossRef] [PubMed]

34. Gagné, F.; Blaise, C.; André, C. Occurrence of pharmaceutical products in a municipal effluent and toxicity to rainbow trout (*Oncorhynchus mykiss*) hepatocytes. *Ecotoxicol. Environ. Saf.* **2006**, *64*, 329–336. [CrossRef] [PubMed]

35. Yu, Z.; Peldszus, S.; Huck, P.M. Optimizing gas chromatographic–mass spectrometric analysis of selected pharmaceuticals and endocrine-disrupting substances in water using factorial experimental design. *J. Chromatogr. A* **2007**, *1148*, 65–77. [CrossRef] [PubMed]

36. Gottschall, N.; Topp, E.; Metcalfe, C.; Edwards, M.; Payne, M.; Kleywegt, S.; Russell, P.; Lapen, D.R. Pharmaceutical and personal care products in groundwater, subsurface drainage, soil, and wheat grain, following a high single application of municipal biosolids to a field. *Chemosphere* **2012**, *87*, 194–203. [CrossRef] [PubMed]

37. Edwards, M.; Topp, E.; Metcalfe, C.D.; Li, H.; Gottschall, N.; Bolton, P.; Curnoe, W.; Payne, M.; Beck, A.; Kleywegt, S.; et al. Pharmaceutical and personal care products in tile drainage following surface spreading and injection of dewatered municipal biosolids to an agricultural field. *Sci. Total Environ.* **2009**, *407*, 4220–4230. [CrossRef] [PubMed]

38. Hummel, D.; Löffler, D.; Fink, G.; Ternes, T.A. Simultaneous determination of psychoactive drugs and their metabolites in aqueous matrices by liquid chromatography mass spectrometry. *Environ. Sci. Technol.* **2006**, *40*, 7321–7328. [CrossRef] [PubMed]

39. Ternes, T.A.; Stüber, J.; Herrmann, N.; McDowell, D.; Ried, A.; Kampmann, M.; Teiser, B. Ozonation: A tool for removal of pharmaceuticals, contrast media and musk fragrances from wastewater? *Water Res.* **2003**, *37*, 1976–1982. [CrossRef]

40. Ternes, T.A. Occurrence of drugs in german sewage treatment plants and rivers. *Water Res.* **1998**, *32*, 3245–3260. [CrossRef]

41. Wiegel, S.; Aulinger, A.; Brockmeyer, R.; Harms, H.; Löffler, J.; Reincke, H.; Schmidt, R.; Stachel, B.; von Tümpling, W.; Wanke, A. Pharmaceuticals in the river elbe and its tributaries. *Chemosphere* **2004**, *57*, 107–126. [CrossRef] [PubMed]

42. Musolff, A.; Leschik, S.; Möder, M.; Strauch, G.; Reinstorf, F.; Schirmer, M. Temporal and spatial patterns of micropollutants in urban receiving waters. *Environ. Pollut.* **2009**, *157*, 3069–3077. [CrossRef] [PubMed]

43. Osenbrück, K.; Gläser, H.-R.; Knöller, K.; Weise, S.M.; Möder, M.; Wennrich, R.; Schirmer, M.; Reinstorf, F.; Busch, W.; Strauch, G. Sources and transport of selected organic micropollutants in urban groundwater underlying the city of halle (saale), Germany. *Water Res.* **2007**, *41*, 3259–3270. [CrossRef] [PubMed]

44. Okuda, T.; Kobayashi, Y.; Nagao, R.; Yamashita, N.; Tanaka, H.; Tanaka, S.; Fujii, S.; Konishi, C.; Houwa, I. Removal efficiency of 66 pharmaceuticals during wastewater treatment process in Japan. *Water Sci. Technol.* **2008**, *57*, 65–71. [CrossRef] [PubMed]

45. Nakada, N.; Komori, K.; Suzuki, Y.; Konishi, C.; Houwa, I.; Tanaka, H. Occurrence of 70 pharmaceutical and personal care products in tone river basin in Japan. *Water Sci. Technol.* **2007**, *56*, 133–140. [CrossRef] [PubMed]

46. Nakada, N.; Kiri, K.; Shinohara, H.; Harada, A.; Kuroda, K.; Takizawa, S.; Takada, H. Evaluation of pharmaceuticals and personal care products as water-soluble molecular markers of sewage. *Environ. Sci. Technol.* **2008**, *42*, 6347–6353. [CrossRef] [PubMed]

47. Kuroda, K.; Murakami, M.; Oguma, K.; Muramatsu, Y.; Takada, H.; Takizawa, S. Assessment of groundwater pollution in tokyo using ppcps as sewage markers. *Environ. Sci. Technol.* **2011**, *46*, 1455–1464. [CrossRef] [PubMed]

48. Choi, K.; Kim, Y.; Park, J.; Park, C.K.; Kim, M.; Kim, H.S.; Kim, P. Seasonal variations of several pharmaceutical residues in surface water and sewage treatment plants of Han River, Korea. *Sci. Total Environ.* **2008**, *405*, 120–128. [CrossRef] [PubMed]

49. Behera, S.K.; Kim, H.W.; Oh, J.-E.; Park, H.-S. Occurrence and removal of antibiotics, hormones and several other pharmaceuticals in wastewater treatment plants of the largest industrial city of Korea. *Sci. Total Environ.* **2011**, *409*, 4351–4360. [CrossRef] [PubMed]

50. Kim, J.-W.; Jang, H.-S.; Kim, J.-G.; Ishibashi, H.; Hirano, M.; Nasu, K.; Ichikawa, N.; Takao, Y.; Shinohara, R.; Arizono, K. Occurrence of pharmaceutical and personal care products (ppcps) in surface water from Mankyung River, South Korea. *J. Health Sci.* **2009**, *55*, 249–258. [CrossRef]

51. Yoon, Y.; Ryu, J.; Oh, J.; Choi, B.-G.; Snyder, S.A. Occurrence of endocrine disrupting compounds, pharmaceuticals, and personal care products in the Han river (Seoul, South Korea). *Sci. Total Environ.* **2010**, *408*, 636–643. [CrossRef] [PubMed]

52. Chen, H.-C.; Wang, P.-L.; Ding, W.-H. Using liquid chromatography–ion trap mass spectrometry to determine pharmaceutical residues in taiwanese rivers and wastewaters. *Chemosphere* **2008**, *72*, 863–869. [CrossRef] [PubMed]

53. Lin, W.-C.; Chen, H.-C.; Ding, W.-H. Determination of pharmaceutical residues in waters by solid-phase extraction and large-volume on-line derivatization with gas chromatography–mass spectrometry. *J. Chromatogr. A* **2005**, *1065*, 279–285. [CrossRef] [PubMed]

54. Kasprzyk-Hordern, B.; Dinsdale, R.M.; Guwy, A.J. The removal of pharmaceuticals, personal care products, endocrine disruptors and illicit drugs during wastewater treatment and its impact on the quality of receiving waters. *Water Res.* **2009**, *43*, 363–380. [CrossRef] [PubMed]

55. Zhang, Z.L.; Zhou, J.L. Simultaneous determination of various pharmaceutical compounds in water by solid-phase extraction–liquid chromatography–tandem mass spectrometry. *J. Chromatogr. A* **2007**, *1154*, 205–213. [CrossRef] [PubMed]

56. Zhou, J.L.; Zhang, Z.L.; Banks, E.; Grover, D.; Jiang, J.Q. Pharmaceutical residues in wastewater treatment works effluents and their impact on receiving river water. *J. Hazard. Mater.* **2009**, *166*, 655–661. [CrossRef] [PubMed]

57. Kasprzyk-Hordern, B.; Dinsdale, R.M.; Guwy, A.J. The occurrence of pharmaceuticals, personal care products, endocrine disruptors and illicit drugs in surface water in south wales, UK. *Water Res.* **2008**, *42*, 3498–3518. [CrossRef] [PubMed]

58. Kasprzyk-Hordern, B.; Dinsdale, R.M.; Guwy, A.J. Multi-residue method for the determination of basic/neutral pharmaceuticals and illicit drugs in surface water by solid-phase extraction and ultra performance liquid chromatography–positive electrospray ionisation tandem mass spectrometry. *J. Chromatogr. A* **2007**, *1161*, 132–145. [CrossRef] [PubMed]

59. Stuart, M.E.; Manamsa, K.; Talbot, J.C.; Crane, E.J. Emerging Contaminants in Groundwater, 2011. Available online: https://nora.nerc.ac.uk/14557/ (accessed on 15 October 2017).

60. Lapworth, D.; Stuart, M.; Hart, A.; Crane, E.; Baran, N. *Emerging Contaminants in Groundwater*; Groundwater Forum: London, UK, 2011; Available online: http://nora.nerc.ac.uk/14093/ (accessed on 15 October 2017).

61. Spongberg, A.L.; Witter, J.D. Pharmaceutical compounds in the wastewater process stream in Northwest Ohio. *Sci. Total Environ.* **2008**, *397*, 148–157. [CrossRef] [PubMed]

62. Vanderford, B.J.; Snyder, S.A. Analysis of pharmaceuticals in water by isotope dilution liquid chromatography/tandem mass spectrometry. *Environ. Sci. Technol.* **2006**, *40*, 7312–7320. [CrossRef] [PubMed]

63. Glassmeyer, S.T.; Furlong, E.T.; Kolpin, D.W.; Cahill, J.D.; Zaugg, S.D.; Werner, S.L.; Meyer, M.T.; Kryak, D.D. Transport of chemical and microbial compounds from known wastewater discharges: Potential for use as indicators of human fecal contamination. *Environ. Sci. Technol.* **2005**, *39*, 5157–5169. [CrossRef] [PubMed]

64. Kolpin, D.W.; Skopec, M.; Meyer, M.T.; Furlong, E.T.; Zaugg, S.D. Urban contribution of pharmaceuticals and other organic wastewater contaminants to streams during differing flow conditions. *Sci. Total Environ.* **2004**, *328*, 119–130. [CrossRef] [PubMed]

65. Conley, J.M.; Symes, S.J.; Kindelberger, S.A.; Richards, S.M. Rapid liquid chromatography–tandem mass spectrometry method for the determination of a broad mixture of pharmaceuticals in surface water. *J. Chromatogr. A* **2008**, *1185*, 206–215. [CrossRef] [PubMed]

66. Katz, B.G.; Griffin, D.W.; Davis, J.H. Groundwater quality impacts from the land application of treated municipal wastewater in a large karstic spring basin: Chemical and microbiological indicators. *Sci. Total Environ.* **2009**, *407*, 2872–2886. [CrossRef] [PubMed]

67. Fram, M.S.; Belitz, K. Occurrence and concentrations of pharmaceutical compounds in groundwater used for public drinking-water supply in california. *Sci. Total Environ.* **2011**, *409*, 3409–3417. [CrossRef] [PubMed]

68. Jelić, A.; Gros, M.; Petrović, M.; Ginebreda, A.; Barceló, D. Occurrence and elimination of pharmaceuticals during conventional wastewater treatment. In *Emerging and Priority Pollutants in Rivers*; Guasch, H., Ginebreda, A., Geiszinger, A., Eds.; Springer: Berlin, Germany, 2012; pp. 1–23. ISBN 9783642257223.

69. Petrovic, M.; de Alda, M.J.L.; Diaz-Cruz, S.; Postigo, C.; Radjenovic, J.; Gros, M.; Barcelo, D. Fate and removal of pharmaceuticals and illicit drugs in conventional and membrane bioreactor wastewater treatment plants

and by riverbank filtration. *Philos. Trans. R. Soc. A Math. Phys. Eng. Sci.* **2009**, *367*, 3979–4003. [CrossRef] [PubMed]

70. Al-Rifai, J.H.; Gabelish, C.L.; Schäfer, A.I. Occurrence of pharmaceutically active and non-steroidal estrogenic compounds in three different wastewater recycling schemes in Australia. *Chemosphere* **2007**, *69*, 803–815. [CrossRef] [PubMed]

71. Clara, M.; Strenn, B.; Gans, O.; Martinez, E.; Kreuzinger, N.; Kroiss, H. Removal of selected pharmaceuticals, fragrances and endocrine disrupting compounds in a membrane bioreactor and conventional wastewater treatment plants. *Water Res.* **2005**, *39*, 4797–4807. [CrossRef] [PubMed]

72. Clara, M.; Strenn, B.; Kreuzinger, N. Carbamazepine as a possible anthropogenic marker in the aquatic environment: Investigations on the behaviour of carbamazepine in wastewater treatment and during groundwater infiltration. *Water Res.* **2004**, *38*, 947–954. [CrossRef] [PubMed]

73. Vieno, N.M.; Tuhkanen, T.; Kronberg, L. Analysis of neutral and basic pharmaceuticals in sewage treatment plants and in recipient rivers using solid phase extraction and liquid chromatography–tandem mass spectrometry detection. *J. Chromatogr. A* **2006**, *1134*, 101–111. [CrossRef] [PubMed]

74. Joss, A.; Keller, E.; Alder, A.C.; Göbel, A.; McArdell, C.S.; Ternes, T.; Siegrist, H. Removal of pharmaceuticals and fragrances in biological wastewater treatment. *Water Res.* **2005**, *39*, 3139–3152. [CrossRef] [PubMed]

75. Vochezer, K. *Modelling of Carbamazepine and Diclofenac in a River Network—Photolytic Degradation in Swiss Rivers*; Swedish University of Agricultural Sciences: Uppsala, Sweden, 2010.

76. Modick, H.; Weiss, T.; Dierkes, G.; Brüning, T.; Koch, H.M. Ubiquitous presence of paracetamol in human urine: Sources and implications. *Reproduction* **2014**, *147*, R105–R117. [CrossRef] [PubMed]

77. Camacho-Muñoz, D.; Martín, J.; Santos, J.L.; Aparicio, I.; Alonso, E. Concentration evolution of pharmaceutically active compounds in raw urban and industrial wastewater. *Chemosphere* **2014**, *111*, 70–79. [CrossRef] [PubMed]

78. Rivera-Jaimes, J.A.; Postigo, C.; Melgoza-Alemán, R.M.; Aceña, J.; Barceló, D.; de Alda, M.L. Study of pharmaceuticals in surface and wastewater from cuernavaca, morelos, mexico: Occurrence and environmental risk assessment. *Sci. Total Environ.* **2018**, *613*, 1263–1274. [CrossRef] [PubMed]

79. Yu, Y.; Wu, L.; Chang, A.C. Seasonal variation of endocrine disrupting compounds, pharmaceuticals and personal care products in wastewater treatment plants. *Sci. Total Environ.* **2013**, *442*, 310–316. [CrossRef] [PubMed]

80. Al Aukidy, M.; Verlicchi, P.; Jelic, A.; Petrovic, M.; Barcelò, D. Monitoring release of pharmaceutical compounds: Occurrence and environmental risk assessment of two wwtp effluents and their receiving bodies in the Po Valley, Italy. *Sci. Total Environ.* **2012**, *438*, 15–25. [CrossRef] [PubMed]

81. Pal, A.; Gin, K.Y.-H.; Lin, A.Y.-C.; Reinhard, M. Impacts of emerging organic contaminants on freshwater resources: Review of recent occurrences, sources, fate and effects. *Sci. Total Environ.* **2010**, *408*, 6062–6069. [CrossRef] [PubMed]

82. Gioia, R.; Dachs, J. The riverine input–output paradox for organic pollutants. *Front. Ecol. Environ.* **2012**, *10*, 405–406. [CrossRef]

83. Kunkel, U.; Radke, M. Fate of pharmaceuticals in rivers: Deriving a benchmark dataset at favorable attenuation conditions. *Water Res.* **2012**, *46*, 5551–5565. [CrossRef] [PubMed]

84. Acuña, V.; von Schiller, D.; García-Galán, M.J.; Rodríguez-Mozaz, S.; Corominas, L.; Petrovic, M.; Poch, M.; Barceló, D.; Sabater, S. Occurrence and in-stream attenuation of wastewater-derived pharmaceuticals in iberian rivers. *Sci. Total Environ.* **2015**, *503*, 133–141. [CrossRef] [PubMed]

85. Writer, J.H.; Antweiler, R.C.; Ferrer, I.; Ryan, J.N.; Thurman, E.M. In-stream attenuation of neuro-active pharmaceuticals and their metabolites. *Environ. Sci. Technol.* **2013**, *47*, 9781–9790. [CrossRef] [PubMed]

86. Gómez, M.J.; Herrera, S.; Solé, D.; García-Calvo, E.; Fernández-Alba, A.R. Spatio-temporal evaluation of organic contaminants and their transformation products along a river basin affected by urban, agricultural and industrial pollution. *Sci. Total Environ.* **2012**, *420*, 134–145. [CrossRef] [PubMed]

87. Wang, C.; Shi, H.; Adams, C.D.; Gamagedara, S.; Stayton, I.; Timmons, T.; Ma, Y. Investigation of pharmaceuticals in missouri natural and drinking water using high performance liquid chromatography-tandem mass spectrometry. *Water Res.* **2011**, *45*, 1818–1828. [CrossRef] [PubMed]

88. Christoffels, E.; Brunsch, A.; Wunderlich-Pfeiffer, J.; Mertens, F.M. Monitoring micropollutants in the swist river basin. *Water Sci. Technol.* **2016**, *74*, 2280–2296. [CrossRef] [PubMed]

89. Kickham, P.; Otton, S.; Moore, M.M.; Ikonomou, M.G.; Gobas, F.A. Relationship between biodegradation and sorption of phthalate esters and their metabolites in natural sediments. *Environ. Toxicol. Chem.* **2012**, *31*, 1730–1737. [CrossRef] [PubMed]

90. Riml, J.; Wörman, A.; Kunkel, U.; Radke, M. Evaluating the fate of six common pharmaceuticals using a reactive transport model: Insights from a stream tracer test. *Sci. Total Environ.* **2013**, *458*, 344–354. [CrossRef] [PubMed]

91. Yamamoto, H.; Nakamura, Y.; Moriguchi, S.; Nakamura, Y.; Honda, Y.; Tamura, I.; Hirata, Y.; Hayashi, A.; Sekizawa, J. Persistence and partitioning of eight selected pharmaceuticals in the aquatic environment: Laboratory photolysis, biodegradation, and sorption experiments. *Water Res.* **2009**, *43*, 351–362. [CrossRef] [PubMed]

92. Weigel, S.; Berger, U.; Jensen, E.; Kallenborn, R.; Thoresen, H.; Hühnerfuss, H. Determination of selected pharmaceuticals and caffeine in sewage and seawater from tromsø/norway with emphasis on ibuprofen and its metabolites. *Chemosphere* **2004**, *56*, 583–592. [CrossRef] [PubMed]

93. Weigel, S.; Bester, K.; Hühnerfuss, H. New method for rapid solid-phase extraction of large-volume water samples and its application to non-target screening of north sea water for organic contaminants by gas chromatography–mass spectrometry. *J. Chromatogr. A* **2001**, *912*, 151–161. [CrossRef]

94. Kinney, C.A.; Furlong, E.T.; Werner, S.L.; Cahill, J.D. Presence and distribution of wastewater-derived pharmaceuticals in soil irrigated with reclaimed water. *Environ. Toxicol. Chem.* **2006**, *25*, 317–326. [CrossRef] [PubMed]

95. Schlenker, G. Pharmaceuticals in the environment. In *Encyclopedia of Quantitative Risk Analysis and Assessment*; Melnick, E.L., Everitt, B.S., Eds.; John Wiley & Sons, Ltd.: Hoboken, NJ, USA, 2008; Volume 3, ISBN 9780470035498.

96. Thacker, P. Pharmaceutical data elude researchers. *Environ. Sci. Technol.* **2005**, *39*, 193A. [PubMed]

97. Postigo, C.; Barceló, D. Synthetic organic compounds and their transformation products in groundwater: Occurrence, fate and mitigation. *Sci. Total Environ.* **2015**, *503*, 32–47. [CrossRef] [PubMed]

98. Chefetz, B.; Mualem, T.; Ben-Ari, J. Sorption and mobility of pharmaceutical compounds in soil irrigated with reclaimed wastewater. *Chemosphere* **2008**, *73*, 1335–1343. [CrossRef] [PubMed]

99. Christou, A.; Agüera, A.; Bayona, J.M.; Cytryn, E.; Fotopoulos, V.; Lambropoulou, D.; Manaia, C.M.; Michael, C.; Revitt, M.; Schröder, P. The potential implications of reclaimed wastewater reuse for irrigation on the agricultural environment: The knowns and unknowns of the fate of antibiotics and antibiotic resistant bacteria and resistance genes—A review. *Water Res.* **2017**, *123*, 448–467. [CrossRef] [PubMed]

100. Ferrari, B.T.; Paxéus, N.; Giudice, R.L.; Pollio, A.; Garric, J. Ecotoxicological impact of pharmaceuticals found in treated wastewaters: Study of carbamazepine, clofibric acid, and diclofenac. *Ecotoxicol. Environ. Saf.* **2003**, *55*, 359–370. [CrossRef]

101. Han, G.H.; Hur, H.G.; Kim, S.D. Ecotoxicological risk of pharmaceuticals from wastewater treatment plants in Korea: Occurrence and toxicity to daphnia magna. *Environ. Toxicol. Chem.* **2006**, *25*, 265–271. [CrossRef] [PubMed]

102. Dussault, E.B.; Balakrishnan, V.K.; Sverko, E.; Solomon, K.R.; Sibley, P.K. Toxicity of human pharmaceuticals and personal care products to benthic invertebrates. *Environ. Toxicol. Chem.* **2008**, *27*, 425–432. [CrossRef] [PubMed]

103. Jos, A.; Repetto, G.; Rios, J.C.; Hazen, M.J.; Molero, M.L.; del Peso, A.; Salguero, M.; Fernández-Freire, P.; Pérez-Martín, J.M.; Cameán, A. Ecotoxicological evaluation of carbamazepine using six different model systems with eighteen endpoints. *Toxicol. In Vitro* **2003**, *17*, 525–532. [CrossRef]

104. Cleuvers, M. Aquatic ecotoxicity of pharmaceuticals including the assessment of combination effects. *Toxicol. Lett.* **2003**, *142*, 185–194. [CrossRef]

105. Li, Z.-H.; Zlabek, V.; Velisek, J.; Grabic, R.; Machovaa, J.; Randaka, T. Physiological condition status and musclebased biomarkers in rainbow trout (*Oncorhynchus mykiss*), after long-term exposure to carbamazepine. *J. Appl. Toxicol.* **2009**, *30*, 197–203.

106. Van den Brandhof, E.-J.; Montforts, M. Fish embryo toxicity of carbamazepine, diclofenac and metoprolol. *Ecotoxicol. Environ. Saf.* **2010**, *73*, 1862–1866. [CrossRef] [PubMed]

107. Hillis, D.G.; Antunes, P.; Sibley, P.K.; Klironomos, J.N.; Solomon, K.R. Structural responses of daucus carota root-organ cultures and the arbuscular mycorrhizal fungus, glomus intraradices, to 12 pharmaceuticals. *Chemosphere* **2008**, *73*, 344–352. [CrossRef] [PubMed]

108. Quinn, B.; Gagné, F.; Blaise, C. An investigation into the acute and chronic toxicity of eleven pharmaceuticals (and their solvents) found in wastewater effluent on the cnidarian, hydra attenuata. *Sci. Total Environ.* **2008**, *389*, 306–314. [CrossRef] [PubMed]

109. Nash, J.P.; Kime, D.E.; Van der Ven, L.T.; Wester, P.W.; Brion, F.; Maack, G.; Stahlschmidt-Allner, P.; Tyler, C.R. Long-term exposure to environmental concentrations of the pharmaceutical ethynylestradiol causes reproductive failure in fish. *Environ. Health Perspect.* **2004**, *112*, 1725. [CrossRef] [PubMed]

110. Corcoran, J.; Winter, M.J.; Tyler, C.R. Pharmaceuticals in the aquatic environment: A critical review of the evidence for health effects in fish. *Crit. Rev. Toxicol.* **2010**, *40*, 287–304. [CrossRef] [PubMed]

111. Snyder, S.A.; Vanderford, B.J.; Drewes, J.; Dickenson, E.; Snyder, E.M.; Bruce, G.M.; Pleus, R.C. *State of Knowledge of Endocrine Disruptors and Pharmaceuticals in Drinking Water*; IWA Publishing: London, UK, 2008; p. 264. ISBN 9781843392415.

112. Chen, P.; Lin, J.-J.; Lu, C.-S.; Ong, C.-T.; Hsieh, P.F.; Yang, C.-C.; Tai, C.-T.; Wu, S.-L.; Lu, C.-H.; Hsu, Y.-C.; et al. Carbamazepine-induced toxic effects and HLA-B* 1502 screening in taiwan. *N. Engl. J. Med.* **2011**, *364*, 1126–1133. [CrossRef] [PubMed]

113. Pereira, F.A.; Mudgil, A.V.; Rosmarin, D.M. Toxic epidermal necrolysis. *J. Am. Acad. Dermatol.* **2007**, *56*, 181–200. [CrossRef] [PubMed]

114. Jentink, J.; Dolk, H.; Loane, A.M.; Morris, K.J.; Wellesley, D.; Garne, E.; de Jong-van den Berg, L. Intrauterine exposure to carbamazepine and specific congenital malformations: Systematic review and case-control study. *Br. Med. J.* **2010**, *341*, 6581–6588. [CrossRef] [PubMed]

115. Cummings, C.; Stewart, M.; Stevenson, M.; Morrow, J.; Nelson, J. Neurodevelopment of children exposed in utero to lamotrigine, sodium valproate and carbamazepine. *Arch. Dis. Child.* **2011**, *96*, 643–647. [CrossRef] [PubMed]

116. Atkinson, E.D.; Brice-Bennett, S.; D'Souza, W.S. Antiepileptic medication during pregnancy: Does fetal genotype affect outcome? *Pediatr. Res.* **2007**, *62*, 120–127. [CrossRef] [PubMed]

117. Virkutyte, J.; Varma, R.S.; Jegatheesan, V. *Treatment of Micropollutants in Water and Wastewater*; IWA Publishing: London, UK, 2010; p. 520. ISBN 9781843393160.

118. Lofgren, H.; Boer, R.D. Pharmaceuticals in Australia: Developments in regulation and governance. *Soc. Sci. Med.* **2004**, *58*, 2397–2407. [CrossRef] [PubMed]

119. MDH. *Carbamazepine in Drinking Water*; Health Risk Assessment Unit, Environmental Health Division: St. Paul, MN, USA, 2011.

120. Cunliffe, D. *Australian Guidelines for Water Recycling: Augmentation of Drinking Water Supplies*; Environment Protection and Heritage Council: Canberra, Australia; National Health and Medical Research Council: Canberra, Australia; Natural Resource Management Ministerial Council: Canberra, Australia, 2008.

121. Garcia-Rodríguez, A.; Matamoros, V.; Fontàs, C.; Salvadó, V. The ability of biologically based wastewater treatment systems to remove emerging organic contaminants—A review. *Environ. Sci. Pollut. Res.* **2014**, *21*, 11708–11728. [CrossRef] [PubMed]

122. Metcalf, E.; EDDY, M. *Wastewater Engineering: Treatment and Resource Recovery*, 5th ed.; McGraw-Hill Professional: New York, NY, USA, 2013; p. 2048. ISBN 9780073401188.

123. Habib, R.; Asif, M.B.; Iftekhar, S.; Khan, Z.; Gurung, K.; Srivastava, V.; Sillanpää, M. Influence of relaxation modes on membrane fouling in submerged membrane bioreactor for domestic wastewater treatment. *Chemosphere* **2017**, *181*, 19–25. [CrossRef] [PubMed]

124. Tran, N.H.; Urase, T.; Ngo, H.H.; Hu, J.; Ong, S.L. Insight into metabolic and cometabolic activities of autotrophic and heterotrophic microorganisms in the biodegradation of emerging trace organic contaminants. *Bioresour. Technol.* **2013**, *146*, 721–731. [CrossRef] [PubMed]

125. Fernandez-Fontaina, E.; Carballa, M.; Omil, F.; Lema, J. Modelling cometabolic biotransformation of organic micropollutants in nitrifying reactors. *Water Res.* **2014**, *65*, 371–383. [CrossRef] [PubMed]

126. Radjenović, J.; Petrović, M.; Barceló, D. Fate and distribution of pharmaceuticals in wastewater and sewage sludge of the conventional activated sludge (cas) and advanced membrane bioreactor (mbr) treatment. *Water Res.* **2009**, *43*, 831–841. [CrossRef] [PubMed]

127. Hai, F.I.; Yamamoto, K.; Fukushi, K. Different fouling modes of submerged hollow-fiber and flat-sheet membranes induced by high strength wastewater with concurrent biofouling. *Desalination* **2005**, *180*, 89–97. [CrossRef]

128. Radjenovic, J.; Petrovic, M.; Barceló, D. Analysis of pharmaceuticals in wastewater and removal using a membrane bioreactor. *Anal. Bioanal. Chem.* **2007**, *387*, 1365–1377. [CrossRef] [PubMed]

129. Nakada, N.; Tanishima, T.; Shinohara, H.; Kiri, K.; Takada, H. Pharmaceutical chemicals and endocrine disrupters in municipal wastewater in tokyo and their removal during activated sludge treatment. *Water Res.* **2006**, *40*, 3297–3303. [CrossRef] [PubMed]

130. Clara, M.; Kreuzinger, N.; Strenn, B.; Gans, O.; Kroiss, H. The solids retention time—A suitable design parameter to evaluate the capacity of wastewater treatment plants to remove micropollutants. *Water Res.* **2005**, *39*, 97–106. [CrossRef] [PubMed]

131. Yan, Q.; Gao, X.; Chen, Y.-P.; Peng, X.-Y.; Zhang, Y.-X.; Gan, X.-M.; Zi, C.-F.; Guo, J.-S. Occurrence, fate and ecotoxicological assessment of pharmaceutically active compounds in wastewater and sludge from wastewater treatment plants in chongqing, the three gorges reservoir area. *Sci. Total Environ.* **2014**, *470*, 618–630. [CrossRef] [PubMed]

132. Subedi, B.; Kannan, K. Occurrence and fate of select psychoactive pharmaceuticals and antihypertensives in two wastewater treatment plants in new york state, USA. *Sci. Total Environ.* **2015**, *514*, 273–280. [CrossRef] [PubMed]

133. Hai, F.I.; Li, X.; Price, W.E.; Nghiem, L.D. Removal of carbamazepine and sulfamethoxazole by MBR under anoxic and aerobic conditions. *Bioresour. Technol.* **2011**, *102*, 10386–10390. [CrossRef] [PubMed]

134. Tadkaew, N.; Hai, F.I.; McDonald, J.A.; Khan, S.J.; Nghiem, L.D. Removal of trace organics by mbr treatment: The role of molecular properties. *Water Res.* **2011**, *45*, 2439–2451. [CrossRef] [PubMed]

135. Asif, M.B.; Habib, R.; Iftekhar, S.; Khan, Z.; Majeed, N. Optimization of the operational parameters in a submerged membrane bioreactor using box behnken response surface methodology: Membrane fouling control and effluent quality. *Desalination* **2017**, *82*, 26–38. [CrossRef]

136. Jumat, M.R.; Hasan, N.A.; Subramanian, P.; Heberling, C.; Colwell, R.R.; Hong, P.-Y. Membrane bioreactor-based wastewater treatment plant in saudi arabia: Reduction of viral diversity, load, and infectious capacity. *Water* **2017**, *9*, 534. [CrossRef]

137. Kimura, K.; Hara, H.; Watanabe, Y. Elimination of selected acidic pharmaceuticals from municipal wastewater by an activated sludge system and membrane bioreactors. *Environ. Sci. Technol.* **2007**, *41*, 3708–3714. [CrossRef] [PubMed]

138. Pérez, S.; Barceló, D. First evidence for occurrence of hydroxylated human metabolites of diclofenac and aceclofenac in wastewater using QqLIT-MS and QqTOF-MS. *Anal. Chem.* **2008**, *80*, 8135–8145. [CrossRef] [PubMed]

139. Cirja, M.; Ivashechkin, P.; Schäffer, A.; Corvini, P. Factors affecting the removal of organic micropollutants from wastewater in conventional treatment plants (CTP) and membrane bioreactors (MBR). *Rev. Environ. Sci. Biotechnol.* **2008**, *7*, 61–78. [CrossRef]

140. Luo, W.; Hai, F.I.; Kang, J.; Price, W.E.; Guo, W.; Ngo, H.H.; Yamamoto, K.; Nghiem, L.D. Effects of salinity build-up on biomass characteristics and trace organic chemical removal: Implications on the development of high retention membrane bioreactors. *Bioresour. Technol.* **2015**, *177*, 274–281. [CrossRef] [PubMed]

141. Jiang, Q.; Ngo, H.H.; Nghiem, L.D.; Hai, F.I.; Price, W.E.; Zhang, J.; Liang, S.; Deng, L.; Guo, W. Effect of hydraulic retention time on the performance of a hybrid moving bed biofilm reactor-membrane bioreactor system for micropollutants removal from municipal wastewater. *Bioresour. Technol.* **2018**, *247*, 1228–1232. [CrossRef] [PubMed]

142. Gurung, K.; Ncibi, M.C.; Sillanpää, M. Assessing membrane fouling and the performance of pilot-scale membrane bioreactor (mbr) to treat real municipal wastewater during winter season in nordic regions. *Sci. Total Environ.* **2017**, *579*, 1289–1297. [CrossRef] [PubMed]

143. Kimura, K.; Hara, H.; Watanabe, Y. Removal of pharmaceutical compounds by submerged membrane bioreactors (mbrs). *Desalination* **2005**, *178*, 135–140. [CrossRef]

144. Abegglen, C.; Joss, A.; McArdell, C.S.; Fink, G.; Schlüsener, M.P.; Ternes, T.A.; Siegrist, H. The fate of selected micropollutants in a single-house mbr. *Water Res.* **2009**, *43*, 2036–2046. [CrossRef] [PubMed]

145. Joss, A.; Andersen, H.; Ternes, T.; Richle, P.R.; Siegrist, H. Removal of estrogens in municipal wastewater treatment under aerobic and anaerobic conditions: Consequences for plant optimization. *Environ. Sci. Technol.* **2004**, *38*, 3047–3055. [CrossRef] [PubMed]

146. Phan, H.V.; Hai, F.I.; Kang, J.; Dam, H.K.; Zhang, R.; Price, W.E.; Broeckmann, A.; Nghiem, L.D. Simultaneous nitrification/denitrification and trace organic contaminant (troc) removal by an anoxic–aerobic membrane bioreactor (mbr). *Bioresour. Technol.* **2014**, *165*, 96–104. [CrossRef] [PubMed]

147. Suarez, S.; Lema, J.M.; Omil, F. Removal of pharmaceutical and personal care products (ppcps) under nitrifying and denitrifying conditions. *Water Res.* **2010**, *44*, 3214–3224. [CrossRef] [PubMed]

148. Yang, S.; Hai, F.I.; Nghiem, L.D.; Price, W.E.; Roddick, F.; Moreira, M.T.; Magram, S.F. Understanding the factors controlling the removal of trace organic contaminants by white-rot fungi and their lignin modifying enzymes: A critical review. *Bioresour. Technol.* **2013**, *141*, 97–108. [CrossRef] [PubMed]

149. Asif, M.B.; Hai, F.I.; Hou, J.; Price, W.E.; Nghiem, L.D. Impact of wastewater derived dissolved interfering compounds on growth, enzymatic activity and trace organic contaminant removal of white rot fungi—A critical review. *J. Environ. Manag.* **2017**, *201*, 89–109. [CrossRef] [PubMed]

150. Hai, F.I.; Yamamoto, K.; Fukushi, K. Development of a submerged membrane fungi reactor for textile wastewater treatment. *Desalination* **2006**, *192*, 315–322. [CrossRef]

151. Hai, F.I.; Yamamoto, K.; Fukushi, K. Hybrid treatment systems for dye wastewater. *Crit. Rev. Environ. Sci. Technol.* **2007**, *37*, 315–377. [CrossRef]

152. Hai, F.I.; Yamamoto, K.; Fukushi, K.; Nakajima, F. Fouling resistant compact hollow-fiber module with spacer for submerged membrane bioreactor treating high strength industrial wastewater. *J. Membr. Sci.* **2008**, *317*, 34–42. [CrossRef]

153. Margot, J.; Maillard, J.; Rossi, L.; Barry, D.A.; Holliger, C. Influence of treatment conditions on the oxidation of micropollutants by trametes versicolor laccase. *New Biotechnol.* **2013**, *30*, 803–813. [CrossRef] [PubMed]

154. Asif, M.B.; Hai, F.I.; Singh, L.; Price, W.E.; Nghiem, L.D. Degradation of pharmaceuticals and personal care products by white-rot fungi—A critical review. *Curr. Pollut. Rep.* **2017**, *3*, 88–103. [CrossRef]

155. Golan-Rozen, N.; Chefetz, B.; Ben-Ari, J.; Geva, J.; Hadar, Y. Transformation of the recalcitrant pharmaceutical compound carbamazepine by pleurotus ostreatus: Role of cytochrome p450 monooxygenase and manganese peroxidase. *Environ. Sci. Technol.* **2011**, *45*, 6800–6805. [CrossRef] [PubMed]

156. Rodarte-Morales, A.; Feijoo, G.; Moreira, M.; Lema, J. Degradation of selected pharmaceutical and personal care products (ppcps) by white-rot fungi. *World J. Microbiol. Biotechnol.* **2011**, *27*, 1839–1846. [CrossRef]

157. Tran, N.H.; Urase, T.; Kusakabe, O. Biodegradation characteristics of pharmaceutical substances by whole fungal culture trametes versicolor and its laccase. *J. Water Environ. Technol.* **2010**, *8*, 125–140. [CrossRef]

158. Nguyen, L.N.; Hai, F.I.; Yang, S.; Kang, J.; Leusch, F.D.L.; Roddick, F.; Price, W.E.; Nghiem, L.D. Removal of pharmaceuticals, steroid hormones, phytoestrogens, uv-filters, industrial chemicals and pesticides by trametes versicolor: Role of biosorption and biodegradation. *Int. Biodeterior. Biodegrad.* **2014**, *88*, 169–175. [CrossRef]

159. Rodarte-Morales, A.; Feijoo, G.; Moreira, M.; Lema, J. Operation of stirred tank reactors (strs) and fixed-bed reactors (fbrs) with free and immobilized phanerochaete chrysosporium for the continuous removal of pharmaceutical compounds. *Biochem. Eng. J.* **2012**, *66*, 38–45. [CrossRef]

160. Jelic, A.; Cruz-Morató, C.; Marco-Urrea, E.; Sarrà, M.; Perez, S.; Vicent, T.; Petrović, M.; Barcelo, D. Degradation of carbamazepine by trametes versicolor in an air pulsed fluidized bed bioreactor and identification of intermediates. *Water Res.* **2012**, *46*, 955–964. [CrossRef] [PubMed]

161. Nguyen, L.N.; Hai, F.I.; Yang, S.; Kang, J.; Leusch, F.D.; Roddick, F.; Price, W.E.; Nghiem, L.D. Removal of trace organic contaminants by an mbr comprising a mixed culture of bacteria and white-rot fungi. *Bioresour. Technol.* **2013**, *148*, 234–241. [CrossRef] [PubMed]

162. Nguyen, L.N.; Hai, F.I.; Price, W.E.; Kang, J.; Leusch, F.D.; Roddick, F.; van de Merwe, J.P.; Magram, S.F.; Nghiem, L.D. Degradation of a broad spectrum of trace organic contaminants by an enzymatic membrane reactor: Complementary role of membrane retention and enzymatic degradation. *Int. Biodeterior. Biodegrad.* **2015**, *99*, 115–122. [CrossRef]

163. Zhang, Y.; Geißen, S.-U. In vitro degradation of carbamazepine and diclofenac by crude lignin peroxidase. *J. Hazard. Mater.* **2010**, *176*, 1089–1092. [CrossRef] [PubMed]

164. Nguyen, L.N.; van de Merwe, J.P.; Hai, F.I.; Leusch, F.D.; Kang, J.; Price, W.E.; Roddick, F.; Magram, S.F.; Nghiem, L.D. Laccase–syringaldehyde-mediated degradation of trace organic contaminants in an enzymatic membrane reactor: Removal efficiency and effluent toxicity. *Bioresour. Technol.* **2016**, *200*, 477–484. [CrossRef] [PubMed]

165. Rodarte-Morales, A.; Feijoo, G.; Moreira, M.; Lema, J. Biotransformation of three pharmaceutical active compounds by the fungus phanerochaete chrysosporium in a fed batch stirred reactor under air and oxygen supply. *Biodegradation* **2012**, *23*, 145–156. [CrossRef] [PubMed]

166. Yang, S.; Hai, F.I.; Nghiem, L.D.; Roddick, F.; Price, W.E. Removal of trace organic contaminants by nitrifying activated sludge and whole-cell and crude enzyme extract of trametes versicolor. *Water Sci. Technol.* **2013**, *67*, 1216–1223. [CrossRef] [PubMed]

167. Espinosa-Ortiz, E.J.; Rene, E.R.; Pakshirajan, K.; van Hullebusch, E.D.; Lens, P.N. Fungal pelleted reactors in wastewater treatment: Applications and perspectives. *Chem. Eng. J.* **2016**, *283*, 553–571. [CrossRef]

168. Mir-Tutusaus, J.; Sarrà, M.; Caminal, G. Continuous treatment of non-sterile hospital wastewater by trametes versicolor: How to increase fungal viability by means of operational strategies and pretreatments. *J. Hazard. Mater.* **2016**, *318*, 561–570. [CrossRef] [PubMed]

169. Mir-Tutusaus, J.A.; Parladé, E.; Llorca, M.; Villagrasa, M.; Barceló, D.; Rodriguez-Mozaz, S.; Martinez-Alonso, M.; Gaju, N.; Caminal, G.; Sarrà, M. Pharmaceuticals removal and microbial community assessment in a continuous fungal treatment of non-sterile real hospital wastewater after a coagulation-flocculation pretreatment. *Water Res.* **2017**, *116*, 65–75. [CrossRef] [PubMed]

170. Badia-Fabregat, M.; Lucas, D.; Pereira, M.A.; Alves, M.; Pennanen, T.; Fritze, H.; Rodríguez-Mozaz, S.; Barceló, D.; Vicent, T.; Caminal, G. Continuous fungal treatment of non-sterile veterinary hospital effluent: Pharmaceuticals removal and microbial community assessment. *Appl. Microbiol. Biotechnol.* **2016**, *100*, 2401–2415. [CrossRef] [PubMed]

171. Cruz-Morató, C.; Lucas, D.; Llorca, M.; Rodriguez-Mozaz, S.; Gorga, M.; Petrovic, M.; Barceló, D.; Vicent, T.; Sarrà, M.; Marco-Urrea, E. Hospital wastewater treatment by fungal bioreactor: Removal efficiency for pharmaceuticals and endocrine disruptor compounds. *Sci. Total Environ.* **2014**, *493*, 365–376. [CrossRef] [PubMed]

172. Srivastava, S.; Goyal, P. *Novel Biomaterials: Decontamination of Toxic Metals from Wasterwater*; Springer: Berlin/Heidelberg, Germany, 2010.

173. Taheran, M.; Brar, S.K.; Verma, M.; Surampalli, R.; Zhang, T.; Valero, J. Membrane processes for removal of pharmaceutically active compounds (phacs) from water and wastewaters. *Sci. Total Environ.* **2016**, *547*, 60–77. [CrossRef] [PubMed]

174. Racar, M.; Dolar, D.; Špehar, A.; Košutić, K. Application of uf/nf/ro membranes for treatment and reuse of rendering plant wastewater. *Process Saf. Environ. Prot.* **2017**, *105*, 386–392. [CrossRef]

175. Schulte-Herbrüggen, H.M.; Semião, A.J.; Chaurand, P.; Graham, M.C. Effect of ph and pressure on uranium removal from drinking water using nf/ro membranes. *Environ. Sci. Technol.* **2016**, *50*, 5817–5824. [CrossRef] [PubMed]

176. Radjenović, J.; Petrović, M.; Ventura, F.; Barceló, D. Rejection of pharmaceuticals in nanofiltration and reverse osmosis membrane drinking water treatment. *Water Res.* **2008**, *42*, 3601–3610. [CrossRef] [PubMed]

177. Bellona, C.; Drewes, J.E.; Oelker, G.; Luna, J.; Filteau, G.; Amy, G. Comparing nanofiltration and reverse osmosis for drinking water augmentation. *Am. Water Work. Assoc. J.* **2008**, *100*, 102–116.

178. Comerton, A.M.; Andrews, R.C.; Bagley, D.M.; Hao, C. The rejection of endocrine disrupting and pharmaceutically active compounds by nf and ro membranes as a function of compound and water matrix properties. *J. Membr. Sci.* **2008**, *313*, 323–335. [CrossRef]

179. Röhricht, M.; Krisam, J.; Weise, U.; Kraus, U.R.; Düring, R.-A. Elimination of carbamazepine, diclofenac and naproxen from treated wastewater by nanofiltration. *CLEAN Soil Air Water* **2009**, *37*, 638–641. [CrossRef]

180. Gur-Reznik, S.; Koren-Menashe, I.; Heller-Grossman, L.; Rufel, O.; Dosoretz, C.G. Influence of seasonal and operating conditions on the rejection of pharmaceutical active compounds by ro and nf membranes. *Desalination* **2011**, *277*, 250–256. [CrossRef]

181. Beier, S.; Köster, S.; Veltmann, K.; Schröder, H.; Pinnekamp, J. Treatment of hospital wastewater effluent by nanofiltration and reverse osmosis. *Water Sci. Technol.* **2010**, *61*, 1691–1698. [CrossRef] [PubMed]

182. Nghiem, L.D.; Coleman, P.J.; Espendiller, C. Mechanisms underlying the effects of membrane fouling on the nanofiltration of trace organic contaminants. *Desalination* **2010**, *250*, 682–687. [CrossRef]

183. Vogel, D.; Simon, A.; Alturki, A.A.; Bilitewski, B.; Price, W.E.; Nghiem, L.D. Effects of fouling and scaling on the retention of trace organic contaminants by a nanofiltration membrane: The role of cake-enhanced concentration polarisation. *Sep. Purif. Technol.* **2010**, *73*, 256–263. [CrossRef]

184. Simon, A.; Price, W.E.; Nghiem, L.D. Effects of chemical cleaning on the nanofiltration of pharmaceutically active compounds (phacs). *Sep. Purif. Technol.* **2012**, *88*, 208–215. [CrossRef]

185. Nghiem, L.D.; Hawkes, S. Effects of membrane fouling on the nanofiltration of pharmaceutically active compounds (phacs): Mechanisms and role of membrane pore size. *Sep. Purif. Technol.* **2007**, *57*, 176–184. [CrossRef]

186. Hajibabania, S.; Verliefde, A.; Drewes, J.E.; Nghiem, L.D.; McDonald, J.; Khan, S.; Le-Clech, P. Effect of fouling on removal of trace organic compounds by nanofiltration. *Drink. Water Eng. Sci.* **2011**, *4*, 117–149. [CrossRef]

187. Vogna, D.; Marotta, R.; Andreozzi, R.; Napolitano, A.; D'Ischia, M. Kinetic and chemical assessment of the uv/h2o2 treatment of antiepileptic drug carbamazepine. *Chemosphere* **2004**, *54*, 497–505. [CrossRef]

188. Mohammad, A.W.; Teow, Y.; Ang, W.; Chung, Y.; Oatley-Radcliffe, D.; Hilal, N. Nanofiltration membranes review: Recent advances and future prospects. *Desalination* **2015**, *356*, 226–254. [CrossRef]

189. Comerton, A.M.; Andrews, R.C.; Bagley, D.M. The influence of natural organic matter and cations on the rejection of endocrine disrupting and pharmaceutically active compounds by nanofiltration. *Water Res.* **2009**, *43*, 613–622. [CrossRef] [PubMed]

190. Kårelid, V.; Larsson, G.; Björlenius, B. Pilot-scale removal of pharmaceuticals in municipal wastewater: Comparison of granular and powdered activated carbon treatment at three wastewater treatment plants. *J. Environ. Manag.* **2017**, *193*, 491–502. [CrossRef] [PubMed]

191. Skouteris, G.; Saroj, D.; Melidis, P.; Hai, F.I.; Ouki, S. The effect of activated carbon addition on membrane bioreactor processes for wastewater treatment and reclamation—A critical review. *Bioresour. Technol.* **2015**, *185*, 399–410. [CrossRef] [PubMed]

192. Moreno-Castilla, C. Adsorption of organic molecules from aqueous solutions on carbon materials. *Carbon* **2004**, *42*, 83–94. [CrossRef]

193. Real, F.J.; Benitez, F.J.; Acero, J.L.; Casas, F. Adsorption of selected emerging contaminants onto pac and gac: Equilibrium isotherms, kinetics, and effect of the water matrix. *J. Environ. Sci. Health Part A* **2017**, 1–8. [CrossRef] [PubMed]

194. Kovalova, L.; Siegrist, H.; Von Gunten, U.; Eugster, J.; Hagenbuch, M.; Wittmer, A.; Moser, R.; McArdell, C.S. Elimination of micropollutants during post-treatment of hospital wastewater with powdered activated carbon, ozone, and UV. *Environ. Sci. Technol.* **2013**, *47*, 7899–7908. [CrossRef] [PubMed]

195. Choi, K.-J.; Kim, S.-G.; Kim, S.-H. Removal of antibiotics by coagulation and granular activated carbon filtration. *J. Hazard. Mater.* **2008**, *151*, 38–43. [CrossRef] [PubMed]

196. Rossner, A.; Snyder, S.A.; Knappe, D.R. Removal of emerging contaminants of concern by alternative adsorbents. *Water Res.* **2009**, *43*, 3787–3796. [CrossRef] [PubMed]

197. Ternes, T.A.; Meisenheimer, M.; McDowell, D.; Sacher, F.; Brauch, H.-J.; Haist-Gulde, B.; Preuss, G.; Wilme, U.; Zulei-Seibert, N. Removal of pharmaceuticals during drinking water treatment. *Environ. Sci. Technol.* **2002**, *36*, 3855–3863. [CrossRef] [PubMed]

198. Pojana, G.; Fantinati, A.; Marcomini, A. Occurrence of environmentally relevant pharmaceuticals in italian drinking water treatment plants. *Int. J. Environ. Anal. Chem.* **2011**, *91*, 537–552. [CrossRef]

199. Huerta-Fontela, M.; Galceran, M.T.; Ventura, F. Occurrence and removal of pharmaceuticals and hormones through drinking water treatment. *Water Res.* **2011**, *45*, 1432–1442. [CrossRef] [PubMed]

200. Kim, S.D.; Cho, J.; Kim, I.S.; Vanderford, B.J.; Snyder, S.A. Occurrence and removal of pharmaceuticals and endocrine disruptors in South Korean surface, drinking, and waste waters. *Water Res.* **2007**, *41*, 1013–1021. [CrossRef] [PubMed]

201. Stackelberg, P.E.; Gibs, J.; Furlong, E.T.; Meyer, M.T.; Zaugg, S.D.; Lippincott, R.L. Efficiency of conventional drinking-water-treatment processes in removal of pharmaceuticals and other organic compounds. *Sci. Total Environ.* **2007**, *377*, 255–272. [CrossRef] [PubMed]

202. Grover, D.P.; Zhou, J.L.; Frickers, P.E.; Readman, J.W. Improved removal of estrogenic and pharmaceutical compounds in sewage effluent by full scale granular activated carbon: Impact on receiving river water. *J. Hazard. Mater.* **2011**, *185*, 1005–1011. [CrossRef] [PubMed]

203. Serrano, D.; Lema, J.M.; Omil, F. Influence of the employment of adsorption and coprecipitation agents for the removal of ppcps in conventional activated sludge (cas) systems. *Water Sci. Technol.* **2010**, *62*, 728–735. [CrossRef] [PubMed]

204. Meinel, F.; Ruhl, A.; Sperlich, A.; Zietzschmann, F.; Jekel, M. Pilot-scale investigation of micropollutant removal with granular and powdered activated carbon. *Water Air Soil Pollut.* **2015**, *226*, 2260. [CrossRef]

205. Snyder, S.A.; Adham, S.; Redding, A.M.; Cannon, F.S.; DeCarolis, J.; Oppenheimer, J.; Wert, E.C.; Yoon, Y. Role of membranes and activated carbon in the removal of endocrine disruptors and pharmaceuticals. *Desalination* **2007**, *202*, 156–181. [CrossRef]

206. Lipp, P.; Groß, H.-J.; Tiehm, A. Improved elimination of organic micropollutants by a process combination of membrane bioreactor (mbr) and powdered activated carbon (pac). *Desalin. Water Treat.* **2012**, *42*, 65–72. [CrossRef]

207. Westerhoff, P.; Yoon, Y.; Snyder, S.; Wert, E. Fate of endocrine-disruptor, pharmaceutical, and personal care product chemicals during simulated drinking water treatment processes. *Environ. Sci. Technol.* **2005**, *39*, 6649–6663. [CrossRef] [PubMed]

208. Soares, A.; Guieysse, B.; Jefferson, B.; Cartmell, E.; Lester, J. Nonylphenol in the environment: A critical review on occurrence, fate, toxicity and treatment in wastewaters. *Environ. Int.* **2008**, *34*, 1033–1049. [CrossRef] [PubMed]

209. Nevskaia, D.; Guerrero-Ruiz, A. Comparative study of the adsorption from aqueous solutions and the desorption of phenol and nonylphenol substrates on activated carbons. *J. Colloid Interface Sci.* **2001**, *234*, 316–321. [CrossRef] [PubMed]

210. Nguyen, L.N.; Hai, F.I.; Kang, J.; Nghiem, L.D.; Price, W.E.; Guo, W.; Ngo, H.H.; Tung, K.-L. Comparison between sequential and simultaneous application of activated carbon with membrane bioreactor for trace organic contaminant removal. *Bioresour. Technol.* **2013**, *130*, 412–417. [CrossRef] [PubMed]

211. Li, X.; Hai, F.I.; Nghiem, L.D. Simultaneous activated carbon adsorption within a membrane bioreactor for an enhanced micropollutant removal. *Bioresour. Technol.* **2011**, *102*, 5319–5324. [CrossRef] [PubMed]

212. Hernández-Leal, L.; Temmink, H.; Zeeman, G.; Buisman, C. Removal of micropollutants from aerobically treated grey water via ozone and activated carbon. *Water Res.* **2011**, *45*, 2887–2896. [CrossRef] [PubMed]

213. Klavarioti, M.; Mantzavinos, D.; Kassinos, D. Removal of residual pharmaceuticals from aqueous systems by advanced oxidation processes. *Environ. Int.* **2009**, *35*, 402–417. [CrossRef] [PubMed]

214. Alharbi, S.K.; Price, W.E. Degradation and fate of pharmaceutically active contaminants by advanced oxidation processes. *Curr. Pollut. Rep.* **2017**, *3*, 268–280. [CrossRef]

215. Andreozzi, R.; Marotta, R.; Pinto, G.; Pollio, A. Carbamazepine in water: Persistence in the environment, ozonation treatment and preliminary assessment on algal toxicity. *Water Res.* **2002**, *36*, 2869–2877. [CrossRef]

216. Hua, W.; Bennett, E.R.; Letcher, R.J. Ozone treatment and the depletion of detectable pharmaceuticals and atrazine herbicide in drinking water sourced from the upper detroit river, ontario, canada. *Water Res.* **2006**, *40*, 2259–2266. [CrossRef] [PubMed]

217. Huber, M.M.; Canonica, S.; Park, G.-Y.; von Gunten, U. Oxidation of pharmaceuticals during ozonation and advanced oxidation processes. *Environ. Sci. Technol.* **2003**, *37*, 1016–1024. [CrossRef] [PubMed]

218. Wert, E.C.; Rosario-Ortiz, F.L.; Snyder, S.A. Effect of ozone exposure on the oxidation of trace organic contaminants in wastewater. *Water Res.* **2009**, *43*, 1005–1014. [CrossRef] [PubMed]

219. Buffle, M.-O.; Schumacher, J.; Salhi, E.; Jekel, M.; von Gunten, U. Measurement of the initial phase of ozone decomposition in water and wastewater by means of a continuous quench-flow system: Application to disinfection and pharmaceutical oxidation. *Water Res.* **2006**, *40*, 1884–1894. [CrossRef] [PubMed]

220. Alharbi, S.K.; Kang, J.; Nghiem, L.D.; van de Merwe, J.P.; Leusch, F.D.; Price, W.E. Photolysis and UV/H_2O_2 of diclofenac, sulfamethoxazole, carbamazepine, and trimethoprim: Identification of their major degradation products by ESI–LC–MS and assessment of the toxicity of reaction mixtures. *Process Saf. Environ. Prot.* **2017**, *112*, 222–234. [CrossRef]

221. Im, J.-K.; Cho, I.-H.; Kim, S.-K.; Zoh, K.-D. Optimization of carbamazepine removal in O_3/UV/H_2O_2 system using a response surface methodology with central composite design. *Desalination* **2012**, *285*, 306–314. [CrossRef]

222. Monteagudo, J.; Durán, A.; González, R.; Expósito, A. In situ chemical oxidation of carbamazepine solutions using persulfate simultaneously activated by heat energy, UV light, Fe^{2+} ions, and H_2O_2. *Appl. Catal. B Environ.* **2015**, *176*, 120–129. [CrossRef]

223. Shirazi, E.; Torabian, A.; Nabi-Bidhendi, G. Carbamazepine removal from groundwater: Effectiveness of the Tio_2/UV, nanoparticulate zero-valent iron, and fenton (NZVI/H_2O_2) processes. *CLEAN Soil Air Water* **2013**, *41*, 1062–1072. [CrossRef]

224. Yang, X.; Sun, J.; Fu, W.; Shang, C.; Li, Y.; Chen, Y.; Gan, W.; Fang, J. Ppcp degradation by uv/chlorine treatment and its impact on dbp formation potential in real waters. *Water Res.* **2016**, *98*, 309–318. [CrossRef] [PubMed]

225. Rosario-Ortiz, F.L.; Wert, E.C.; Snyder, S.A. Evaluation of UV/H_2O_2 treatment for the oxidation of pharmaceuticals in wastewater. *Water Res.* **2010**, *44*, 1440–1448. [CrossRef] [PubMed]

226. Sichel, C.; Garcia, C.; Andre, K. Feasibility studies: UV/chlorine advanced oxidation treatment for the removal of emerging contaminants. *Water Res.* **2011**, *45*, 6371–6380. [CrossRef] [PubMed]

227. Pereira, V.J.; Linden, K.G.; Weinberg, H.S. Evaluation of uv irradiation for photolytic and oxidative degradation of pharmaceutical compounds in water. *Water Res.* **2007**, *41*, 4413–4423. [CrossRef] [PubMed]

228. Doll, T.E.; Frimmel, F.H. Photocatalytic degradation of carbamazepine, clofibric acid and iomeprol with p25 and hombikat UV100 in the presence of natural organic matter (nom) and other organic water constituents. *Water Res.* **2005**, *39*, 403–411. [CrossRef] [PubMed]

229. Bernabeu, A.; Palacios, S.; Vicente, R.; Vercher, R.F.; Malato, S.; Arques, A.; Amat, A.M. Solar photo-fenton at mild conditions to treat a mixture of six emerging pollutants. *Chem. Eng. J.* **2012**, *198–199*, 65–72. [CrossRef]

230. Klamerth, N.; Rizzo, L.; Malato, S.; Maldonado, M.I.; Agüera, A.; Fernández-Alba, A.R. Degradation of fifteen emerging contaminants at μg/L initial concentrations by mild solar photo-fenton in mwtp effluents. *Water Res.* **2010**, *44*, 545–554. [CrossRef] [PubMed]

231. Liu, N.; Zheng, M.; Sijak, S.; Tang, L.; Xu, G.; Wu, M. Aquatic photolysis of carbamazepine by UV/H_2O_2 and UV/Fe(ii) processes. *Res. Chem. Interm.* **2015**, *41*, 7015–7028. [CrossRef]

232. Dai, C.-m.; Zhou, X.-F.; Zhang, Y.-L.; Duan, Y.-P.; Qiang, Z.-M.; Zhang, T.C. Comparative study of the degradation of carbamazepine in water by advanced oxidation processes. *Environ. Technol.* **2011**, *33*, 1101–1109. [CrossRef] [PubMed]

233. Hollender, J.; Zimmermann, S.G.; Koepke, S.; Krauss, M.; McArdell, C.S.; Ort, C.; Singer, H.; von Gunten, U.; Siegrist, H. Elimination of organic micropollutants in a municipal wastewater treatment plant upgraded with a full-scale post-ozonation followed by sand filtration. *Environ. Sci. Technol.* **2009**, *43*, 7862–7869. [CrossRef] [PubMed]

234. Reungoat, J.; Macova, M.; Escher, B.; Carswell, S.; Mueller, J.; Keller, J. Removal of micropollutants and reduction of biological activity in a full scale reclamation plant using ozonation and activated carbon filtration. *Water Res.* **2010**, *44*, 625–637. [CrossRef] [PubMed]

235. Comninellis, C.; Kapalka, A.; Malato, S.; Parsons, S.A.; Poulios, I.; Mantzavinos, D. Advanced oxidation processes for water treatment: Advances and trends for R&D. *J. Chem. Technol. Biotechnol.* **2008**, *83*, 769–776.

236. Hai, F.I.; Nguyen, L.N.; Nghiem, L.D.; Liao, B.-Q.; Koyuncu, I.; Price, W.E. Trace organic contaminants removal by combined processes for wastewater reuse. In *Advanced Treatment Technologies for Urban Wastewater Reuse*; Fatta-Kassinos, D., Dionysiou, D.D., Kümmerer, K., Eds.; Springer: Cham, Switzerland, 2014; pp. 39–77. ISBN 9783319238869.

237. Nguyen, L.N.; Hai, F.I.; Kang, J.; Price, W.E.; Nghiem, L.D. Removal of emerging trace organic contaminants by mbr-based hybrid treatment processes. *Int. Biodeterior. Biodegrad.* **2013**, *85*, 474–482. [CrossRef]

238. Laera, G.; Chong, M.N.; Jin, B.; Lopez, A. An integrated MBR–Tio2 photocatalysis process for the removal of carbamazepine from simulated pharmaceutical industrial effluent. *Bioresour. Technol.* **2011**, *102*, 7012–7015. [CrossRef] [PubMed]

239. Wang, S.; Wang, J. Carbamazepine degradation by gamma irradiation coupled to biological treatment. *J. Hazard. Mater.* **2017**, *321*, 639–646. [CrossRef] [PubMed]

240. Ivancev-Tumbas, I.; Hobby, R. Removal of organic xenobiotics by combined out/in ultrafiltration and powdered activated carbon adsorption. *Desalination* **2010**, *255*, 124–128. [CrossRef]

241. Serrano, D.; Suárez, S.; Lema, J.M.; Omil, F. Removal of persistent pharmaceutical micropollutants from sewage by addition of pac in a sequential membrane bioreactor. *Water Res.* **2011**, *45*, 5323–5333. [CrossRef] [PubMed]

242. Nguyen, L.N.; Hai, F.I.; Kang, J.; Price, W.E.; Nghiem, L.D. Removal of trace organic contaminants by a membrane bioreactor–granular activated carbon (mbr–gac) system. *Bioresour. Technol.* **2012**, *113*, 169–173. [CrossRef] [PubMed]

243. Hübner, U.; Seiwert, B.; Reemtsma, T.; Jekel, M. Ozonation products of carbamazepine and their removal from secondary effluents by soil aquifer treatment–indications from column experiments. *Water Res.* **2014**, *49*, 34–43. [CrossRef] [PubMed]

244. Kleywegt, S.; Pileggi, V.; Yang, P.; Hao, C.; Zhao, X.; Rocks, C.; Thach, S.; Cheung, P.; Whitehead, B. Pharmaceuticals, hormones and bisphenol a in untreated source and finished drinking water in Ontario, Canada—Occurrence and treatment efficiency. *Sci. Total Environ.* **2011**, *409*, 1481–1488. [CrossRef] [PubMed]

245. Hu, L.; Martin, H.M.; Arce-Bulted, O.; Sugihara, M.N.; Keating, K.A.; Strathmann, T.J. Oxidation of carbamazepine by Mn(vii) and Fe(vi): Reaction kinetics and mechanism. *Environ. Sci. Technol.* **2008**, *43*, 509–515. [CrossRef]

246. Kosjek, T.; Andersen, H.R.; Kompare, B.; Ledin, A.; Heath, E. Fate of carbamazepine during water treatment. *Environ. Sci. Technol.* **2009**, *43*, 6256–6261. [CrossRef] [PubMed]

247. Prado, M.; Borea, L.; Cesaro, A.; Liu, H.; Naddeo, V.; Belgiorno, V.; Ballesteros, F. Removal of emerging contaminant and fouling control in membrane bioreactors by combined ozonation and sonolysis. *Int. Biodeterior. Biodegrad.* **2017**, *119*, 577–586. [CrossRef]

248. Suarez, S.; Lema, J.M.; Omil, F. Pre-treatment of hospital wastewater by coagulation-flocculation and flotation. *Bioresour. Technol.* **2009**, *100*, 2138–2146. [CrossRef] [PubMed]

249. Vieno, N.; Tuhkanen, T.; Kronberg, L. Removal of pharmaceuticals in drinking water treatment: Effect of chemical coagulation. *Environ. Technol.* **2006**, *27*, 183–192. [CrossRef] [PubMed]

250. Vieno, N.M.; Härkki, H.; Tuhkanen, T.; Kronberg, L. Occurrence of pharmaceuticals in river water and their elimination in a pilot-scale drinking water treatment plant. *Environ. Sci. Technol.* **2007**, *41*, 5077–5084. [CrossRef] [PubMed]

251. Wert, E.C.; Gonzales, S.; Dong, M.M.; Rosario-Ortiz, F.L. Evaluation of enhanced coagulation pretreatment to improve ozone oxidation efficiency in wastewater. *Water Res.* **2011**, *45*, 5191–5199. [CrossRef] [PubMed]

252. Ciardelli, G.; Ranieri, N. The treatment and reuse of wastewater in the textile industry by means of ozonation and electroflocculation. *Water Res.* **2001**, *35*, 567–572. [CrossRef]

253. McDowell, D.C.; Huber, M.M.; Wagner, M.; von Gunten, U.; Ternes, T.A. Ozonation of carbamazepine in drinking water: Identification and kinetic study of major oxidation products. *Environ. Sci. Technol.* **2005**, *39*, 8014–8022. [CrossRef] [PubMed]

254. Marco-Urrea, E.; Radjenović, J.; Caminal, G.; Petrović, M.; Vicent, T.; Barceló, D. Oxidation of atenolol, propranolol, carbamazepine and clofibric acid by a biological fenton-like system mediated by the white-rot fungus trametes versicolor. *Water Res.* **2010**, *44*, 521–532. [CrossRef] [PubMed]

255. Bu, H.-Z.; Zhao, P.; Dalvie, D.K.; Pool, W.F. Identification of primary and sequential bioactivation pathways of carbamazepine in human liver microsomes using liquid chromatography/tandem mass spectrometry. *Rapid Commun. Mass Spectrom.* **2007**, *21*, 3317–3322. [CrossRef] [PubMed]

256. Leclercq, M.; Mathieu, O.; Gomez, E.; Casellas, C.; Fenet, H.; Hillaire-Buys, D. Presence and fate of carbamazepine, oxcarbazepine, and seven of their metabolites at wastewater treatment plants. *Arch. Environ. Contam. Toxicol.* **2009**, *56*, 408–415. [CrossRef] [PubMed]

257. Mohapatra, D.P.; Brar, S.K.; Tyagi, R.D.; Picard, P.; Surampalli, R.Y. A comparative study of ultrasonication, fenton's oxidation and ferro-sonication treatment for degradation of carbamazepine from wastewater and toxicity test by yeast estrogen screen (yes) assay. *Sci. Total Environ.* **2013**, *447*, 280–285. [CrossRef] [PubMed]

258. Vom Eyser, C.; Börgers, A.; Richard, J.; Dopp, E.; Janzen, N.; Bester, K.; Tuerk, J. Chemical and toxicological evaluation of transformation products during advanced oxidation processes. *Water Sci. Technol.* **2013**, *68*, 1976–1983. [CrossRef] [PubMed]

259. Li, Z.; Fenet, H.; Gomez, E.; Chiron, S. Transformation of the antiepileptic drug oxcarbazepine upon different water disinfection processes. *Water Res.* **2011**, *45*, 1587–1596. [CrossRef] [PubMed]

water

MDPI

Article

Sulfide Precipitation in Wastewater at Short Timescales

Bruno Kiilerich [1,2,*], Wilbert van de Ven [2], Asbjørn Haaning Nielsen [1] and Jes Vollertsen [1]

[1] Department of Civil Engineering, Aalborg University, Thomas Manns Vej 23, DK-9220 Aalborg Ø, Denmark; ahn@civil.aau.dk (A.H.N.); jv@civil.aau.dk (J.V.)
[2] Grundfos Holding A/S, Poul Due Jensens Vej 7, DK-8850 Bjerringbro, Denmark; wvandeven@grundfos.com
* Correspondence: bkiilerich@grundfos.com; Tel.: +45-2463-0165

Received: 3 July 2017; Accepted: 1 September 2017; Published: 5 September 2017

Abstract: Abatement of sulfides in sewer systems using iron salts is a widely used strategy. When dosing at the end of a pumping main, the reaction kinetics of sulfide precipitation becomes important. Traditionally the reaction has been assumed to be rapid or even instantaneous. This work shows that this is not the case for sulfide precipitation by ferric iron. Instead, the reaction time was found to be on a timescale where it must be considered when performing end-of-pipe treatment. For real wastewaters at pH 7, a stoichiometric ratio around 14 mol Fe(II) (mol S(−II))$^{-1}$ was obtained after 1.5 s, while the ratio dropped to about 5 mol Fe(II) (mol S(−II))$^{-1}$ after 30 s. Equilibrium calculations yielded a theoretic ratio of 2 mol Fe(II) (mol S(−II))$^{-1}$, indicating that the process had not equilibrated within the span of the experiment. Correspondingly, the highest sulfide conversion only reached 60%. These findings differed significantly from what has been demonstrated in previous studies and what is attained from theoretical equilibrium conditions.

Keywords: ferrous iron; ferrous sulfide; hydrogen sulfide; odor control; pumping mains; sewerage

1. Introduction

Biogenic sulfide formation in sewers is directly related to problems such as corrosion of sewer assets, health impacts, and malodors [1]. Sulfide is formed in submerged biofilms, and the formation requires anaerobic conditions, which are found in all parts of the network. Pumping mains are typically sulfide formation hotspots and, in temperate climates, sulfide generation is almost solely related to these parts of the network [2].

A common and well-documented practice to manage sulfide related problems is addition of iron salts. Ferrous iron (Fe^{2+}) reacts with sulfide (S^{2-}) and precipitates as ferrous sulfide (FeS) [3,4]. The low solubility constant of ferrous sulfide (3.7×10^{-19} g·mol^2·L^{-2} at 18 °C) implies that this method should be very effective, and it is unlikely that sulfides will be released back into solution after the precipitate has formed [4,5].

According to [6], addition of iron salts is the most common method used worldwide for abatement of sulfide in wastewater. This was also the finding from a more recent survey performed in Australia on methods applied for chemical sulfide control [7]. However, when applying iron salts for sulfide control in pumping mains, there are several challenges to consider.

Due to practical issues, such as available space for storage of chemicals and access to mains power, the dosing point is most often positioned at the start of the pumping main. This necessitates that the amount of iron added must correspond to the amount of sulfide that later will be generated in the pumping main during transport. However, on a short time scale, wastewater flows and characteristics can be highly variable and unpredictable, and many wastewater characteristics change on the time scale of minutes [8]. The result is that retention time in the pumping main, concentration of biodegradable matter, pH, etc. of the wastewater varies substantially. All of these parameters influence sulfide formation [2], and their high variability and inherent unpredictability makes it impossible to consistently add the correct amount of iron to a specific wastewater plug. Further difficulties occur when start-of-pipe treatment is used for branched pressurized sewer systems. These systems are, for example, used to collect wastewater from dispersed houses in the countryside. Due to the hydraulics of the system, wastewater from side branches does not get mixed into the wastewater of the main pipe, but continues herein as distinctive plugs. Treating only the flow in the main pipe with iron will not have the desired effect, and treatment would be needed at each and every of the many pumping stations in the network.

Injection at the start of a pumping main is still the most common strategy for iron dosing. In many practical applications, however, injection at the end of the main can be the only practical solution. Additionally, it offers the benefit of potentially knowing exactly how much sulfide was formed and hence must be managed. At the end of a main, it is in principle possible to measure the sulfide formed in the upstream pipe, and inject iron or other chemicals at the optimum dosage rate. End-of-pipe treatment using nitrate was shown to be possible by [9,10], and that dosage could be optimized compared to the conventional start-of-pipe strategy.

Injecting iron at the end of a pipe entails that the rate of the precipitation reaction must be sufficiently fast to ensure all hydrogen sulfide is precipitated when the sewage depressurizes and releases unreacted H_2S into the headspace. Ferrous iron is preferred over ferric iron, as precipitation of ferric iron with sulfide might not be a significant process under sewer conditions [2,11]. While ferric iron can precipitate with sulfide, this process is comparatively slow, and sulfide precipitation might depend on a biological reduction to ferrous iron, which might simultaneously oxidize sulfide [12]. The biologically formed ferrous iron can then precipitate with sulfide [6]. Ferrous iron (Fe^{2+}) reacts with bisulfide (HS^-) to produce ferrous sulfide (FeS) as shown in Equation (1):

$$Fe^{2+} + HS^- \rightarrow FeS(s) \downarrow + H^+ \tag{1}$$

Ferrous sulfide precipitation in pure water has been reported to occur with fast kinetics, and the process is often assumed to equilibrate within seconds [13,14]. For the reaction rates in wastewater, the literature gives no conclusive answer. Even though the stoichiometry of ferrous sulfide formation has been extensively studied for decades (e.g. [3,11,15]), the exact kinetics of sulfide precipitation with ferrous iron is not known [2]. Some authors state that the precipitation process is rapid without being more specific [16]. On the contrary, it was found that ferrous sulfide formation because of ferric chloride addition to anaerobic wastewater continued for a few hours [12]. Even though it has been stated that the initial reduction of ferric iron to ferrous iron was quick the subsequent kinetics of the ferrous sulfide formation were not specified. Furthermore, most of the experiments reported on sulfide precipitation using ferrous iron do not specify the exact time allowed for the chemical reaction before analysis. Those who did, reported times of 10–40 min [6,11,15]. However, these previous studies did not state the conversion of the reaction.

The reaction rates at short reaction times is not well understood, which can cause problems when performing end-of-pipe abatement of sulfide, as described previously. In the present work, the short-term reaction of ferrous sulfide precipitation is investigated in buffered water and wastewater by measuring stoichiometric ratios of sulfide precipitation at 1.5 and 30 s, to give an estimate of the minimum time needed for sulfide precipitation in practical applications. The measured stoichiometric ratios and conversions, i.e. how much sulfide had reacted with iron within the tested reaction times, are compared to the results of the precipitation process, as predicted by a theoretically determined chemical equilibrium.

2. Materials and Methods

Stoichiometry was studied by mixing iron solutions into waters containing sulfides in a plug-flow reactor and measuring the aqueous H_2S-concentration after 1.5 and 30 s of reaction time. The 1.5 s was chosen to reveal whether precipitation from a practical H_2S management point of view could be seen as nearly instantaneous, while 30 s was chosen to confirm or reject that equilibrium of the reaction was reached after the 1.5 s. H_2S was measured applying an amperometric sensor with a response time of approximate 5 s. To ensure that the sensor response time did not affect the measurements, a continuous plug-flow system was designed where the desired reaction time in a plug equaled the travel time from the point of mixing to the position of the sensor. Two types of experiments were conducted: One addressing the influence of pH and one where the differences between different wastewaters was tested.

2.1. Continuous Flow System

The continuous plug-flow system allowed stable precipitation conditions down to a reaction time of 1.5 s. The setup consisted of a compressible reservoir for the liquid holding sulfide and a flask for holding the iron stock solution (Figure 1). The liquids were pumped separately into a three-way valve followed by an in-line static mixer where mixing occurred rapidly. A Clark type sensor measuring aqueous H_2S was placed at a set distance from the junction, allowing the two reaction times to be obtained. pH was measured continuously just next to the sulfide sensor. The H_2S sensor was a Unisense Sulfide Gas Minisensor 500 μm (H_2S-500) coupled to a Unisense Multimeter (Unisense A/S, Aarhus, Denmark). H_2S concentrations could be converted to dissolved sulfide concentrations applying measured pH. For the latter, a Radiometer Analytical PHC2001-8 Combination Red-Rod pH Electrode coupled with a Radiometer Analytical PHM210 standard pH meter, MeterLab®(Radiometer Analytical SAS, Villeurbanne Cedex, France) was used together with a proprietary data acquisition program for the setup. Prior to use, sensors were calibrated as described in the manuals.

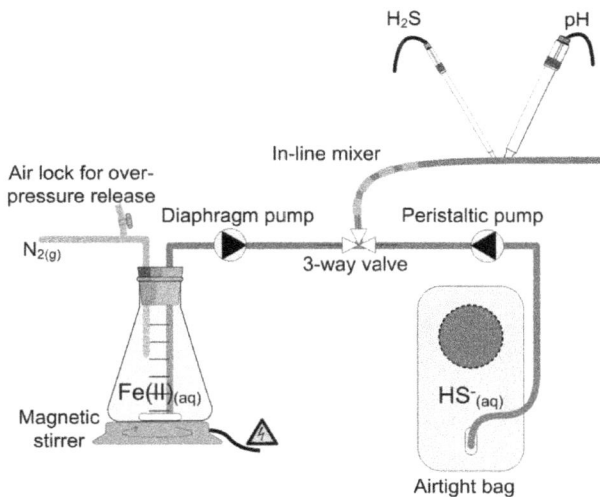

Figure 1. Experimental setup for measurement of sulfide precipitation with ferrous iron. Sulfide and iron solutions were pumped in separate lines to a mixing point and injected to the reaction tubing where sulfide and pH were measured online. The reaction time of the precipitation reaction was altered by varying the length of the reaction tube.

The liquid holding sulfide was fed by a Masterflex L/S 100 rpm peristaltic pump equipped with a Masterflex L/S Easy-Load II Head (Cole&Parmer, Vernon Hills, Illinois, US) running at a high speed, to yield as equal a flow in the setup as possible. To mimic a dosing situation in real sewer system applications, the ferrous iron stock was dosed at a small flow (13 mL·h^{-1}) compared to the liquid holding the sulfide (2532 mL·h^{-1}), using a DDC 6-10 digital dosing diaphragm pump mounted with a multifunction valve (Grundfos A/S, Bjerringbro, Denmark). The digital dosing diaphragm pump is characterized by having a fast and short suction stroke, followed by a long discharge stroke. However, even though the suction stroke was short, the abruption in ferrous iron supply could be seen on the measured dissolved sulfide levels. These parts of the time series were omitted from the sulfide and pH measurements.

To minimize oxygen ingress into the system, liquids were conveyed in Masterflex Tygon® (Vernon Hills, Illinois, US) chemical tubing (internal diameter 3 mm). Prior to the experiments, oxygen ingress was measured and found to be negligible over the course of the experiments. The pump rate, in combination with the distance between the three-way valve and the sensors, determined the specific reaction time of sulfide precipitation, and the stoichiometry could then be determined from the measured dissolved sulfide. A magnetic stirrer kept the ferrous iron stock mixed. To avoid oxidation to ferric iron, the stock was kept under a nitrogen atmosphere.

The sulfide solution was prepared in an airtight bag prior to use by dissolving di-sodium sulfide crystals ($Na_2S \cdot 9H_2O$) directly in deoxygenated buffered water or wastewater. Deoxygenation was done by flushing with high-purity nitrogen gas (5.0). A PreSens Fibox 3 fiber optic oxygen meter and an oxygen sensitive optode (PreSens GmbH, Regensburg, Germany) were used to certify oxygen-free conditions. After addition of di-sodium sulfide (to concentrations between 168–299 µM), the headspace of the bag was evacuated, sealed, and pH was subsequently adjusted. The ferrous iron stock (99 mM) was prepared freshly by dissolving anhydrous ferrous chloride ($FeCl_2$) in 2 mM deoxygenated HCl. Ferrous iron and sulfide were mixed in the three-way valve at a constant flow ratio of 1:195.

Stoichiometric ratios were determined as triplicates. At each measuring cycle, a baseline of the H_2S concentration and pH was determined (Figure 2). The baseline was obtained by solely feeding the sulfide solution through the experimental setup. After a steady baseline was established, the ferrous iron stock was fed into the setup, leading to new steady levels for pH and H_2S at the measuring point. Before the next measurement cycle was conducted, the system was flushed with 5 mM·HCl. The specific levels of H_2S and pH before and after addition of ferrous iron for each measurement cycle were found by root mean square error fitting of data. Subsequently, the stoichiometric ratio on a mole-to-mole basis of ferrous iron to total sulfide could be determined. Ferrous iron could not be measured online, and losses to ferrous iron side reactions were hence included in the stoichiometric ratio and the sulfide conversion.

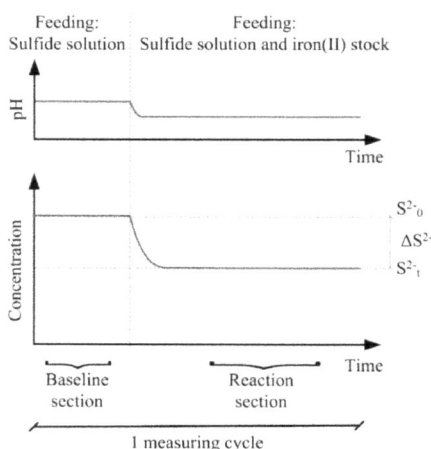

Figure 2. Schematic drawing of one measurement cycle of sulfide precipitation using ferrous iron.

The initial molar ratio of an experiment is defined by Equation (2), where $[Fe(II)_0]$ and $[S(-II)_0]$ are the molar concentrations of dissolved ferrous iron and dissolved sulfide at time t = 0. The stoichiometric ratio of the precipitation reaction is defined by Equation (3), where $[S(-II)_t]$ is the molar concentration of dissolved sulfide at time t = 1.5 or 30 s; consequently $[\Delta S(-II)]$ is the amount of dissolved sulfide that has been removed by precipitation at the specified reaction time. In Equation (3), $[Fe(II)_0]$ is used as a substitution for $[\Delta Fe(II)]$, as it is not possible to measure $[Fe(II)_t]$ online at t = 1.5 and 30 s. Sulfide conversion at a specific reaction time is defined by Equation (4).

Concentrations were used for these calculations instead of activities of the species, even though the ionic strength of the buffered water was 0.01 M and the wastewater from Frejlev was estimated to be in the range of 0.02 M, based on conductivity measurements, and following the conversion according to [17]. Using concentrations was justified, as the activity coefficient of divalent cations and anions was approximately equal at these ionic strengths [17], and the activity coefficients in the following calculations hence cancelled out.

$$\text{Initial molar ratio} = \frac{[Fe(II)_0]}{[S(-II)_0]} \tag{2}$$

$$\text{Stoichiometric ratio} = \frac{[Fe(II)_0]}{[S(-II)_0] - [S(-II)_t]} = \frac{[Fe(II)_0]}{[\Delta S(-II)]} \text{ for t > 0} \tag{3}$$

$$\text{Sulfide conversion} = \frac{[S(-II)_0] - [S(-II)_t]}{[S(-II)_0]} = \frac{[\Delta S(-II)]}{[S(-II)_0]} \text{ for t > 0} \tag{4}$$

2.2. Conducted Experiments

The study comprised two sets of experiments. The first set addressed the pH dependency of the reaction stoichiometry by applying buffered MilliQ water. The second set addressed differences in reaction stoichiometry for wastewaters of different characteristics and buffered water. The latter experiments were all run at pH 7.

For the experiments on pH dependency, buffered water was prepared freshly from MilliQ water (18 $M\Omega\cdot cm^{-1}$) before every experimental run. To reflect a typical wastewater composition of approximately 4 meq·L^{-1} [18], Na_2CO_3 was added to a concentration of 2 mM and $NaH_2PO_4\cdot 2H_2O$ to a concentration of 0.105 mM. The pH of the water was adjusted between runs by adding hydrochloric acid and sodium hydroxide.

For the experiments at pH 7, raw municipal wastewater was collected in sewers of the towns Frejlev and Svenstrup, and in the inlet of Aalborg West wastewater treatment plant (WWTP), Denmark. The drainage area of Frejlev is small, steep, and fully aerobic with approx. half an hour conveyance time from the sources to the sampling point. The drainage area of Svenstrup is larger, shallower, and collects wastewater from several small towns. The wastewater at the sampling location is hence of mixed age and has been less oxygenated. Some of it has furthermore been transported in intercepting pump mains. The drainage area of Aalborg West WWTP is large, and in places rather flat. It receives wastewater from outlying towns as well as the city of Aalborg itself. This wastewater is hence the oldest of the sampled waters and has been the least oxygenated. In combination, the three wastewaters consequently represent fresh, medium-fresh, and old wastewater. Prior to use, the wastewaters were settled for approx. half an hour to eliminate gross particles. Total and dissolved chemical oxygen demand (COD) was measured using Hach Lange COD cuvette kits. Samples for dissolved COD were filtered through a Sartorius GF + CA 0.45 μm filter. Phosphate was measured using the protocol described in [19]. Carbonate alkalinity was measured according to [20]. Prior to the experiments, the pH of the three wastewaters was adjusted to be close to 7 by addition of hydrochloric acid and sodium hydroxide. For comparison, the wastewater experiments included the results from the previous runs of buffered water at pH 7.

2.3. Theoretical Equilibrium Calculations

Visual MINTEQ ver. 3.1. (Kungliga Tekniska Högskolan, Stockholm, Sweden) was used to predict the equilibrium conditions of sulfide precipitation in the buffered water. Equilibrium conditions for the wastewaters were not predicted as not all species affecting such equilibrium were determined. All scenarios were modelled at 20 °C, with fixed pH and amorphous ferrous sulfide, as well as crystalline mackinawite, set as possible solids of the reaction. Initial calculations showed that inclusion of siderite and vivianite as possible end products did not affect the stoichiometry in the pH range of interest and were hence omitted from the simulations. Chemical species added to the reaction were identical in concentration to those used for making the buffered water.

Addressing the stoichiometric ratio and conversion versus pH, the iron to sulfide ratio in Visual MINTEQ was set to 1.75 mol Fe $(mol·S)^{-1}$, which corresponds to the average ratio of the samples at a reaction time of 1.5 s. When addressing the differences in reaction stoichiometry and conversion for wastewaters of different characteristics and buffered water at pH 7, the iron to sulfide ratio in Visual MINTEQ was set to 1.85 mol Fe $(mol·S)^{-1}$. This value was chosen as it corresponded to the average value of the buffered waters used for comparison. In both cases, the ferrous iron concentration was fixed and sulfide concentrations were varied to yield the desired ratios.

3. Results and Discussion

Characteristics of buffered water and wastewater used for the experiments are shown in Table 1 The COD and phosphate contents of the wastewaters from Frejlev and Svenstrup can be characterized as low to medium strength wastewater [18]. The wastewater from Aalborg West WWTP was sampled the day after a storm event; thus it was somewhat diluted as can be seen from the concentrations of phosphate and dissolved COD. The concentration of COD in this sample was only around half of what can be characterized as low strength wastewater, and for phosphate, the concentration was around five times lower [18].

Table 1. Average values (standard deviation) of key parameters characterizing the liquids used. Buffered water was prepared in the laboratory and thus not measured.

Liquid	Alkalinity before Adjusting pH to Approx. 7 $(meq·L^{-1})$	PO_4^{2-} $(mg·L^{-1})$	$COD_{dissolved}$ $(mg·L^{-1})$	COD_{total} $(mg·L^{-1})$
Buffered water	4.05	10	0	0
Frejlev	9.04 (0.04)	7.04 (0.09)	312 (74)	587 (22)
Svenstrup	8.65 (0.02)	7.26 (0.04)	347 (33)	438 (169)
Aalborg West WWTP	6.12 (0.07)	2.31 (0.04)	88 (9)	306 (8)

The buffered water was prepared to have a calcium carbonate alkalinity reflecting that of typical wastewater, as reported in the literature. Its pH was adjusted to approx. 6.5, 7.0 and 7.5 (Table 2), which is typical for wastewater that has been subject to anaerobic transformation and hence sulfide formation [2]. The three wastewaters had calcium carbonate alkalinities that were 3–4.5 times higher than the buffered water. This variation from typical values was due to the public water supply of the region being based on groundwater extracted from limestone aquifers of high carbonate content. pH of the wastewaters was adjusted to approx. 7 prior to experiments, in order to allow comparable precipitation conditions. The sulfide solution and the iron stock were mixed in the three-way valve (Figure 1), achieving initial molar ratios ranging from 1.75–2.84 mol Fe $(mol·S)^{-1}$ (Table 2).

Table 2. Initial molar ratios of ferrous iron to sulfide after mixing in the 3-way valve (Figure 1).

Liquid	Reaction Time 1.5 s		Reaction Time 30 s	
	pH	[Fe(II)] [S(−II)]$^{-1}$	pH	[Fe(II)] [S(−II)]$^{-1}$
Buffered water	6.46	1.8	6.4	1.92
Buffered water	7.01	1.68	6.95	2.05
Buffered water	7.47	1.75	7.49	2.02
Aalborg West	6.96	2.48	7.02	2.21
Svenstrup	7.04	2.44	6.97	2.84
Frejlev	7.07	2.69	6.97	2.61

3.1. Precipitation Stoichiometry Versus pH in Buffered Water

Figure 3 shows the precipitation ratios and the sulfide conversion in buffered water (Table 1) at 1.5 and 30 s of reaction time. A comparison with the ratio of a fully equilibrated reaction is included, which has been calculated from a simulation hereof using Visual MINTEQ. It is evident from the figure that the stoichiometric ratios and sulfide conversions depended on pH, where a lower pH resulted in higher stoichiometric ratio and lower conversion. For the buffered water, the stoichiometric ratios were not greatly affected by the reaction time. The difference in the ratio between 1.5 and 30 s reaction time at pH 6.5 was believed to be due to a difference in pH between the two runs. Although the pH only differed by 0.05 pH-units, the trend of the data indicates that this slight variation was likely to have caused the difference in stoichiometric ratio. Similarly, [1] stated that below pH 6.5, addition of iron will only have little effect, and a slight change in pH might consequently be expected to result in much higher demands for ferrous iron and thus a higher stoichiometric ratio. An increase in ratio when lowering the pH was also observed for the modelled equilibrium conditions, albeit less distinct.

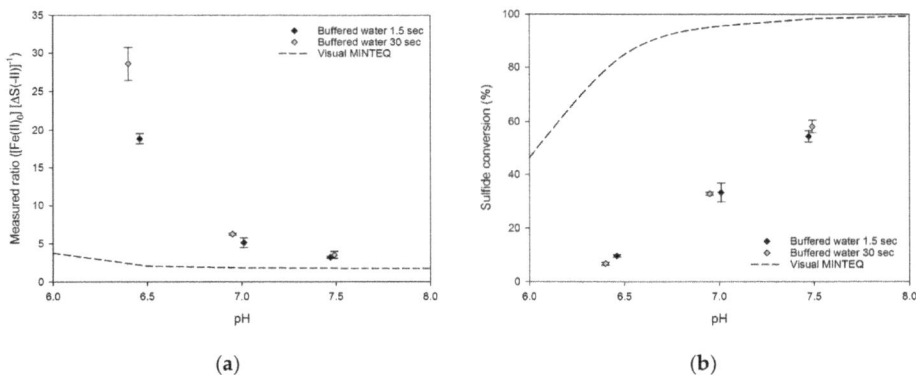

(a)   (b)

Figure 3. (a) The stoichiometric ratio of sulfide precipitation using ferrous iron in buffered water as a function of pH. (b) Sulfide conversion in buffered water as a function of pH. Each point represents the average of the three individual measurements and error bars indicate their standard deviation. Visual MINTEQ was used to calculate equilibrium conditions.

The stoichiometric ratios at the lowest pH-values were up to 30 times higher than the stoichiometric requirement (1 mol Fe (mol·S)$^{-1}$) and 15 times higher than the equilibrium condition modelled by Visual MINTEQ. At pH 7.5, the obtained stoichiometric ratios were around 3.5 times higher than the stoichiometric requirement and two-fold higher than the modeled ratio. Even though the stoichiometric ratio came closer to the modeled equilibrium ratio as pH increased, it was still higher than what has been reported in literature (e.g. [3,11,15,21]). In the analysis performed by Visual MINTEQ, the concentrations of the different chemical species were kept equal to the buffered water experiments. This implies that the inorganic ligands for ferrous iron in the buffered water were also considered during modelling of equilibrium conditions. The 2–15 fold differences between measured and modeled ratios were consequently expected to account for the fact that the ferrous sulfide reaction had not fully equilibrated. The validity of this statement was indirectly supported by a study by [22] where reaction times of 5–7 minutes agreed with the analysis performed in Visual MINTEQ.

The fact that the difference between modeled and measured ratios became less as pH increased, might be related to speciation of sulfide and iron. Such a phenomenon was observed by [14,23] which showed that the precipitation mechanism at ambient temperature depends on pH and sulfide concentration, and thus indirectly also on the speciation of sulfide. It was reported that at sulfide concentrations below 10^{-3} M, a H_2S-pathway of FeS formation dominated in environments up to pH 8, with FeS forming directly, and that the rate of FeS formation was greater in neutral and acidic environments. At concentrations above 10^{-3} M, the rate was instead greater at neutral to alkaline conditions where a bisulfide pathway dominated. For this concentration range, $Fe(HS)_2$ formed as an intermediate before further transformation to FeS. However, even in the concentration range above 10^{-3} M, the H_2S-pathway took over in acidic environments to yield FeS directly.

The findings of the present study, where sulfide was added to concentrations of 10^{-6} M, are somewhat contradictory to those of [14,23]. In contrast here, a higher conversion was observed at increasing pH even though sulfide in the present study was added to concentrations $< 10^{-3}$ M. Thereby it seems that the bisulfide pathway, with an intermediate forming as suggested by [14], could be dominant even at sulfide concentrations as low as 10^{-6} M.

Similarly, [24] suggested that an intermediate species in the reaction pathway is present. By stopped-flow spectrophotometry they found that within the first few seconds of the reaction, an intermediate formed and a subsequent conversion of this intermediate took place. They suggested that this initial product of the reaction between Fe^{2+} and HS^- was $Fe(HS)^+$. Also, [25] found that different types of FeS form at neutral pH compared to slightly acidic conditions, indicating that different reaction pathways are followed.

As reported in literature, it is not completely agreed upon whether the FeS that forms under the different pH conditions of these earlier studies is in the form of an intermediate reaction product, amorphous FeS, nanocrystallineor microcrystalline mackinawite, or greigite (e.g. [23,25,26]). This discrepancy might be because ferrous sulfide salts appear in nature in various different forms and crystal structures, and the mechanisms leading to the different forms are complicated [25]. The specific form of FeS generated, and hence also the resulting reaction kinetics, might thus vary with e.g. pH, redox conditions, ionic strength, and available ligands.

FeS readily precipitates, and hence plays a role in controlling the concentrations of aqueous ferrous and sulfide concentrations [27,28]. The equilibrium simulations in Visual MINTEQ indicate that there might have been free sulfide at low pH, as the removal of sulfide was not complete even though ferrous iron was in excess (Figure 3). The concentration of free sulfide decreased at higher pH, indicating that more FeS was formed (Figure 3). Depending on the resulting structure of the FeS complex the values of the solid/liquid-partitioning coefficient, pKs, are reported to be in the range of 2.95–5.25, with the crystalline forms having the highest values [27]. This agrees well with the values used in Visual MINTEQ for the simulations with pKs values of 2.95 for amorphous FeS and 3.60 for mackinawite.

The crystalline forms of ferrous sulfide, such as mackinawite and greigite [26], are probably not the first to be formed in the reaction. In shallow and deep natural water bodies, [29] found that aging of the FeS precipitate might play a role in the solubility and that a metastable phase of FeS is transformed on aging to crystalline and less soluble forms. Also, [30] found that amorphous FeS at room temperature transformed into mackinawite and greigite; however, this was found to occur on a timescale of days and months. Whether the process of aging is pH-dependent is not reported in these works, but the pH-values tested in the present study are within the same range as those of natural waters. The tested reaction times of 1.5 and 30 s were very different from those addressed by [29] and would not induce an aging effect, where amorphous FeS transforms into its crystalline forms. This further supports that the precipitation of sulfide did not reach equilibrium at the reaction times tested.

The overall trend of sulfide conversion in the buffered water at different pH showed that higher conversions could be obtained at higher pH (Figure 3). The highest conversions were around 60% at pH 7.5, and the sulfide conversion decreased almost linearly with decreasing pH and reached around 10% at pH 6.5. This trend in decreasing conversion level with decreasing pH is in line with findings of [11], who in the range of pH 5–10.5 obtained sulfide conversions of 10% and 90%, respectively. At reaction times of 1.5 and 30 s, no significant differences in the sulfide conversion were observed for the samples. Even though the absolute level of sulfide conversion was much lower in the experiments compared to the model results, this tendency was still in line with the overall trend for conversion as predicted by Visual MINTEQ, where a higher sulfide conversion was obtained at higher pH values.

Nevertheless, using buffered water for the reaction is a simplification of the real wastewater system, where the actual precipitation must take place during abatement of sulfide. In the wastewater matrix, organic or inorganic ligands could also be of importance to the reaction, complexing with sulfide or iron and impeding the reaction.

3.2. Precipitation Stoichiometry versus Water Type

The impact of water type on the precipitation was studied for three wastewaters and the buffered water previously discussed (Table 1). The pH was kept close to 7 and the reaction times were 1.5 and 30 s. MINTEQ simulations were done to estimate equilibrium conditions corresponding to an infinite reaction time, also at pH 7 (Figure 4). The wastewaters had an average initial molar ratio of ferrous iron to sulfide of 2.54 ± 0.11 and 2.55 ± 0.26 mol Fe $(\text{mol·S})^{-1}$ for reaction times of 1.5 and 30 s, respectively. The buffered water samples had initial molar ratios of 1.68 and 2.05 mol Fe $(\text{mol·S})^{-1}$ for the two reaction times (Table 2).

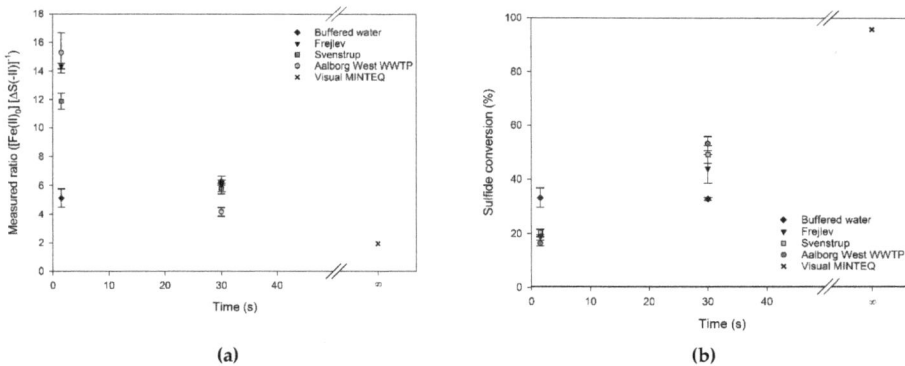

Figure 4. (a) The stoichiometric ratio of sulfide precipitation by ferrous iron in different waters at pH 7 (average of three individual measurements and their standard deviation). (b) Sulfide conversion of the same samples. The value at t = ∞ is a simulation of equilibrium conditions applying Visual MINTEQ.

It is evident from Figure 4 that the stoichiometric ratios and sulfide conversions of the three wastewaters depended on reaction time, with a longer reaction time resulting in a lower stoichiometric ratio and thereby a higher conversion. This trend was not observed for the buffered water, where no time-dependency could be documented (two-sided t-test, $\alpha = 5\%$). However, this lack of time-dependency might be caused by the slight dissimilarities in pH and initial molar ratios between the experimental runs. Nevertheless, the ratio in buffered water did not reach the values predicted by equilibrium simulations, indicating that the reaction might continue at a slow rate for longer time.

The stoichiometric ratios for wastewater at a reaction time of 1.5 s were 2.5–3 times higher than those for the buffered water. However, at a reaction time of 30 s, this difference disappeared and the ratios showed values in the same range as for the buffered water. Compared to the 1 mol Fe $(\text{mol·S})^{-1}$ theoretically needed according to Equation (1), and the 1.94 mol Fe $(\text{mol·S})^{-1}$ predicted by the equilibrium modeling, the obtained ratios at 1.5 and 30 s reaction times were 5–15 and 2.5–7.5 times higher, respectively.

The results for the three wastewaters showed comparable stoichiometric ratios at the two reaction times, despite the fact that wastewater is a heterogeneous medium and the variation in COD between the wastewaters was large (Table 1). A one-way analysis of variance (ANOVA) test ($\alpha = 5\%$) revealed that the mean of the samples, including the buffered water, were not equal and a multiple comparison procedure using the Holm-Sidiak method showed that at 1.5 s the stoichiometric ratio of the buffered water and Svenstrup wastewater differed from the Aalborg and Frejlev wastewaters. However, at 30 s, the Aalborg wastewater differed as the only one from the three other samples.

Previous studies have reported the stoichiometric ratios for iron sulfide precipitation in wastewater to vary between a better than stoichiometric ratio and up to a ratio of 5.7 (e.g. [3,11,15,21]). This variation could be due to the fact that many studies are site- and wastewater-specific, as both pH, initial sulfide concentration, and other ligands for ferrous iron are known to influence the stoichiometric ratios [3,6,11]. Furthermore [3,6] reported that the stoichiometric ratio depends on the initial sulfide concentration. A near stoichiometric ratio was achieved by [3] at high initial sulfide concentrations, and it was observed that the ratio increased drastically at lower initial concentrations. The initial sulfide concentrations in the present study were in the high end of what is typical for septic wastewater [2]; however, the stoichiometric ratios obtained in the present study were higher than what those studies led to expect. Also, compared to the ratios found by [11], which used an almost equal initial sulfide concentration of around 0.3 mM, the obtained stoichiometric ratios were high. The above indicates that the reaction most likely had not run to completion within the 30 s of reaction time.

The amount of iron needed to precipitate sulfide in wastewater will exceed the stoichiometric amount of Equation (1) as stated by [1,11]. According to those studies, near stoichiometric ratios can be expected at pH around 8. For pH around 6.5, a ratio of around 4.6 mol Fe $(\text{mol·S})^{-1}$ can be expected, and below this pH, addition of iron will only have little effect, thus the stoichiometric ratio will increase substantially. According to the above, the present experiments conducted at pH 7 should have had a stoichiometric ratio somewhat better than 4.6 mol Fe $(\text{mol·S})^{-1}$. They were, however, on average 13.8 ± 1.43 for reaction times of 1.5 s and of 5.3 ± 0.82 for reaction times of 30 s, indicating that reaction times on this timescale are important for the achieved stoichiometric ratio.

The difference in stoichiometric ratios observed for wastewater samples (Figure 4) at the two reaction times might be ascribed to the fact that ferrous iron initially reacted with other inorganic or organic constituents in the wastewater before it subsequently precipitated dissolved sulfides as previously discussed by [11,12]. This implies that the reaction between ferrous iron and sulfide was inhibited, and thus the stoichiometric ratio observed at low reaction times became higher than that for longer reaction times. However, the stoichiometric ratios observed for buffered water showed time-independent behavior, and thus the carbonate and phosphate content by themselves did not seem to influence the precipitation reaction to any significant extent, and thus explained the difference observed for wastewater at the two reaction times. It hence seems reasonable to assume that the retardation in precipitation was caused by organic wastewater constituents.

3.3. Conversion of Sulfide in Different Water Types

The mean of conversions (Figure 4) was tested statistically using a one-way ANOVA ($\alpha = 5\%$) and a subsequent multiple comparison with the Holm-Sidiak method to find statistical differences. This showed that there was no difference in conversion between the three wastewaters, and that they all differed significantly from the conversion mean of the buffered water at both 1.5 and 30 s of reaction time. The conversion at 1.5 s was found to be greater in buffered water compared to the wastewater samples, and reversed at 30 s, where conversion was greater in the wastewaters. This might again be due to the interaction between iron and organic matter in the samples, and thus ultimately the differences in matrices between buffered water and wastewater.

The sulfide conversion observed in this study differed considerably from previously reported numbers where [31] in real wastewater installations in Florida at a stoichiometric ratio of 1.43–2.86 mol Fe $(mol \cdot S)^{-1}$ attained a conversion of more than 80%. For a trunk sewer in California, [32] showed that a 95% reduction of initial sulfide levels could be attained with a stoichiometric ratio of around 1.4 mol Fe $(mol \cdot S)^{-1}$ when adding a mix of ferrous and ferric iron. The pH range of the precipitation in these full-scale installations was not reported. However, under laboratory conditions in a setup using wastewater at pH around 8, [11] showed an 80% conversion of sulfide. But when increasing the ratio from 0.8 to 1.3 mol Fe $(mol \cdot S)^{-1}$, thus adding iron in excess, they only experienced a 90% conversion of sulfide using wastewater. They moreover observed that when lowering pH below 7, conversion in some cases decreased and even attained values below 40%. This trend of decreasing conversion level is in line with both what was observed in this study and predicted theoretically by Visual MINTEQ (Figure 3).

Overall, the differences in sulfide conversion in wastewater between the two reaction times, as well as the differences found comparing to literature values and theoretical equilibrium conversions, most likely was caused by the reaction not having reached completion within the maximum reaction time tested.

3.4. Influence of Organic Matter

The organic content of the wastewaters seemed to influence the precipitation reaction. This influence was most pronounced at a reaction time of 1.5 s where a clear difference in stoichiometric ratios as well as conversions was observed (Figure 4). Between the two different criteria evaluated for the precipitation reaction (stoichiometric ratio and conversion), a consistent picture of a certain type of wastewater differing statistically significant from the others, either due to wastewater age or a specific reaction time, could not be established. This indicates that gross wastewater characteristics such as COD or age were poor indicators for the precipitation reaction.

The observed differences in stoichiometry between the three wastewaters must hence have been caused by specific organic and inorganic substances of the waters. This hypothesis is supported by, for example, [33], who showed that ferrous and ferric iron in aqueous solution at neutral pH and under oxic conditions interacted with organic matter and that both iron species were held in solution in concentrations in excess of what would be theoretically expected. Furthermore, [34] discovered that under reduced conditions, ferrous iron in the form of ferrous hydroxide can interact with organic matter, resulting in co-precipitation of the species. They showed that it is primarily the proteinaceous fraction of the organic matter which participates in these interactions. Those findings substantiate the conclusion of the present study that a gross parameter like COD is a poor indicator for the precipitation reaction in itself, and that a more detailed analysis of both organic and inorganic substances, as well as their interrelations, is needed to allow the prediction of stoichiometric requirements and reactions rates. Which specific substances are of importance and how their importance and interrelations depend on conditions like pH and redox, is still an unsolved issue.

The finding of this study implies that practitioners should take reaction time into account when managing H_2S issues in sewer networks and at wastewater treatment plants. Depending on the actual wastewater characteristics, several minutes of reaction time might be needed to achieve optimal precipitation. Such reaction times can though be difficult to achieve in practice. The issue of sufficient reaction time must hence be considered when choosing the best suited H_2S management strategies.

4. Conclusions

In agreement with other studies on sulfide precipitation by ferrous iron, it was found that the stoichiometric ratio of the precipitation process and the conversion of sulfide became poorer with decreasing pH. However, the study also showed that the precipitation of dissolved sulfide by ferrous iron was not as instantaneous as commonly assumed. At a reaction time of 1.5 s and a pH of 7, the achieved stoichiometric ratio was as high as 5–15 mol (mol·S)$^{-1}$, while it only dropped to 4–6 mol Fe (mol·S)$^{-1}$ at a reaction time of 30 s. Sulfide conversions were consequently poor, and equilibrium calculations indicated that the precipitation had not run to completion during the first 30 s. Another finding was that the precipitation at 1.5 s reaction time was slower in wastewater than in buffered water, while this was not seen at a reaction time of 30 s. Even though there was no simple relationship with the wastewater COD, it was hypothesized that organic substances in the wastewater influenced this.

The findings show that under conditions where short reaction times are adamant, ignoring the rate of the precipitation process will lead to inefficient sulfide management. Locations where precipitation rates must be considered, are for example, the end of a pumping main or the inlet to a wastewater treatment plant, where sufficient distance between dosing point and depressurization must be ensured to allow the reaction to equilibrate. The study furthermore makes clear that the precipitation process is not yet understood at a level where exact predictions of required stoichiometric ratio or sulfide conversion can be made based on knowledge of the chemical composition of the wastewater.

Acknowledgments: This work is partly funded by the Innovation Fund Denmark (IFD) under File No. 4135-00076B.

Author Contributions: B.K., A.H. and J.V. conceived and designed the experiments; B.K. performed the experiments; B.K. analyzed the data; B.K. wrote the paper, W.V., A.H. and J.V. revised the paper.

Conflicts of Interest: The authors declare no conflict of interest. The founding sponsors had no role in the design of the study; in the collection, analyses, or interpretation of data; in the writing of the manuscript, and in the decision to publish the results.

References

1. Boon, A.G. Septicity in sewers: Causes, consequences and containment. *Water Sci. Technol.* **1995**, *31*, 237–253. [CrossRef]
2. Hvitved-Jacobsen, T.; Vollertsen, J.; Nielsen, A.H. *Sewer Processes-Microbial and Chemical Process Engineering of Sewer Networks*; CRC press: Boca Raton, FL, USA, 2013.
3. Pomeroy, R.; Bowlus, F.D. Progress Report on Sulfide Control Research. *Sewage Work. J.* **1946**, *18*, 597–640. [PubMed]
4. Jameel, P. The Use of Ferrous Chloride To Control Dissolved Sulfides in Interceptor Sewers. *J. Water Pollut. Control Fed.* **1989**, *61*, 230–236.
5. ASCE. *Sulfide in Wastewater Collection and Treatment Systems*; American Society of Civil Engineers: New York, NY, USA, 1989.
6. Firer, D.; Friedler, E.; Lahav, O. Control of sulfide in sewer systems by dosage of iron salts: Comparison between theoretical and experimental results, and practical implications. *Sci. Total Environ.* **2008**, *392*, 145–156. [CrossRef] [PubMed]
7. Ganigue, R.; Gutierrez, O.; Rootsey, R.; Yuan, Z. Chemical dosing for sulfide control in Australia: An industry survey. *Water Res.* **2011**, *45*, 6564–6574. [CrossRef] [PubMed]
8. Gudjonsson, G.; Vollertsen, J.; Hvitved-Jacobsen, T. Dissolved oxygen in gravity sewers—Measurement and simulation. *Water Sci. Technol.* **2002**, *45*, 35–44. [PubMed]
9. Gutierrez, O.; Sutherland-Stacey, L.; Yuan, Z. Simultaneous online measurement of sulfide and nitrate in sewers for nitrate dosage optimisation. *Water Sci. Technol.* **2010**, *61*, 651–658. [CrossRef] [PubMed]
10. Auguet, O.; Pijuan, M.; Guasch-Balcells, H.; Borrego, C.M.; Gutierrez, O. Implications of Downstream Nitrate Dosage in anaerobic sewers to control sulfide and methane emissions. *Water Res.* **2015**, *68*, 522–532. [CrossRef] [PubMed]
11. Nielsen, A.H.; Hvitved-Jacobsen, T.; Vollertsen, J. Effects of pH and iron concentrations on sulfide precipitation in wastewater collection systems. *Water Environ. Res.* **2008**, *80*, 380–384. [CrossRef] [PubMed]
12. Nielsen, A.H.; Lens, P.; Vollertsen, J.; Hvitved-Jacobsen, T. Sulfide-iron interactions in domestic wastewater from a gravity sewer. *Water Res.* **2005**, *39*, 2747–2755. [CrossRef]

13. Rickard, D. Experimental concentration-time curves for the iron(II) sulphide precipitation process in aqueous solutions and their interpretation. *Chem. Geol.* **1989**, *78*, 315–324. [CrossRef]

14. Rickard, D. Kinetics of FeS precipitation: Part 1. Competing reaction mechanisms. *Geochim. Cosmochim. Acta* **1995**, *59*, 4367–4379. [CrossRef]

15. Tomar, M.; Abdullah, T.H.A. Evaluation of chemicals to control the generation of malodorous hydrogen sulfide in waste water. *Water Res.* **1994**, *28*, 2545–2552. [CrossRef]

16. Zhang, L.; De Schryver, P.; De Gusseme, B.; De Muynck, W.; Boon, N.; Verstraete, W. Chemical and biological technologies for hydrogen sulfide emission control in sewer systems: A review. *Water Res.* **2008**, *42*, 1–12. [CrossRef] [PubMed]

17. Snoeyink, V.L.; Jenkins, D. *Water Chemistry*; John Wiley & Sons: Chichester, England, 1980.

18. Henze, M.; Comeau, Y. *Chapter 3—Wastewater Characterization*; Henze, M., Loosdrecht, M.C.M., van Ekama, G.A., Brdjanovic, D., Eds.; IWA Publishing: London, UK, 2008.

19. APHA. *Standard Methods for the Examination of Water and Wastewater*, 19th ed.; American Public Health Association (APHA): Washington, DC, USA, 1995.

20. *Water quality—Determination of alkalinity—Part 2: Determination of carbonate alkalinity, Dansk Standard DS/EN ISO 9963-2-1996*; Fonden Dansk Standard: Copenhagen, Denmark, 1996.

21. Zhang, L.; Keller, J.; Yuan, Z. Ferrous Salt Demand for Sulfide Control in Rising Main Sewers: Tests on a Laboratory-Scale Sewer System. *J. Environ. Eng.* **2010**, *136*, 1180–1187. [CrossRef]

22. Oviedo, E.R.; Johnson, D.; Shipley, H. Evaluation of hydrogen sulphide concentration and control in a sewer system. *Environ. Technol.* **2012**, *33*, 1207–1215. [CrossRef] [PubMed]

23. Rickard, D. The solubility of FeS. *Geochim. Cosmochim. Acta* **2006**, *70*, 5779–5789. [CrossRef]

24. Wei, D.; Osseo-Asare, K. Formation of Iron Monosulfide: A Spectrophotometric Study of the Reaction between Ferrous and Sulfide Ions in Aqueous Solutions. *J. Colloid Interface Sci.* **1995**, *174*, 273–282. [CrossRef]

25. Harmandas, N.G.; Koutsoukos, P.G. The formation of iron(II) sulfides in aqueous solutions. *J. Cryst. Growth* **1996**, *167*, 719–724. [CrossRef]

26. Morse, J.W.; Millero, F.J.; Cornwell, J.C.; Rickard, D. The chemistry of the hydrogen sulfide and iron sulfide systems in natural waters. *Earth-Science Rev.* **1987**, *24*, 1–42. [CrossRef]

27. Davison, W. The Solubility of Iron Sulphides in Synthetic and Natural-Waters at Ambient-Temperature. *Aquat. Sci.* **1991**, *53*, 309–329. [CrossRef]

28. Bågander, L.E.; Carman, R. In situ determination of the apparent solubility product of amorphous iron sulphide. *Appl. Geochemistry* **1994**, *9*, 379–386. [CrossRef]

29. Davison, W. A critical comparison of the measured solubilities of ferrous sulphide in natural waters. *Geochim. Cosmochim. Acta* **1980**, *44*, 803–808. [CrossRef]

30. Csakberenyi-Malasics, D.; Rodriguez-Blanco, J.D.; Kis, V.K.; Recnik, A.; Benning, L.G.; Posfai, M. Structural properties and transformations of precipitated FeS. *Chem. Geol.* **2012**, *294–295*, 249–258. [CrossRef]

31. Bowker, R.P.G.; Smith, J.M.; Webster, N.A. *Design Manual Odor and Corrosion Control in Sanitary Sewerage Systems and Treatment Plants*; United States Environmental Protection Agency: Cincinnati, OH, USA, 1985.

32. Padival, N.A.; Kimbell, W.A.; Redner, J.A. Use of Iron Salts to Control Dissolved Sulfide in Trunk Sewers. *J. Environ. Eng.* **1995**, *121*, 824–829. [CrossRef]

33. Akiyama, T. Interactions of ferric and ferrous irons and organic matter in water environment. *Geochem. J.* **1973**, *7*, 167–177. [CrossRef]

34. Theis, T.L.; Singer, P.C. Complexation of Iron(II) by Organic Matter and Its Effect on Iron(II) Oxygenation. *Environ. Sci. Technol.* **1974**, *8*, 569–573. [CrossRef]

Article

Effective Removal of Lead Ions from Aqueous Solution Using Nano Illite/Smectite Clay: Isotherm, Kinetic, and Thermodynamic Modeling of Adsorption

Juan Yin [1,2], Chaobing Deng [1,3,*], Zhen Yu [4], Xiaofei Wang [3] and Guiping Xu [1,3]

[1]   College of Light industry and Food Engineering, Guangxi University, Nanning 530004, China;
      yinjuan101@163.com (J.Y.); xuguiping@126.com (G.X.)

[2]   Department of Management Science and Engineering, Guangxi University of Finance and Economics,
      Nanning 530003, China

[3]   Guangxi Zhuang Autonomous Region Enviromental Monitoring Center, Nanning 530028, China;
      wangxiaofei26@163.com

[4]   College of Resources, Enviroment and Materials, Guangxi University, Nanning 530004, China;
      yz2465@hotmail.com

*   Correspondence: dcb715@sina.com; Tel.: +86-771-532-5805

Received: 14 January 2018; Accepted: 13 February 2018; Published: 16 February 2018

Abstract: Illite-smectite clay is a new mixed mineral of illite and montmorillonite. The ability of nano illite/smectite clay to remove Pb(II) from slightly polluted aqueous solutions has been investigated. The effects of pH, contact time, initial concentration of Pb(II), nano illite/smectite clay dosage, and temperature on the adsorption process were studied. The nano illite/smectite clay was characterized by X-Ray Diffraction (XRD), Fourier transform infrared spectrometry (FTIR), and Scanning electron microscopy (SEM). The results showed that Pb(II) was adsorbed efficiently by nano illite/smectite clay in aqueous solution. The pseudo-second-order kinetic model best described the kinetic of the adsorption, and the adsorption capacity of nano illite/smectite (I-S_m) clay was found to be 256.41 $\mu g \cdot g^{-1}$ for Pb(II). The adsorption patterns followed the Langmuir isotherm model. Thermodynamic parameters, including the Gibbs free energy (ΔG), enthalpy (ΔH), and entropy (ΔS) changes, indicated that the present adsorption process was feasible, spontaneous, and endothermic in the temperature range of 298–333 K.

Keywords: Pb(II); nano illite/smectite clay; adsorption kinetics; adsorption thermodynamic

---

1. Introduction

Water is a source of life. In recent years, a large number of studies have indicated that the water, especially of rivers, in urban areas has been seriously contaminated by heavy metals [1–3]. Because heavy metals are not readily degradable in nature and accumulate in animals as well as human bodies, people who drink water or eat food containing heavy metals for a long time are susceptible to disease. Therefore, heavy metal contamination in the water environment has attracted great concern owing to its environmental toxicity and persistence.

Lead is a widely distributed and accumulative pollutant, and is the third-most common toxic element in the heavy metal toxicity list. It is also one of the 10 chemicals that the World Health Organization (WHO) has set out as a cause for significant public health concerns. Once the lead in the environment through various ways enters the human body and accumulates, the nerve, digestive, immune, and reproductive systems will be compromised and the health of human beings will be threatened, especially that of children [4,5]. The permissible limit for lead in potable water is

0.01 mg·L^{-1} [6]. The removal of Pb(II) has become a great concern globally due to these toxic effects of Pb(II) on living beings. In the past few years, various techniques have been used, such as chemical precipitation, membrane filtration, ion exchange, and biological treatment, for lead removal [7–10]. Among these methods, adsorption has the advantages of being easy to perform and having a low cost and high efficiency. Thus, it has been used commonly in the heavy metal pollution treatment of water [11].

Clay as an adsorbent is widely used for the removal of heavy metals and has great applicability due to its being economical and having an environment-friendly nature, a high adsorption capacity, and a wide pH range [12,13]. In recent years, many kinds of clay, i.e., bentonite, kaolin, and montmorillonite, have been reported for the removal of high-concentration heavy metals from water [13–17]. Illite-smectite ($I\text{-}S_m$) clay is a new mixed mineral of illite and montmorillonite, and a transition mineral from montmorillonite to illite, belonging to the typical 2:1-type layered silicate mineral. In 2014, nearly 30 billion tons of $I\text{-}S_m$ mineral was discovered in Shangsi County, Guangxi. Because $I\text{-}S_m$ mineral has the characteristics of high purity, fine particles, and a large surface area, $I\text{-}S_m$ has been studied for use as a rubber modifier [18] and as an adsorbent to remove high concentrations of heavy metals from aqueous solution [19,20]. Only a few studies have focused on the adsorption of low concentrations of heavy metals in contaminated water using $I\text{-}S_m$.

In this study, the $I\text{-}S_m$ mineral was first made nano-size, then nano $I\text{-}S_m$ was used to adsorb Pb(II) in a solution close to that of real polluted water. The effects of various analytical conditions, such as initial pH of the solution, contact time, and initial adsorbate and adsorbent concentration, were evaluated in detail on the removal performance of nano $I\text{-}S_m$. Isotherm, kinetic, and thermodynamic modeling of the adsorption process was analyzed.

2. Materials and Methods

2.1. Preparation of Nano I-S$_m$

$I\text{-}S_m$ mineral was provided by Sino-nanotech Holdings Co., Ltd. (Shangsi, Guangxi, China). The main steps of nano $I\text{-}S_m$ preparation included crushing, soaking, dispersing, sieving and purification, and ultrafine grinding [18,21]. In detail, the preparation was as follows. Firstly, $I\text{-}S_m$ mineral was crushed into small pieces (d < 2 cm) and then soaked in a certain amount of water (the water-to-clay ratio was 9:11) for about 10 h. Secondly, the soaked $I\text{-}S_m$ mud was dispersed for 30 min with a mixer beater and then passed through a 100-mesh and 325-mesh vibrating screen to obtain the primary nano $I\text{-}S_m$ slurry (d < 45 µm). Thirdly, the slurry was ground by a high energy density medium stirring mill (FPML OML-H/V, Buhler Group Co., Uzwil, Switzerland) for 2 h. Finally, the slurry was dried by azeotropic distillation and then dispersed by high-speed mill.

2.2. Characterization of Nano I-S$_m$

The Fourier transform infrared (FT-IR) spectra of nano $I\text{-}S_m$ were recorded with the Fourier transform infrared spectrophotometer (Nicolet 380, Thermo Fisher, Waltham, MA, USA). The X-ray diffraction (XRD) analysis was determined using a MiniFlex X-ray diffractometer (Miniflex 600, Rigaku, Tokyo, Japan), and the scanning regions of the diffraction were 5–80° on the 2θ angle. The morphology of nano $I\text{-}S_m$ was analyzed by a scanning electron microscope (SEM) (ProX, Phenom World, Shanghai, China), The Brunauer-Emmett-Teller (BET) surface area and pore properties of nano $I\text{-}S_m$ were determined via N_2 adsorption–desorption isotherms using a Micromeritics analyzer (ASAP 2460, Micromeritics, Norcross, GA, USA). The cation exchange capacity (CEC) of nano $I\text{-}S_m$ was determined by the ammonium acetate method [22]. The slurry's pH was determined by soaking 1 g of nano $I\text{-}S_m$ in 50 mL distilled water, stirring the solution for 24 h, filtering it, and then measuring the final pH [23].

2.3. Batch Adsorption Experiments

Stock solutions of Pb(II) were prepared by dissolving appropriate amounts of $(CH_3COO)_2Pb\cdot3H_2O$ in distilled water. Batch adsorption experiments were carried out in a series of centrifuge tubes by mixing a constant amount of nano I-S_m with 40 mL of the aqueous solution of Pb(II) at varying concentration and different temperatures. Then, the centrifuge tubes were put in a shaker incubator at 150 rpm for a certain time interval, the nano I-S_m was separated from the aqueous solutions by centrifugation at 3000 rpm for 5 min (TDZ5-WS, Cence, Changsha, China), and the supernatant was filtered through a 0.45 μm filter membrane. Pb(II) concentration in the solutions was measured by inductively coupled plasma mass spectrometry (7700 e, Agilent Technologies, Santa Clara, CA, USA).

The adsorption capacity of Pb(II) on nano I-S_m in the batch test was calculated using Equations (1) and (2).

$$R_{ratio} = \frac{C_0 - C_{eq}}{C_0} \times 100\% \tag{1}$$

$$Q_e = \frac{(C_0 - C_{eq})V}{m} \tag{2}$$

where R_{ratio} is the Pb(II) removal rate; Q_e is the equilibrium capacity of lead on the nano I-S_m, mg·g^{-1}; C_0 is the initial concentration of the Pb(II) solution, mg·L^{-1}; C_{eq} is the equilibrium concentration of the Pb(II) solution, mg·L^{-1}; V is the solution volume, L; and m is the mass of nano I-S_m, g. All assays were carried out in triplicate and only mean values are presented.

The effects of process variables, including pH of the solution, initial concentration of Pb(II), contact time, adsorbent dosage, and temperature, on the adsorption were studied. The pH of the solution at the start of the experiments was adjusted with 0.1 mg·L^{-1} HCl or 0.1 mg·L^{-1} NaOH. Adsorption isotherms studies were conducted at 298, 308, 313, 323, and 333 K, whereby 0.1 g of nano I-S_m was kept in contact with 40 mL of Pb(II) solution of varying concentrations (0.25, 0.50, 1.50, 2.50, 3.50, and 5 mg·L^{-1}) at pH 5. The kinetic experiments were performed using a Pb(II) concentration of 1 mg·L^{-1} with 0.1 g nano I-S_m at different time intervals (5, 10, 20, 30, and 60 min) at pH 5.

2.4. Theoretical Model

2.4.1. Adsorption Kinetics Model

The equations of the pseudo-first-order [24] and the pseudo-second-order kinetic model [25] were used to fit experiment data obtained from the batch experiments. The formulas of the pseudo-first-order and the pseudo-second-order kinetic model are expressed as Equations (3) and (4), respectively.

$$\ln(Q_e - Q_t) = \ln Q_e - k_1 t \tag{3}$$

$$\frac{t}{Q_t} = \frac{1}{k_2 Q_e^2} + \frac{1}{Q_e}t \tag{4}$$

where Q_t is the amount of Pb(II) adsorbed at time t, mg·g^{-1}, k_1 is the pseudo-first-order rate constant adsorption rate, min^{-1}; and k_2 is the adsorption rate constant in the pseudo-second-order kinetic rate constant, g·μg^{-1}·min^{-1}.

2.4.2. Adsorption Equilibrium

The isotherm models of Langmuir [26] and Freundlich [27] were tested to analyze the equilibrium data. The Langmuir isotherm model and Freundlich isotherm model equations are expressed by Equations (5) and (6).

$$\frac{C_{eq}}{Q_e} = \frac{1}{Q_{max}}C_{eq} + \frac{1}{Q_{max}K_L} \tag{5}$$

$$\ln Q_e = \ln K_f + \frac{1}{n}\ln C_{eq} \tag{6}$$

where Q_{max} is the monolayer capacity of nano I-S_m, mg·g^{-1}; K_L is the Langmuir constant, L·µg^{-1}; K_f is the Freundlich constant, µg·g^{-1}; and n is the heterogeneity.

2.4.3. Adsorption Thermodynamics

The thermodynamic parameters can be determined using the equilibrium constant and temperature [28,29]. The change in the Gibbs free energy (ΔG), enthalpy (ΔH), and entropy (ΔS) in the adsorption process was calculated using Equations (7) and (8).

$$\Delta G = -RT \ln K_d \tag{7}$$

$$\ln K_d = \frac{\Delta S}{R} - \frac{\Delta H}{RT} \tag{8}$$

where R is the universal gas constant, 8.314 J·mol^{-1}·K^{-1}; T is the absolute temperature, K; and K_d is the distribution coefficient of nano I-S_m, $K_d = Q_e/C_{eq}$.

3. Results

3.1. Characterization of Nano I-S_m

The chemical composition and physicochemical properties of nano I-S_m are presented in Table 1. The XRD patterns of nano I-S_m are given in Figure 1 A. Nano I-S_m is mainly composed of quartz, mixed-layer illite/smectite, illite, and kaolinite, and the characteristic diffraction peak of nano I-S_m was observed between 5 and 10° (2θ) [30]. Furthermore, an FT-IR analysis was applied to identify the functional groups on the nano I-S_m sample's surface. The FT-IR spectra of the nano I-S_m sample are shown in Figure 1B. The absorption bands at 3698.96 and 3620.60 cm^{-1} represent the inner surface OH stretching vibration, while the absorption band at 3423.76 cm^{-1} represents the outer surface OH stretching vibration. These OH groups function as an active site for the binding of positively charged cations. The absorption band at 1629.97 cm^{-1} represents the OH bending of water retained in the silica matrix [23]. The absorption bands at 1031.99 and 470.19 cm^{-1} represent the Si–O–Si stretching vibration [31]. The absorption band at 912.4 cm^{-1} represents the Al–OH bending vibrations [29], while those at 798.04 and 694.4 cm^{-1} represent the Si–O stretching vibration [23].

Table 1. Chemical composition and physicochemical properties of nano illite-smectite (I-S_m) clay.

Parameter	Value
SiO$_2$ (wt %)	64.29
Al$_2$O$_3$ (wt %)	20.38
Fe$_2$O$_3$ (wt %)	2.95
K$_2$O (wt %)	2.74
MgO (wt %)	1.82
TiO$_2$ (wt %)	0.82
Na$_2$O (wt %)	0.19
Loss of ignition (wt %)	6.46
BET surface area (m^2·g^{-1})	39.46
Micropore area (m^2·g^{-1})	10.46
External surface area (m^2·g^{-1})	28.99
Total pore volume (cm^3·g^{-1})	0.011
Micropore volume (cm^3·g^{-1})	0.0055
Adsorption average pore diameter (4 V/A by BET)	1.07
CEC (meg/100 g)	2.11
Slurry pH	6.75

CEC: cation exchange capacity. BET: Brunauer-Emmett-Teller.

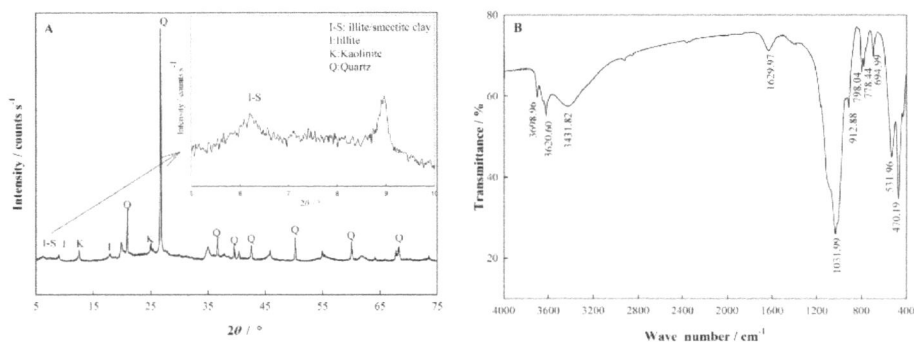

Figure 1. XRD spectra (**A**) and FT-IR spectrum (**B**) of nano I-S$_m$.

SEM analysis is another important tool used in the determination of the surface morphology of an adsorbent. In this study, SEM was used to probe the change in morphological features of nano I-S$_m$ and Pb-adsorbed nano I-S$_m$ (Figure 2).

Figure 2. (**A**) SEM micrograph of nano I-S$_m$ (before adsorption); (**B**) after Pb(II) adsorption.

3.2. Effect of Adsorption Conditions

3.2.1. Effect of pH

The effect of pH on Pb(II) removal rate was investigated at 298 K for 60 min as shown in Figure 3A. It was observed that the levels of adsorption efficiency of Pb(II) increased significantly with increasing pH. The removal rate of Pb(II) on nano I-S$_m$ was only 41.25 % at pH 2.0. In addition, the removal rate of Pb(II) tended to equilibrate at pH 4.0. When the solution had a pH > 6.0, the solution of Pb(II) gradually formed Pb(OH)$_2$ precipitate, and the solution system became relatively complex.

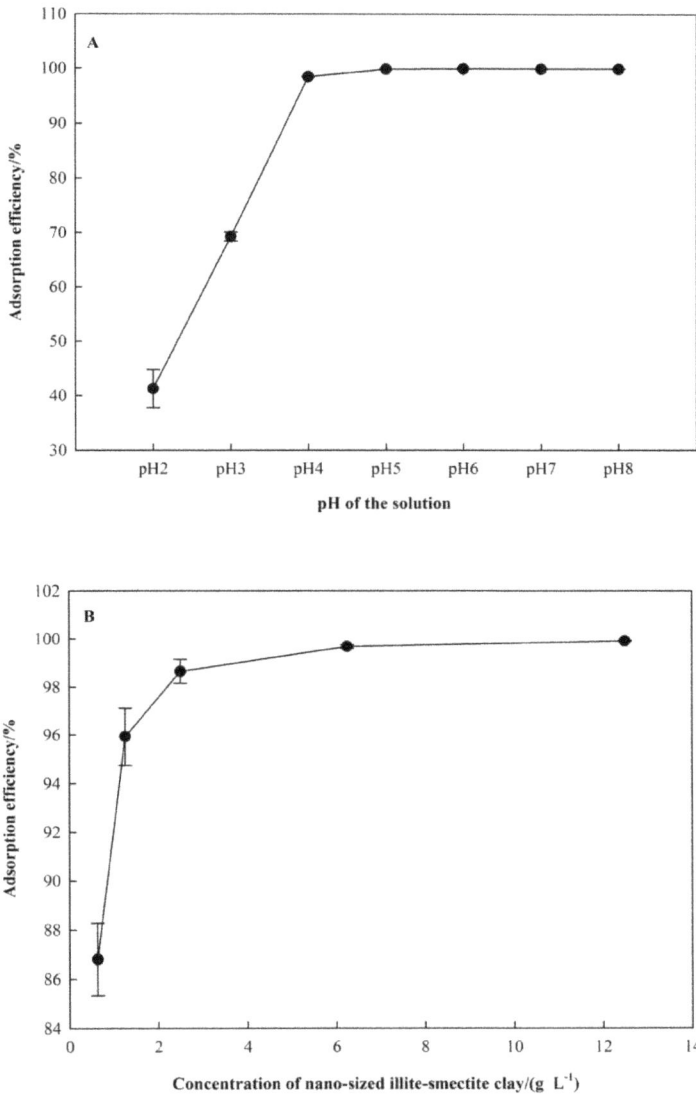

Figure 3. *Cont.*

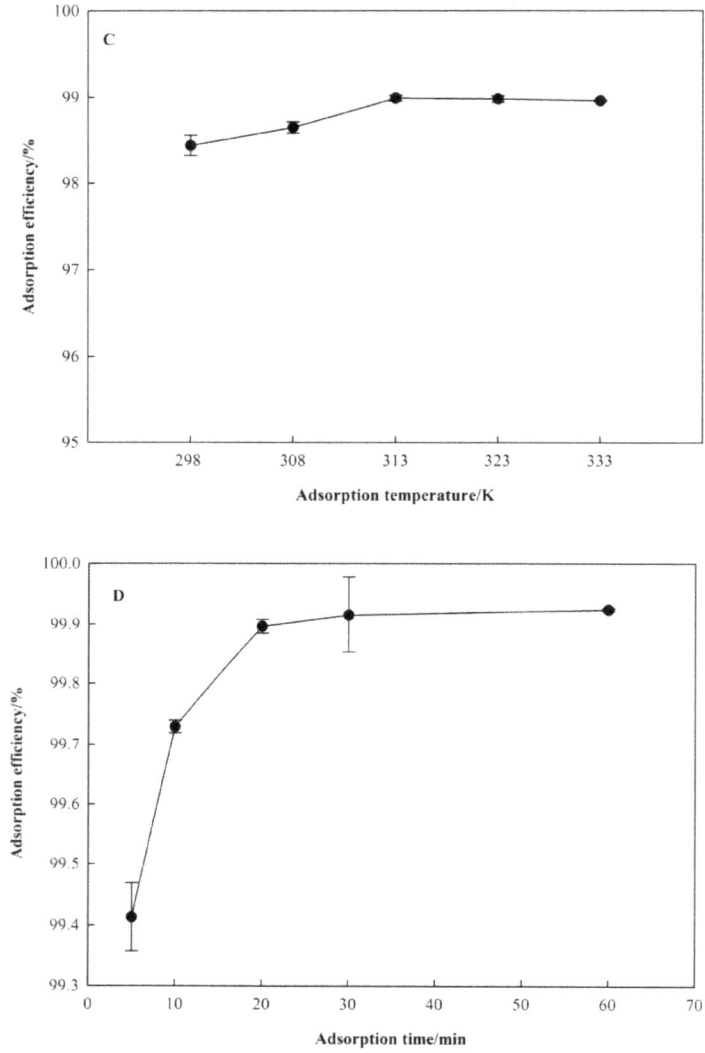

Figure 3. *Cont.*

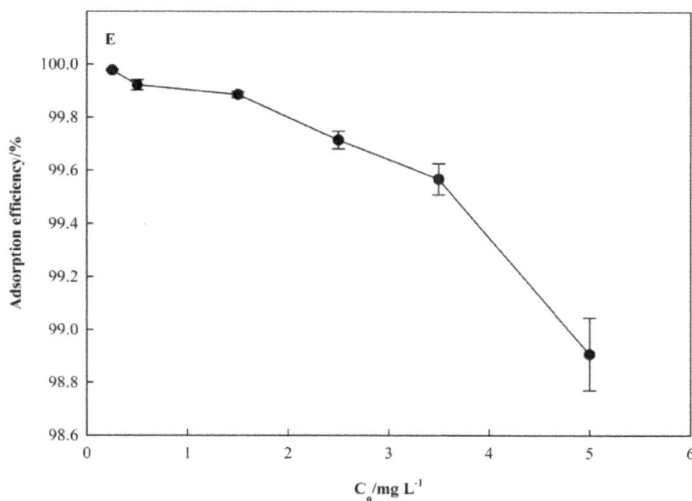

Figure 3. Effect of adsorption conditions on the removal rate of Pb(II). (**A**) for pH, (**B**) for dosage of I-S_m, (**C**) for adsorption temperature, (**D**) for adsorption time, (**E**) for Pb(II) initial concentration.

3.2.2. Effect of Nano I-S_m Dosage

The effect of nano I-S_m dosage on Pb(II) removal rate is shown in Figure 3B. The nano I-S_m dosage varied from 0.625 to 12.5 g·L^{-1} with a constant initial Pb(II) concentration of 1 mg·L^{-1} for 60 min at 298 K. Figure 3B shows the effect of nano I-S_m dosage on the removal rate of Pb(II). It was observed that the removal rate of Pb(II) increased with an increase in the nano I-S_m dosage from 0.625 to 2.5 g·L^{-1}. A further increase in the nano I-S_m dosage, however, did not result in a sufficient increase in the removal rate of Pb(II).

3.2.3. Effect of Adsorption Temperature

The effect of temperature on Pb(II) removal rate is shown in Figure 3C. It was observed that the removal rate of Pb(II) was 98.44–98.99% when the temperature was set at 298, 308, 313, 323, and 333 K. The trend of the removal rate with the increase of temperature is not obvious.

3.2.4. Effect of Adsorption Time

The effect of adsorption time on the removal rate of Pb(II) is shown in Figure 3D. In a Pb(II) solution with a low initial concentration, the removal rate of Pb(II) in solution reached 99.41% when the adsorption time was 5 min, and the removal rate of Pb(II) tended to be stable after 20 min.

3.2.5. Effect of Initial Concentration of Pb(II)

The effect of initial concentration on the removal rate of Pb(II) adsorbed by nano I-S_m is shown in Figure 3E. The removal rate of Pb(II) decreased with the increase of initial Pb(II) concentration. When the initial concentration of Pb(II) increased from 0.25 to 5 mg·L^{-1}, the removal rate of Pb(II) decreased from 99.45% to 98.90%.

3.3. Kinetic Parameters of the Adsorption

The kinetic of adsorption of Pb(II) on nano I-S_m was fitted by pseudo-first-order and pseudo-second-order kinetic equations. The results are shown in Table 2 and Figure 4. The correlation coefficient of the linear plots of t/Q_t against t for the pseudo-first-order model and the pseudo-second-order model was 0.985 and 1, respectively. The Q_e of the pseudo-first-order kinetics model was 2.603 µg·g^{-1}, and the Q_e of pseudo-second-order dynamic model was 256.410 µg·g^{-1}.

Table 2. The predicted parameters by pseudo-first-order and pseudo-second-order kinetic models and experimental data.

Pseudo-First-Order Kinetic Model			Pseudo-Second-Order Kinetic Model			Experimental Data
k_1/min^{-1}	$Q_e/(\text{µg·g}^{-1})$	R^2	$K_2/(\text{µg·g}^{-1}\cdot\text{min}^{-1})$	$Q_e/(\text{µg·g}^{-1})$	R^2	$Q_e/(\text{µg·g}^{-1})$
0.380	2.603	0.985	0.251	256.410	1.000	254.680

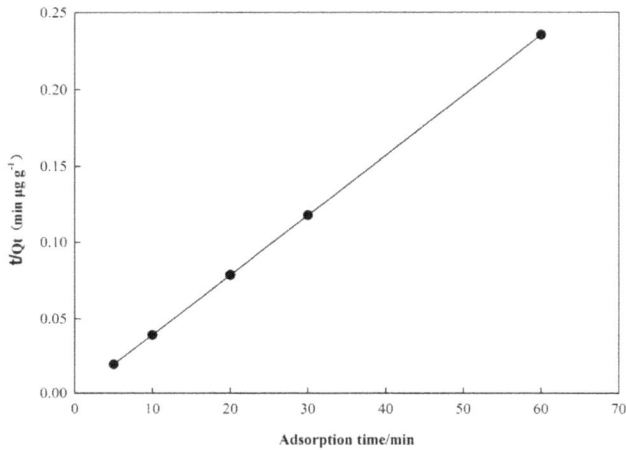

Figure 4. Pseudo-second-order plots for the adsorption of Pb(II) at 298 K.

3.4. Equilibrium Parameters of the Adsorption

The Langmuir and Freundlich isotherm models were used to analyze the adsorption of Pb(II) on nano I-S$_m$. All of the isotherm constants and correlation coefficients were calculated from the linear forms of the isotherm model equations and are provided in Table 3 and Figure 5.

Table 3. Parameters calculated by the Langmuir and Freundlich isotherm models for the adsorption of Pb(II) on nano I-S$_m$.

Langmuir			Freundlich		
$Q_{max}/(\text{mg·g}^{-1})$	$K_L/(\text{L·µg}^{-1})$	R^2	$K_f/(\text{µg·g}^{-1})$	$1/n$	R^2
2.104	0.216	0.985	8.825	0.457	0.980

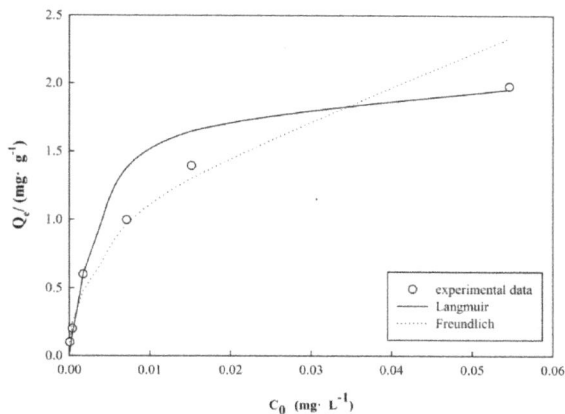

Figure 5. Comparison of equilibrium isotherms between the experimental data and the theoretical data.

3.5. Thermodynamic Parameters of the Adsorption

The results of the analysis of the thermodynamic parameters of adsorption are shown in Table 4 and Figure 6.

Table 4. Thermodynamic parameters for adsorption of Pb^{2+} on nano I-S_m.

ΔS	ΔH	$\Delta G/(kJ \cdot mol^{-1})$				
$J/mol^{-1} \cdot K^{-1}$	$kJ \cdot mol^{-1}$	298 K	308 K	313 K	323 K	333 K
9.658	4.844	-2.541	-2.637	-2.695	-2.788	-2.876

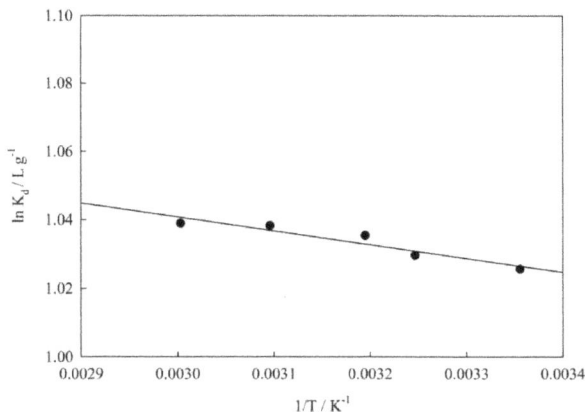

Figure 6. Relationship between $1/T$ and $\ln K_d$ for nano I-S_m.

4. Discussion

Among the adsorption conditions, the pH of the aqueous solution is an important variable for the adsorption of metals onto the adsorbents [32]. In this study, the adsorption efficiency was significantly inhibited when the pH of the aqueous solution was low. At a low pH, the number of H^+ ions exceeds that of Pb(II) ions several times and the surface of nano I-S_m is most likely covered with H^+ ions, which account for less Pb(II) adsorbed [33]. As the pH increases, more and more H^+ ions leave the nano $I-S_m$ surface, making the sites available to the Pb(II), which could increasingly bind to the nano $I-S_m$ surface through a mechanism similar to that of exchange interactions (H(I)/Pb(II)) [28]. In the meantime, Yuan et al. [19] have found that the Zeta potential decreased from 3.29 to -69.95 mV when the pH value increased from 2 to 7 in a nano $I-S_m$ solution, indicating that the surface charge of the nano $I-S_m$ changed from positive to negative, and further verifying the deprotonation processes of nano- $I-S_m$ with the increase of pH. It was observed that nano $I-S_m$ was suitable for removing Pb(II) from waste water under an acidic condition, which was similar to the results of some clay adsorbing heavy metals [28,29]. The removal rate of Pb(II) is also related to the $I-S_m$ dosage. At a lower nano $I-S_m$ dosage, Pb(II) ions compete for the limited adsorption sites in the nano $I-S_m$. As the quantity of nano $I-S_m$ increased, more available sites promoted a greater percentage removal of Pb(II) [33]. When the amount of nano $I-S_m$ was 6 $g \cdot L^{-1}$, 1 $mg \cdot L^{-1}$ Pb(II) in solution would be reduced to 0.01 $mg \cdot L^{-1}$, reaching the potable water standard. In addition, nano $I-S_m$ showed a rapid adsorption effect in the temperature range of 298–333 K. The above results revealed an important advantage of high efficiency removal of Pb(II) by the nano $I-S_m$.

The Q_e of the pseudo-second-order dynamic model was much closer to the experimental result and the correlation coefficients were found to be relatively high. The pseudo-second-order adsorption mechanism was predominant for the adsorption of Pb(II) on nano $I-S_m$. The pseudo-second-order model assumes that two reactions are happening: the first one is fast and reaches equilibrium quickly, whereas the second one is a slower reaction [15]. Accordingly, the following mechanism may be proposed [33]:

$$Clay + Pb(II) = Clay \cdots Pb(II)$$

in which the number of adsorption sites on the nano $I-S_m$ surface and the number of Pb(II) ions in the liquid phase determine the kinetics. Depending on pH, different Pb-species may be held to the clay surface at appropriate ion-exchange sites [25,33].

The correlation coefficient of the Langmuir isotherm model was higher than that of the Freundlich isothermal model, from which we can conclude that the Langmuir isotherm model was more suitable for nano $I-S_m$ removal of Pb(II) in aqueous solutions. Similar results were also reported in earlier studies in which the adsorption of heavy metal ions fitted well to the Langmuir isotherm [15,19,33]. Furthermore, the n values of the Freundlich isothermal model relate to the adsorption properties of the adsorbent, where values of n between 2 and 10 represent good adsorption [34], which is an indication of the good adsorption of Pb(II) by nano $I-S_m$.

The Gibbs free energy (ΔG) was -2.541, -2.637, -2.695, -2.788, and -2.876 kJ mol^{-1}($\Delta G < 0$) when the temperature was set at 298, 308, 313, 323, and 333 K, respectively. These indicate that the adsorption of Pb(II) on nano $I-S_m$ is a spontaneous process [23]. The ΔH was 4.844 kJ mol^{-1}($0 < \Delta H < 16$), which indicates that the adsorption of Pb(II) on nano $I-S_m$ is an endothermic process [19,34].

The SEM results showed that the surface morphology of Pb-adsorbed nano $I-S_m$ is different from that of natural nano $I-S_m$. The natural nano $I-S_m$ showed loose aggregates with a porous structure. After adsorption, the surface of nano $I-S_m$ demonstrates compacted aggregates. The surface morphology of the natural nano $I-S_m$ changed evidently during the adsorption process, indicating that significant interaction at the lead–clay interface occurred during the experiment. Similar SEM results were reported by other researchers [28,35].

5. Conclusions

As a new adsorbent, nano $I-S_m$ can be used for depth treatment in lead-contaminated water. The pseudo-second-order adsorption mechanism was predominant for the adsorption of Pb(II) on nano $I-S_m$. The saturated adsorption capacity of Pb(II) on nano $I-S_m$ in the aqueous solution was 256.41 $\mu g \cdot g^{-1}$. The adsorption patterns followed the Langmuir isotherm model. The adsorption of Pb(II) on nano $I-S_m$ is a thermodynamically feasible, spontaneous, and endothermic process.

Acknowledgments: This work was supported by the Guangxi Natural Science Foundation (grant No. 2015GXNSFEA139001) from the Guangxi science and Technology Department, the project of International Scientific Exchange Program (grant No. 7–1, 2017) from the Ministry of Science and Technology of the People's Republic of China, and the key project of Guangxi Social Sciences. (grant No. gxsk201605)

Author Contributions: Juan Yin and Chaobing Deng designed the study; Juan Yin and Zhen Yu analyzed the data; Juan Yin wrote the article, and Xiaofei Wang and Guiping Xu provided access to the experimental site.

Conflicts of Interest: The authors declare no conflict of interest.

References

1. Dipak, P. Research on heavy metal pollution of river Ganga: A review. *Ann. Agrar. Sci.* **2017**, *15*, 278–286. [CrossRef]

2. IslamabMd, M.S.; Ahmed, M.K.; Raknuzzaman, M.; Mamun, M.H.-A.; Islam, M.K. Heavy metal pollution in surface water and sediment: A preliminary assessment of an urban river in a developing country. *Ecol. Indic.* **2015**, *48*, 282–291. [CrossRef]

3. Xie, X.J.; Wang, F.Y.; Wang, G.J.; Mei, R.W.; Wang, C.Z. Study on Heavy Metal Pollution in Surface Water in China. *Envron. Sci. Manag.* **2017**, *42*, 31–34. [CrossRef]

4. Zhang, X.W.; Yang, L.S.; Li, Y.H.; Li, H.R.; Wang, W.Y.; Ye, B.X. Impacts of lead /zinc mining and smelting on the environment and human health in China. *Environ. Monit. Assess.* **2012**, *184*, 2261–2273. [CrossRef] [PubMed]

5. Li, Z.Y.; Ma, Z.W.; Kuijp, T.J.; Yuan, Z.W.; Huang, L. A review of soil heavy metal pollution from mines in China: Pollution and health risk assessment. *Sci. Total Environ.* **2014**, *468–469*, 843–853. [CrossRef] [PubMed]

6. Ahmaruzzaman, M.; Gupta, V.K. Rice husk and its ash as low-cost adsorbents in water and wastewater treatment. *Ind. Eng. Chem. Res.* **2011**, *50*, 13589–13613. [CrossRef]

7. Zhao, D.D.; Yu, Y.; Chen, P. Treatment of lead contaminated water by a PVDF membrane that is modified by zirconium, phosphate and PVA. *Water Res.* **2016**, *101*, 564–573. [CrossRef] [PubMed]

8. Vergili, I.; Gönder, Z.B.; Kaya, Y.; Gürdağ, G.; Çavuş, S. Sorption of Pb (II) from battery industry wastewater using a weak acid cation exchange resin. *Process Saf. Environ. Prot.* **2017**, *107*, 498–507. [CrossRef]

9. Milind, M.N.; Santosh, K.D. Lead resistant bacteria: Lead resistance mechanisms, their applications in lead bioremediation and biomonitoring. *Ecotox. Environ. Saf.* **2013**, *98*, 1–7. [CrossRef]

10. Abreham, T.B.; Abaynesh, Y.G.; Ramato, A.T.; Dawit, N.; Efrem, C.; Lidietta, G. Removal of emerging micropollutants by activated sludge process and membrane bioreactors and the effects of micropollutants on membrane fouling: A review. *J. Environ. Chem. Eng.* **2017**, *5*, 2395–2414. [CrossRef]

11. Zhao, G.X.; Li, J.X.; Ren, X.M.; Chen, C.L.; Wang, X.K. Few-layered graphene oxide nanosheets as superior sorbents for heavy metal ion pollution management. *Environ. Sci. Technol.* **2011**, *45*, 10454–10462. [CrossRef] [PubMed]

12. Burakov, A.E.; Galunin, E.V.; Burakova, I.V.; Kucherova, A.E.; Agarwal, S.; Tkachev, A.G.; Gupta, V.K. Adsorption of heavy metals on conventional and nanostructured materials for wastewater treatment purposes: A review. *Ecotoxicol. Environ. Saf.* **2018**, *148*, 702–712. [CrossRef] [PubMed]

13. Mohammad, K.U. A review on the adsorption of heavy metals by clay minerals, with special focus on the past decade. *Chem. Eng. J.* **2017**, *308*, 438–462. [CrossRef]

14. Masindi, V.; Gitari, W.M. Simultaneous removal of metal species from acidic aqueous solutions using cryptocrystalline magnesite/bentonite clay composite: An experimental and modelling approach. *J. Clean. Prod.* **2016**, *112*, 1077–1085. [CrossRef]

15. Joziane, G.M.; Murilo, P.M.; Thirugnanasambandham, K.; Sergio, H.B.F.; Marcelino, L.G.; Maria, A.S.D.B.; Sivakumar, V. Preparation and characterization of calcium treated bentonite clay and its application for the removal of lead and cadmium ions: Adsorption and thermodynamic modeling. *Process Saf. Environ.* **2017**, *111*, 244–252. [CrossRef]

16. Sari, A.; Tuzen, M. Cd(II) adsorption from aqueous solution by raw and modified kaolinite. *Appl. Clay Sci.* **2014**, *88–89*, 63–72. [CrossRef]

17. Atta, A.M.; Al-Lohedan, H.A.; ALOthman, Z.A.; Abdel-Khalek, A.A.; Tawfeek, A.M. Characterization of reactive amphiphilic montmorillonite nanogels and its application for removal of toxic cationic dye and heavy metals water pollutants. *J. Ind. Eng. Chem.* **2015**, *31*, 374–384. [CrossRef]

18. Qiu, J.Y. Preparation of Illite/Smectite Clay Nano-Powder and Its Application as Rubber Filler. Master's Thesis, South China University of Technology, Guangzhou, China, 4 June 2014.

19. Yuan, S.S.; Li, Z.Y.; Pan, Z.D.; Wang, Y.M. Removal of Copper and Cadmium Ions in Aqueous Solution via Adsorption by Nano-sized Illite-Smectite Clay. *J. Chin. Ceram. Soc.* **2016**, *44*, 43–49. [CrossRef]

20. Zhang, L.H.; Yuan, Y.H.; Yan, Z.G.; Zhou, Y.Y.; Zhang, C.Y.; Huang, Y.; Xu, M. Application of nano illite/smectite clay for adsorptive removal of metals in water. *Res. Environ. Sci.* **2016**, *29*, 115–123. [CrossRef]

21. Sakthivel, S.; Venkatesan, V.; Krishnan, B.; Pitchumani, B. Influence of suspension stability on wet grinding for production of mineral nanoparticles. *Particuology* **2008**, *6*, 120–124. [CrossRef]

22. Bao, S.D. *Soil Agricultural Chemistry Analysis*, 3rd ed.; China Agricultural Press: Beijing, China, 2000; pp. 152–176, ISBN 9787109066441.

23. Dawodu, F.A.; Akpomie, K.G. Simultaneous adsorption of Ni(II) and Mn(II) ions from aqueous solution unto a Nigerian kaolinite. *J. Mater. Res. Technol.* **2014**, *3*, 129–141. [CrossRef]

24. Lagergren, S. *Zur theorie der sogenannten adsorption gelöster stoffe (about the theory of so-called adsorption of soluble substances)*; Kungliga Svenska Vetenskapsakademiens (Royal Swedish Academy of Sciences): Stockholm, Sweden, 1898; pp. 1–39.

25. Ho, Y.S.; McKay, G. Pseudo-second order model for sorption processes. *Process Biochem.* **1999**, *34*, 451–465. [CrossRef]

26. Langmuir, I. The adsorption of gases on plane surfaces of glass, mica and platinum. *J. Am. Chem. Soc.* **1918**, *40*, 1361–1403. [CrossRef]

27. Freundlich, H.M.F. Über die adsorption in lösungen. *Z. Phys. Chem.* **1906**, *57*, 385–470. [CrossRef]

28. Jiang, M.Q.; Wang, Q.P.; Jin, X.Y.; Chen, Z.L. Removal of Pb(II) from aqueous solution using modified and unmodified kaolinite clay. *J. Hazard. Mater.* **2009**, *170*, 332–339. [CrossRef] [PubMed]

29. Olgun, A.; Atar, N. Equilibrium, thermodynamic and kinetic studies for the adsorption of lead (II) and nickel (II) onto clay mixture containing boron impurity. *J. Ind. Eng. Chem.* **2012**, *18*, 1751–1757. [CrossRef]

30. Campo, M.D.; Bauluz, B.; Nieto, F.; Papa, C.D.; Hongn, F. SEM and TEM evidence of mixed-layer illite-smectite formed by dissolution- crystallization processes in continental Paleogene sequences in northwestern Argentina. *Clay Miner.* **2016**, *51*, 723–740. [CrossRef]

31. Li, Y.J.; Zeng, L.; Zhou, Y.; Wang, T.F.; Zhang, Y.J. Preparation and Characterization of Montmorillonite Intercalation Compounds with Quaternary Ammonium Surfactant: Adsorption Effect of Zearalenone. *J. Nanomater.* **2014**, *2014*. [CrossRef]

32. Deng, L.L.; Yuan, P.; Liu, D.; Liu, Z.W. Effects of microstructure of clay minerals, montmorillonite, kaolinite and halloysite, on their benzene adsorption behaviors. *Appl. Clay Sci.* **2017**, *143*, 184–191. [CrossRef]

33. Krishna, G.B.; Susmita, S.G. Pb(II) uptake by kaolinite and montmorillonite in aqueous medium: Influence of acid activation of the clays. *Colloid Surf. A* **2006**, *277*, 191–200. [CrossRef]

34. Nir, S.; Undabeytia, T.; Dana, Y.; Yasser, E.; Polubesova, T.; Serban, C.; Rytwo, G.; Lagaly, G.; Rubin, B. Optimization of adsorption of hydrophobic herbicides on montmorillonite presorbed by monovalent organic cations: Interaction between phenyl rings. *Environ. Sci. Technol.* **2000**, *34*, 1269–1274. [CrossRef]

35. Li, J.S.; Xue, Q.; Wang, P.; Li, Z.Z. Effect of lead (II) on the mechanical behavior and microstructure development of a Chinese clay. *Appl. Clay Sci.* **2015**, *105–106*, 192–199. [CrossRef]

Article

Lignin Biodegradation in Pulp-and-Paper Mill Wastewater by Selected White Rot Fungi

Stefania Costa [1], Davide Gavino Dedola [1], Simone Pellizzari [2], Riccardo Blo [2], Irene Rugiero [1], Paola Pedrini [1] and Elena Tamburini [1,*]

[1] Department of Life Science and Biotechnology, University of Ferrara, Via L. Borsari, 44121 Ferrara, Italy; stefania.costa@unife.it (S.C.); davidegavino.dedola@student.unife.it (D.G.D.); irene.rugiero@unife.it (I.R.); pdp@unife.it (P.P.)
[2] NCR-Biochemical SpA, Via dei Carpentieri, 40050 Castello d'Argile (BO), Italy; S.Pellizzari@ncr-biochemical.it (S.P.); R.Blo@ncr-biochemical.it (R.B.)
* Correspondence: tme@unife.it; Tel.: +39-0532-455-329

Received: 20 October 2017; Accepted: 29 November 2017; Published: 2 December 2017

Abstract: An investigation has been carried out to explore the lignin-degrading ability of white rot fungi, as *B. adusta* and *P. crysosporium*, grown in different media containing (i) glucose and mineral salts; (ii) a dairy residue; (iii) a dairy residue and mineral salts. Both fungi were then used as inoculum to treat synthetic and industrial pulp-and-paper mill wastewater. On synthetic wastewater, up to 97% and 74% of lignin degradation by *B. adusta* and *P. crysosporium*, respectively, have been reached. On industrial wastewater, both fungal strains were able to accomplish 100% delignification in 8–10 days, independent from pH control, with a significant reduction of total organic carbon (TOC) of the solution. Results have confirmed the great biotechnological potential of both *B. adusta* and *P. crysosporium* for complete lignin removal in industrial wastewater, and can open the way to next industrial applications on large scale.

Keywords: lignin; delignification; pulp-and-paper-mill c; wastewater; white rot fungi; *B. adusta*; *P. crysosporium*

1. Introduction

The pulp and paper industry in Europe accounts for about a quarter of world manufacturing, producing more than 90 million tons of paper and board, and more than 36 million tons of pulp annually [1]. The manufacture of paper generates significant quantities of wastewater, as high as 60 m^3/ton of paper produced [2]. These raw wastewaters—sometimes called black liquor—can be potentially very polluting [3]. Pulp-and-paper mill wastewater contains a considerable amount of pollutants characterized by high biochemical oxygen demand (BOD), chemical oxygen demand (COD), and high dissolved solids, mainly due to alkali–lignin and polysaccharide degradation residues [4]. The environmental impact of pulp-and-paper mill wastewater depends not only on its chemical nature, but also on its dark coloration that negatively affects aquatic fauna and flora [5]. The primary contributors to the color and toxicity of wastewater are high-molecular-weight lignin and its derivatives. Lignin is the generic term for a large group of aromatic rigid and impervious polymers resulting from the oxidative coupling of 4-hydroxyphenylpropanoids, present predominantly in woody plants [6]. The chemical or biological degradation of lignin is very difficult due to the presence of recalcitrant and not-hydrolysable carbon-carbon linkages and aryl ether bonds [7]. Notwithstanding, pulp-and-paper mills are now facing challenges to comply with stringent environmental regulations [8]. For years, various methods have been developed and attempted for wastewater treatment and organic pollutants removal, including incineration [9], photochemical UV/TiO$_2$ oxidation [10], adsorption of organic compounds on activated carbon and polymer resin [11], chemical coagulation/flocculation of lignin using synthetic or natural coagulants [12], and catalytic wet air oxidation [13]. However, all these processes are expensive, environmentally overburdening, and often not very efficient [14]. Furthermore, in these processes lignin is not really degraded, but transferred from a water-suspended state into a solid or absorbed state, only moving the problem [15]. A valid alternative to remove organic pollutants from pulp-and-paper wastewater is now represented by biological treatments. In nature, various ligninolytic organisms and enzymes including fungi, actinomycetes, and bacteria are implicated in lignin biodegradation, and can have potential application in wastewater treatments [16]. Several studies have been carried out on biological delignification of pulp-and-paper mill wastewater using pure bacterial strains [17]: about 70–80% of lignin degradation and COD removal have been achieved with *Pseudomonas putida* and *Acinetobacter calcoaceticus* [18], *Aeromonas formicans* [19], and *Bacillus* sp. [20]. In this field, white-rot fungi have also received increasing attention due to their powerful lignin-degrading enzyme system [21]. White-rot wood fungi use the cellulose fraction as a carbon source and are able to completely degrade the lignin in order to have access to the cellulose. Basidiomycetes species are extensively studied due to the high degradation ability of the extracellular oxidative enzymes (i.e., laccase, peroxidase) that need low-molecular weight cofactors [22]. Recent developments of new technologies and/or improvements of existing ones for the treatment of effluents from the pulp and paper industries include the use of the white rot fungi *Aspergillus foetidus*, *Phanerochaete chrysosporium*, and *Trametes versicolor* [23], but scarce industrial experience is available concerning the degradation of highly-contaminated pulp-and-paper mill wastewater by fungi. In particular, *Phanerochaete chrysosporium* is a well-known white-rot fungus and a strong degrader of various xenobiotics [24]. It has been extensively investigated as a model organism for fungal lignin and organopollutant degradation, since it was the first fungus found to produce lignin peroxidase and manganese peroxidase [25]. *Bjerkandera adusta* is a wood-rotting basidiomycete belonging to the white-rot fungi commonly found in Europe. Its capability to degrade aromatic xenobiotics [26] and extractives [27] has progressively increased its biotechnological interest in wastewater treatments for lignin degradation [28]. Due to its laccase and manganese peroxidase activity [29,30], applications of *B. adusta* to the biomineralization of lignin in soils [31] and to the decoloration of industrial dye effluents [32] has been already attempted, but to date not at an industrial level. This study reports the lignin removal capability and effectiveness of *B. adusta* and *P. crysosporium*, grown in different culture media containing lignin, on synthetic and industrial pulp-and-paper mill wastewater.

2. Materials and Methods

2.1. Fungal Strain Master Cell Bank and Working Cell Bank

Bierkandera adusta and *Phenarochete crysosporium* were purchased from Leibniz Institute DSMZ–German Collection of Microorganisms and Cell Cultures (Braunschweig, Germany). The strains have been stored as a master cell bank (MCB), maintained at −20 °C in 3% malt extract and 3% peptone cryovials (1 mL) with added glycerol (0.5 mL). Cells from the MCB were expanded to form the working cell bank (WCB), using an identical procedure. Prior to being used in the process, the fungal strains from WCB were maintained for 7 days in 3% malt extract agar Petri dishes.

2.2. Standard Media and Pulp-and-Paper Mill Wastewater

Three growth media have been prepared for this study: (i) a medium (standard glucose medium, SGM) containing glucose (10 g/L), KH_2PO_4 (1 g/L), yeast extract (0.5 g/L), $MgSO_47H_2O$ (0.5 g/L), and KCl (0.5 g/L) was adjusted to pH 5.5 with 1 M HCl and autoclaved; (ii) a medium (standard lactose medium, SLM) where glucose has been replaced with 50 mL of a dairy residue from cheese processing containing 50 g/L lactose, supplied by Granarolo S.p.A. (Bologna, Italy) (Table 1); (iii) a medium made up with the sole dairy by-product (standard dairy medium, SDM). Before inoculation, SGM medium was added with 5 g/L of standard lignin. Spore and mycelium suspensions obtained from agar Petri dishes were used to inoculate a 250 mL Erlenmeyer flask containing 100 mL of SGM. Cell cultures were all incubated at 24 °C without pH control for 10 days under mild stirring rate (60 rpm) and samples were withdrawn at 1–3 day intervals for residual lignin content analysis.

Table 1. Dairy residue chemical composition.

Constituent	%
Total solids	6.0
Lactose	5.0
Proteins	0.6
Non-protein N *	0.2
Lipids	0.05
Ash	0.5

Note: * N = Nitrogen.

A synthetic pulp-and-paper mill wastewater was prepared by dissolving 5 g/L of standard lignin in distilled water. Three 1-L Erlenmeyer flasks containing 500 mL of the synthetic wastewater were inoculated with 50 mL of cell cultures grown in the SGM, SLM, and SDM media, respectively, all added with standard lignin (5 g/L) and incubated for 10 days at 24 °C and mild agitation (60 rpm).

The industrial pulp-and-paper mill wastewater utilized for this study was supplied by a local pulp-and-paper firm, collected in a closed container and stored in darkness at 4 °C until use. The concentration of soluble and insoluble lignin was determined, as well as total organic carbon (TOC), as described in Section 2.3.

Two 1-L Erlenmeyer flasks containing 500 mL of wastewater were inoculated with 50 mL of cell cultures grown in the SLM added with lignin (5 g/L), and incubated for 10 days at 24 °C and mild agitation (60 rpm). In one flask, pH was adjusted to 5.5 with 1 M HCl, in the other pH was left at the original value measured for industrial wastewater of 6.5 without control.

All of the above experiments were conducted in triplicate. The data in subsequent sections are based on the average and standard deviation of the three measurements.

2.3. Chemicals and Analysis

All chemicals were reagent grade or better. Unless specified otherwise, they were obtained from Sigma-Aldrich Chemical Co (Saint Louis, MO, USA). The concentration of lignin was measured using the INNVENTIA—Biorefinery Test Methods L 2:2016 [33], specific for the determination of lignin isolated from a Kraft pulping process. The procedure is based on the sulphuric acid hydrolysis of the samples. This method makes it possible to determine concentrations of total lignin content, measured as the sum of the amount of acid-insoluble matter (AIM) and acid-soluble matter (ASM) after sulphuric acid hydrolysis, down to 10 mg/g oven-dry sample.

TOC was determined with a Carbon Analyzer TOC-V-CSM (Shimadzu, Tokio, Japan) after acidification with 2 M HCl to remove dissolved carbonate [34]. The instrument has a detection limit of 5 µg/L and a measurement accuracy expressed as coefficient of variation (CV) 1.5%. Biomass concentration (dry weight, DW) was determined gravimetrically after drying overnight at 105 °C on a pre-weighed 0.2 µm filter.

3. Results

3.1. B. adusta and P. crysosporium Growth on SGM

Lignin was added to the standard medium SGM before inoculation of *B. adusta* and *P. crysosporium* because several studies describe that the presence of lignin in the liquid medium exerts an influence on the expression profile of lignin peroxidase, manganese peroxidase, and laccase—all enzymes held responsible for the lignin degradation of natural lignocellulosic residue [35,36]. Under the condition maintained on 100 mL-scale, in 10 days *B. adusta* was able to uptake and metabolize lignin up to 67%, while *P. crysosporium* only 30% (Figure 1).

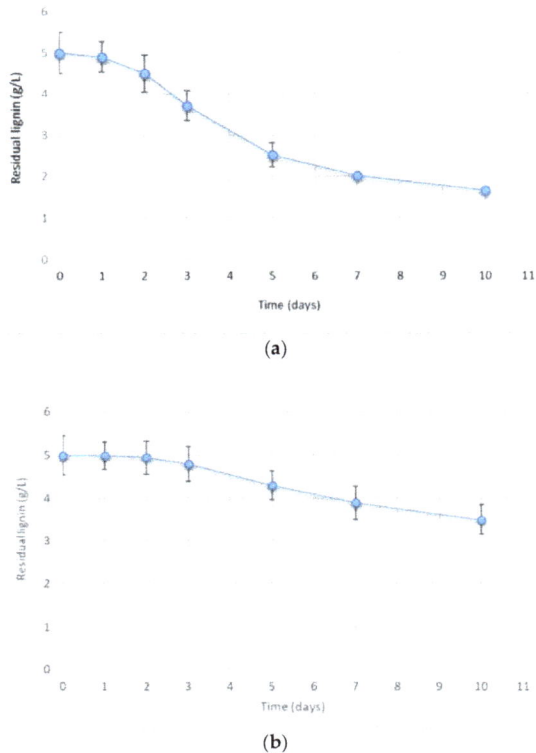

(a)

(b)

Figure 1. Lignin removal from 100 mL of standard glucose medium (SGM) with lignin 5 g/L added by (**a**) *B. adusta* and (**b**) *P. crysosporium*.

As described by Girard et al. [37], in both cases the expression of delignifying enzymes only initiated after 2–3 days from inoculation, corresponding to complete glucose depletion (data not shown).

3.2. Lignin Removal Efficiency on Synthetic Pulp-and-Paper Mill Wastewater

The addition of agro-food by-products to fungal cultures may reflect complex growth conditions close to nature, and could stimulate the secretion of various enzymes required for degradation or detoxification processes [38]. This, in addition to the evidence that the production of lignin peroxidase and manganese peroxidase in *B. adusta* is stimulated by the presence of organic nitrogen (N) source (unlike *P. chrysosporium*, which produces ligninolytic peroxidases in response to N limitation [39]), has driven the study towards the possibility of integrating the growth medium with a dairy by-product, usually rich in protein and amino acids, apart from sugar. Furthermore, in view of industrial application, the use of a by-product instead of pure substrates could permit the considerable reduction of operational investments, among which chemicals required for fungal growth are the most relevant. The use of cheese whey has been previously proposed by Feijoo et al. [40] as an inexpensive substrate for fungal growth. *B. adusta* and *P. chrysosporium* have been incubated in SGM, SLM, and only dairy residue with no addition of other nutrients or mineral salts (SDM). The largest amount of fungal biomass was obtained when dairy residue was present in the media (Table 2).

Table 2. Fungal cells dry weight (g/L) obtained from growth in SGM, standard lactose medium (SLM), and standard dairy medium (SDM) media.

Strain	SGM	SLM	SDM
B. adusta	2.5 ± 0.4	3.6 ± 0.5	3.5 ± 0.4
P. crysosporium	2.7 ± 0.3	4.3 ± 0.5	3.8 ± 0.6

In both cases, the results seem to confirm the correlation between organic N source and fungal cell growth. Identical amounts of cells of *B. adusta* and *P. crysosporium* grown in the three media were used as inoculum for synthetic wastewater, in order to verify if cell cultures developed in different media would express different enzymatic patterns or different enzyme activities. Figure 2a shows that *B. adusta* grown in the SGM medium was able to remove 73% of lignin, whilst *B. adusta* grown in the presence of a source of protein and amino acids in both cases reached delignification yields of 97% with SLM and 86% with SDM. On the other hand, *P. crysosporium* in all three cases obtained yields not higher than 74% when grown in SLM (54% in SGM and 69% in SDM, respectively).

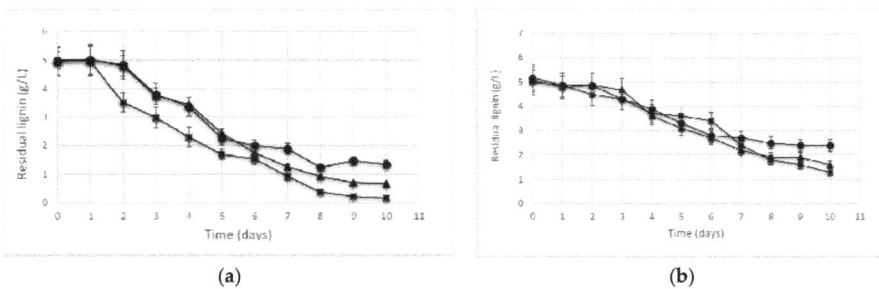

Figure 2. Lignin removal from synthetic pulp-and-paper mill wastewater during 10 days by (**a**) *B. adusta* and (**b**) *P. crysosporium* grown in SGM (dots), SDM (triangles), and SLM (squares) media, all with the addition of 5 g/L lignin.

The time courses of delignification in 10 days were quite similar in all three cases for *B. adusta*, having a 1-day reduced lag phase cell culture grown in SLM. The interesting point is that the slopes of the three curves are similar in the 3–8 days' interval, but from day-8 on, cell culture grown in SGM seemed to miss the lignin removal capacity, even though residual lignin was still present in the fermentation broth. This could be due to the decline of lignin peroxidase activity caused by the appearance of extracellular protease activity that has been observed after day 6–10 in cultures of *P. chrysosporium* grown on glucose [41]. This also confirmed what was reported by Nakamura et al. [42], whereby in glucose-based media, enzymes produced by *B. adusta* can only degrade part of the chemical structure of lignin. Otherwise, in order to maximize peroxidase activity, lactose has already been identified as a good carbon source for *Bierkandera* spp. when the nitrogen source was organic [43], as in SLM and SDM media. *P. crysosporium* was found to be surprisingly less active than *B. adusta* in lignin removal effectiveness in all three growth conditions (Figure 2b). Moreover, it showed a longer lag phase before starting to degrade lignin. According to Keyser et al. [44], lignin metabolism in *P. crysosporium* did not reflect the depletion of glucose, as in *B. adusta*, but instead appeared to be a response to nitrogen starvation. The prolonged lag phase could be induced by the need to wait for the partial or complete depletion of the N source transferred with inoculum.

3.3. Lignin Removal Efficiency on Industrial Pulp-and-Paper Mill Wastewater

B. adusta grown on SLM with lignin 5 g/L added has demonstrated to be effective for almost complete lignin biodegradation in synthetic wastewater. Based on these promising results, an application on industrial wastewater has been attempted, in comparison with *P. crysosporium* grown in the same conditions. The industrial wastewater supplied by the local pulp-and-paper mill for these tests (pH 6.5 and with a 100 g/L lignin content on dry weight basis) was diluted (12% dry weight). The ability of both fungal strains to biodegrade lignin has been tested, verifying the effect of pH on their enzymatic activities. In one case, the pH of wastewater was adjusted to the optimum value for fungi cell growth (pH 5.5), and in the other the process was allowed to proceed without correction (pH 6.5). From the perspective of industrial application, the possibility of avoiding costs deriving from the use of acids as correction agent could be very relevant. The results of the tests carried out using an inoculum of *B. adusta* and *P. crysosporium* grown on SLM medium on industrial wastewater with and without pH correction are reported in Figure 3.

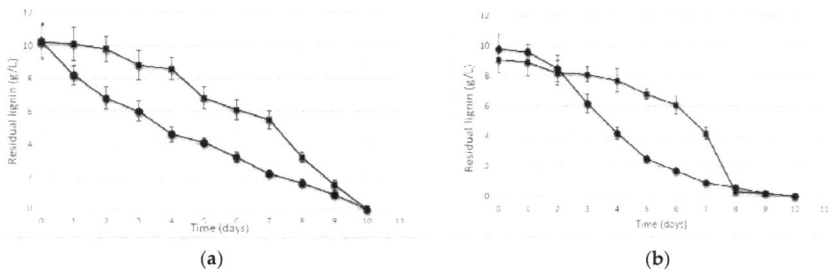

Figure 3. Lignin removal from industrial pulp-and-paper mill wastewater with pH correction (dots) and without pH correction (squares) during 10 days by (**a**) *B. adusta* and (**b**) *P. crysosporium*, grown in SLM medium added with lignin 5 g/L.

As expected, at optimum pH condition *B. adusta* started to biodegrade lignin without any lag phase and maintained an almost constant biodegradation rate of about 1 g/L of lignin per day over the entire test course. In contrast, without pH control fungal cells needed 1–2 days for adapting, before starting biodegradation. This leads to a variable delignification rate during the process—slower at the beginning (0.9 g/L × day) and higher from five days on (1.7 g/L × day). The final result in both cases was complete lignin removal, with an efficiency of 100%. One hundred percent delignification was also obtained when treating the pulp-and-paper mill wastewater with *P. crysosporium*, almost complete in 8 days. At a first glance, the time courses seemed to confirm the previous results obtained on synthetic pulp-and-paper mill wastewater, regarding the need of a longer lag phase compared with *B. adusta*. Otherwise, a sharp decline of residual lignin was observed from day 6. These results appeared to be particularly promising, compared with an average delignification yield of 70–80% reported for white rot fungi: both *P. crysosporium* and *B. adusta* were competitive against the 71% delignification yield on pulp-and-paper mill residues obtained by *Pseudomonas putida* [45], 78% by *Aeromonas formicans* [19], and 80% by *Acinetobacter calcoaceticus* [46].

To confirm the overall organic C removal, TOC analysis of samples was carried out. It is usually reported that lignin represents about 30–45% of the total organics in pulp-and-paper mill wastewater [47], so a corresponding decrease of TOC was expected (Figure 4). In both cases, an overall reduction of about 35% of organic charge of wastewater was obtained, reasonably due to lignin uptake for fungal metabolism.

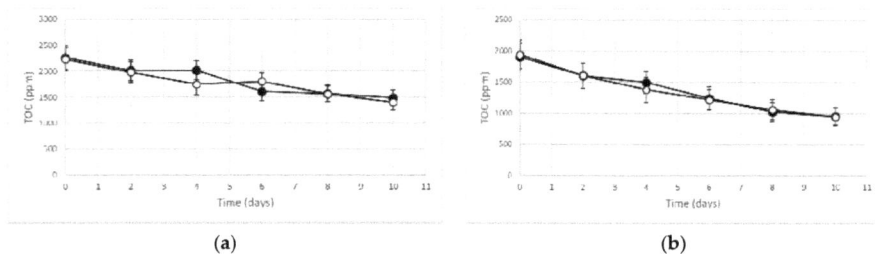

Figure 4. Total organic carbon (TOC) decrease of industrial pulp-and-paper mill wastewater with pH correction (white dots) and without pH correction (black dots) during 10 days by (**a**) *B. adusta* and (**b**) *P. crysosporium* grown in SLM medium with 5 g/L lignin added.

4. Conclusions

This study opens new perspectives for the bioremediation of industrial effluents such as pulp-and-paper mill wastewater using white rot fungi. In particular both *B. adusta* and *P. crysosporium* were found able to growth on non-conventional media, better than on glucose as sole carbon source, and to improve the delignifying activity in the presence of organic N and mineral salts. Moreover, they can survive on synthetic wastewater and proved to be effective for the complete degradation of lignin. The biotechnological potential of these strains was also confirmed on industrial wastewater, being active up to the total depletion of lignin. No operational problem was detected at 500 mL scale, as a first confirmation of the robustness and applicability of this system. The results obtained lay the ground for further scaling up to pilot plant level.

Author Contributions: Davide Gavino Dedola, Riccardo Blo and Irene Rugiero performed all the experiments and carried out all the analytical assays, also giving a great contribution to the discussion. Simone Pellizzari conceived and designed the experiments, together with Stefania Costa and Elena Tamburini, who wrote the manuscript. As supervisor of the research group, Paola Pedrini defined the general research statement.

Conflicts of Interest: The authors declare no conflict of interest.

References

1. Key Statistics 2015. CEPI-Confederation of European Paper Industries. Available online: http://www.cepi.org/statistics/keystatistics2015 (accessed on 9 October 2017).
2. Pokhrel, D.; Viraraghavan, T. Treatment of pulp and paper mill wastewater—A review. *Sci. Total Environ.* **2004**, *333*, 37–58. [CrossRef] [PubMed]
3. Thompson, G.; Swain, J.; Kay, M.; Forster, C.F. The treatment of pulp and paper mill effluent: A review. *Bioresour. Technol.* **2011**, *77*, 275–286. [CrossRef]
4. Lara, M.A.; Rodríguez-Malaver, A.J.; Rojas, O.J.; Holmquist, O.; González, A.M.; Bullón, J.; Araujo, E. Black liquor lignin biodegradation by *Trametes elegans*. *Int. Biodeterior. Biodegrad.* **2003**, *52*, 167–173. [CrossRef]
5. Ali, M.; Sreekrishnan, T.R. Aquatic toxicity from pulp and paper mill effluents: A review. *Adv. Environ. Res.* **2001**, *5*, 175–196. [CrossRef]
6. Vanholme, R.; Demedts, B.; Morreel, K.; Ralph, J.; Boerjan, W. Lignin biosynthesis and structure. *Plant Physiol.* **2010**, *153*, 895–905. [CrossRef] [PubMed]
7. Minu, K.; Jiby, K.K.; Kishore, V.V.N. Isolation and purification of lignin and silica from the black liquor generated during the production of bioethanol from rice straw. *Biomass Bioenergy* **2012**, *39*, 210–217. [CrossRef]
8. Kamali, M.; Khodaparast, Z. Review on recent developments on pulp and paper mill wastewater treatment. *Ecotoxicol. Environ. Saf.* **2015**, *114*, 326–342. [CrossRef] [PubMed]
9. Harila, P.; Kivilinna, V.A. Biosludge incineration in a recovery boiler. *Water Sci. Technol* **1999**, *40*, 195–200.
10. Chang, C.N.; Ma, Y.S.; Fang, G.C.; Chao, A.C.; Tsai, M.C.; Sung, H.F. Decolorizing of lignin wastewater using the photochemical UV/TiO$_2$ process. *Chemosphere* **2004**, *56*, 1011–1017. [CrossRef] [PubMed]

11. Zhang, Q.; Chuang, K.T. Adsorption of organic pollutants from effluents of a Kraft pulp mill on activated carbon and polymer resin. *Adv. Environ. Res.* **2001**, *5*, 251–258. [CrossRef]

12. Wang, J.P.; Chen, Y.Z.; Wang, Y.; Yuan, S.J.; Yu, H.Q. Optimization of the coagulation-flocculation process for pulp mill wastewater treatment using a combination of uniform design and response surface methodology. *Water Res.* **2011**, *45*, 5633–5640. [CrossRef] [PubMed]

13. Sales, F.G.; Abreu, C.A.M.; Pereira, J.A.F.R. Catalytic wet-air oxidation of lignin in a three-phase reactor with aromatic aldehyde production. *Braz. J. Chem. Eng.* **2004**, *21*, 211–218. [CrossRef]

14. Wu, J.; Xiao, Y.Z.; Yu, H.Q. Degradation of lignin in pulp mill wastewaters by white-rot fungi on biofilm. *Bioresour. Technol.* **2005**, *96*, 1357–1363. [CrossRef] [PubMed]

15. Puyol, D.; Batstone, D.J. Resource Recovery from wastewater by biological technologies. *Front. Microbiol.* **2017**, *8*. [CrossRef] [PubMed]

16. Ruiz-Dueñas, F.J.; Martínez, Á.T. Microbial degradation of lignin: How a bulky recalcitrant polymer is efficiently recycled in nature and how we can take advantage of this. *Microb. Biotechnol.* **2009**, *2*, 164–177. [CrossRef] [PubMed]

17. Brown, M.E.; Chang, M.C. Exploring bacterial lignin degradation. *Curr. Opin. Chem. Biol.* **2014**, *19*, 1–7. [CrossRef] [PubMed]

18. Jain, N.; Shrivastava, A.K.; Srivastava, S.K. Treatment of black liquor by Pseudomonas putida and Acinetobacter calcoaceticus in continuous reactor. *Environ. Technol.* **1996**, *17*, 903–907. [CrossRef]

19. Gupta, V.K.; Minocha, A.K.; Jain, N. Batch and continuous studies on treatment of pulp mill wastewater by *Aeromonas formicans*. *J. Chem. Technol. Biotechnol.* **2001**, *76*, 547–552. [CrossRef]

20. Raj, A.; Reddy, M.K.; Chandra, R. Identification of low molecular weight aromatic compounds by gas chromatography–mass spectrometry (GC–MS) from kraft lignin degradation by three *Bacillus* sp. *Int. Biodeterior. Biodegrad.* **2007**, *59*, 292–296. [CrossRef]

21. Leonowicz, A.; Matuszewska, A.; Luterek, J.; Ziegenhagen, D.; Wojtaś-Wasilewska, M.; Cho, N.S.; Rogalski, J. Biodegradation of lignin by white rot fungi. *Fungal Genet. Biol.* **1999**, *27*, 175–185. [CrossRef] [PubMed]

22. Kirk, T.K.; Farrell, R.L. Enzymatic "combustion": The microbial degradation of lignin. *Ann. Rev. Microbiol.* **1987**, *41*, 465–501. [CrossRef] [PubMed]

23. Fu, Y.; Viraraghavan, T. Fungal decolorization of dye wastewaters: A review. *Bioresour. Technol.* **2001**, *79*, 251–262. [CrossRef]

24. Sağlam, N.; Say, R.; Denizli, A.; Patır, S.; Arıca, M.Y. Biosorption of inorganic mercury and alkylmercury species on to Phanerochaete chrysosporium mycelium. *Proc. Biochem.* **1999**, *34*, 725–730. [CrossRef]

25. Faison, B.D.; Kirk, T.K. Factors involved in the regulation of a ligninase activity in Phanerochaete chrysosporium. *App. Environ. Microbiol.* **1985**, *49*, 299–304.

26. Sodaneath, H.; Lee, J.I.; Yang, S.O.; Jung, H.; Ryu, H.W.; Cho, K.S. Decolorization of textile dyes in an air-lift bioreactor inoculated with Bjerkandera adusta OBR105. *J. Environ. Sci. Health Part A* **2017**, *52*, 1099–1111. [CrossRef] [PubMed]

27. Kinnunen, A.; Maijala, P.; JArvinen, P.; Hatakka, A. Improved efficiency in screening for lignin-modifying peroxidases and laccases of basidiomycetes. *Curr. Biotechnol.* **2017**, *6*, 105–115. [CrossRef]

28. Peláez, F.; Martínez, M.J.; Martinez, A.T. Screening of 68 species of basidiomycetes for enzymes involved in lignin degradation. *Mycol. Res.* **1995**, *99*, 37–42. [CrossRef]

29. Rodriguez, E.; Pickard, M.A.; Vazquez-Duhalt, R. Industrial dye decolorization by laccases from ligninolytic fungi. *Curr. Microbiol.* **1999**, *38*, 27–32. [CrossRef] [PubMed]

30. Wang, Y.; Vazquez-Duhalt, R.; Pickard, M.A. Purification, characterization, and chemical modification of manganese peroxidase from *Bjerkandera adusta* UAMH 8258. *Curr. Microbiol.* **2002**, *45*, 77–87. [CrossRef] [PubMed]

31. Tuomela, M.; Oivanen, P.; Hatakka, A. Degradation of synthetic 14 C-lignin by various white-rot fungi in soil. *Soil Biol. Biochem.* **2002**, *34*, 1613–1620. [CrossRef]

32. Anastasi, A.; Spina, F.; Prigione, V.; Tigini, V.; Giansanti, P.; Varese, G.C. Scale-up of a bioprocess for textile wastewater treatment using Bjerkandera adusta. *Bioresour. Technol.* **2010**, *101*, 3067–3075. [CrossRef] [PubMed]

33. Kraft Lignins—Lignin and Carbohydrate Content—Acid Hydrolysis Method. Available online: http://www.innventia.com (accessed on 12 October 2017).

34. Potter, B.B.; Wimsatt, J.C. *Method 415.3. Determination of Total Organic Carbon and Specific UV Absorbance at 254 nm in Source Water and Drinking Water*; EPA/600/R-05/055; US Environmental Protection Agency: Cincinnati, OH, USA, 2005.

35. Reina, R.; Kellner, H.; Jehmlich, N.; Ullrich, R.; García-Romera, I.; Aranda, E.; Liers, C. Differences in the secretion pattern of oxidoreductases from *Bjerkandera adusta* induced by a phenolic olive mill extract. *Fungal Genet. Biol.* **2014**, *72*, 99–105. [CrossRef] [PubMed]

36. Arora, D.S.; Chander, M.; Gill, P.K. Involvement of lignin peroxidase, manganese peroxidase and laccase in degradation and selective ligninolysis of wheat straw. *Int. Biodeterior. Biodegrad.* **2002**, *50*, 115–120. [CrossRef]

37. Girard, V.; Dieryckx, C.; Job, C.; Job, D. Secretomes: The fungal strike force. *Proteomics* **2013**, *13*, 597–608. [CrossRef] [PubMed]

38. Schützendübel, A.; Majcherczyk, A.; Johannes, C.; Hüttermann, A. Degradation of fluorene, anthracene, phenanthrene, fluoranthene, and pyrene lacks connection to the production of extracellular enzymes by *Pleurotus ostreatus* and *Bjerkandera adusta*. *Int. Biodeterior. Biodegrad.* **1999**, *43*, 93–100. [CrossRef]

39. Bonnarme, P.; Asther, M.; Asther, M. Influence of primary and secondary proteases produced by free or immobilized cells of the white-rot fungus *Phanerochaete chrysosporium* on lignin peroxidase activity. *J. Biotechnol.* **1993**, *30*, 271–282. [CrossRef]

40. Feijoo, G.; Moreira, M.T.; Roca, E.; Lema, J.M. Use of cheese whey as a substrate to produce manganese peroxidase by *Bjerkandera* sp. BOS55. *J. Ind. Microbiol. Biotechnol.* **1999**, *23*, 86–90. [CrossRef] [PubMed]

41. Dosoretz, C.G.; Chen, H.C.; Grethlein, H.E. Effect of environmental conditions on extracellular protease activity in lignolytic cultures of *Phanerochaete chrysosporium*. *Appl. Environ. Microbiol.* **1990**, *56*, 395–400. [PubMed]

42. Nakamura, Y.; Sungusia, M.G.; Sawada, T.; Kuwahara, M. Lignin-degrading enzyme production by *Bjerkandera adusta* immobilized on polyurethane foam. *J. Biosci. Bioeng.* **1999**, *88*, 41–47. [CrossRef]

43. Taboada-Puig, R.; Lú-Chau, T.; Moreira, M.T.; Feijoo, G.; Martínez, M.J.; Lema, J.M. A new strain of *Bjerkandera* sp. production, purification and characterization of versatile peroxidase. *World J. Microbiol. Biotechnol.* **2011**, *27*, 115–122. [CrossRef]

44. Keyser, P.; Kirk, T.K.; Zeikus, J.G. Ligninolytic enzyme system of Phanaerochaete chrysosporium: Synthesized in the absence of lignin in response to nitrogen starvation. *J. Bacteriol.* **1978**, *135*, 790–797. [PubMed]

45. Srivastava, S.K.; Shrivastava, A.K.; Jain, N. Degradation of black liquor, a pulp mill effluent by bacterial strain *Pseudomonas putida*. *Ind. J. Exp. Biol.* **1995**, *33*, 962–966.

46. Jain, N.; Shrivastava, A.K.; Srivastava, S.K. Degradation of black liquor, a pulp mill effluent by bacterial strain *Acinetobacter calcoaceticus*. *Ind. J. Exp. Biol.* **1997**, *35*, 139–143.

47. Alekhina, M.; Ershova, O.; Ebert, A.; Heikkinen, S.; Sixta, H. Softwood kraft lignin for value-added applications: Fractionation and structural characterization. *Ind. Crops Prod.* **2015**, *66*, 220–228. [CrossRef]

water MDPI

Article

Adsorption Capacity of a Volcanic Rock—Used in Constructed Wetlands—For Carbamazepine Removal, and Its Modification with Biofilm Growth

Allan Tejeda [1], Arturo Barrera [2] and Florentina Zurita [1,*]

[1] Quality Environmental Laboratory, Centro Universitario de la Ciénega, University of Guadalajara, Ocotlán, Jalisco 47820, Mexico; allanteor@hotmail.com

[2] Laboratory of Catalytic Nanomaterials, Centro Universitario de la Ciénega, University of Guadalajara, Ocotlán, Jalisco 47820, Mexico; arturo.barrera@cuci.udg.mx

* Correspondence: fzurita2001@yahoo.com; Tel.: +52-392-925-9400

Received: 26 August 2017; Accepted: 15 September 2017; Published: 20 September 2017

Abstract: In this study, the aim was to evaluate the adsorption capacity of a volcanic rock commonly used in Mexico as filter medium in constructed wetlands (locally named *tezontle*) for carbamazepine (CBZ) adsorption, as well as to analyze the change in its capacity with biofilm growth. Adsorption essays were carried out under batch conditions by evaluating two particle sizes of *tezontle*, two values of the solution pH, and two temperatures; from these essays, optimal conditions for carbamazepine adsorption were obtained. The optimal conditions (pH 8, 25 °C and 0.85–2.0 mm particle-size) were used to evaluate the adsorption capacity of *tezontle* with biofilm, which was promoted through *tezontle* exposition to wastewater in glass columns, for six months. The maximum adsorption capacity of clean *tezontle* was 3.48 µg/g; while for the *tezontle* with biofilm, the minimum value was 1.75 µg/g (after the second week) and the maximum, was 3.3 µg/g (after six months) with a clear tendency of increasing over time. The adsorption kinetic was fitted to a pseudo-second model for both *tezontle* without biofilm and with biofilm, thus indicating a chemisorption process. On clean *tezontle*, both acid active sites (AAS) and basic active sites (BAS) were found in 0.087 and 0.147 meq/g, respectively. The increase in the adsorption capacity of *tezontle* with biofilm, along the time was correlated with a higher concentration of BAS, presumably from a greater development of biofilm. The presence of biofilm onto *tezontle* surface was confirmed through FTIR and FE-SEM. These results confirm the essential role of filter media for pharmaceutical removal in constructed wetlands (CWs).

Keywords: pharmaceuticals; micropollutant removal; *tezontle*; filter media; active sites

1. Introduction

Carbamazepine (CBZ) is a drug extensively used worldwide in the treatment of a variety of mental disorders, neuralgia, and seizure disorders [1,2]. In the last two decades, this drug has been found along with a huge diversity of pharmaceuticals in aquatic environments. Moreover, according to different studies, CBZ is the pharmaceutical most frequently detected in water bodies around the world [3,4] so that it has been proposed as an anthropogenic marker in such environments [5]. This situation is mainly a consequence of CBZ's poor removal in conventional wastewater treatment plants (WWTP), which is generally lower than 10%. The effluents of municipal WWTPs are considered the main route for pharmaceutical release into the environment [6,7].

In recent years, the removal of pharmaceutical compounds in constructed wetlands (CW) have been the aim in many studies worldwide [8]. Although the removal efficiencies for many drugs in such systems have been found to be similar or much higher than those removals achieved in conventional WWTPs, the removal of CBZ has been very low, varying in the range of 20–30% with an average efficiency of lower than 30% [6]. However, in a recent study carried out by this research group, removal efficiencies of 62.5% and 59% were found in two two-stage hybrid constructed wetlands configured with horizontal subsurface flow wetland (HSSFW), followed by stabilization ponds and HSSFW, then followed by vertical subsurface flow wetland, respectively [9]. These results were achieved under a subtropical climate by using ornamental species as emergent vegetation in conjunction with a porous-local ground volcanic rock as filter medium, commonly named *tezontle*. In Mexico, this is the substrate most frequently used in CWs.

On the other hand, it has been reported that the main mechanisms for CBZ removal in constructed wetlands, include microbial degradation, plant uptake, and adsorption [10], which are not completely elucidated due to the fact that most studies have been focused only on the inlet and outlet loads [8]. In general, adsorption as a removal process for pharmaceuticals has been extensively assessed through the use of different adsorbents; among those adsorbents recently evaluated are commercial activated carbons [11], silica-based materials [12], chitosan-based magnetic composite [13], ion-exchange resins [14], clay minerals [15], etc. Nevertheless, commonly used filter media in CWs have been scarcely evaluated as adsorbents for pharmaceutical removal [8,16]; as a result, their contribution for each specific drug removal, such as CBZ, is almost unknown.

Among the few reported studies is that performed by Dordio et al. [2] who evaluated light expanded clay aggregates (LECA) for CBZ removal as a single compound and as a mixture of three compounds (CBZ, ibuprofen, and clofibric acid), obtaining higher removal efficiencies for CBZ in all of the tested conditions in comparison to the other two drugs. In another study, Matamoros et al. [17] quantified sorption on gravel used as substrate in HSSFWs, finding a higher sorption of CBZ in comparison to clofibric acid and ibuprofen; they attributed such results to the presence of biofilm covering the gravel bed. From these few studies, it is evident the relevance of sorption as a mechanism for CBZ removal and also the necessity of evaluating filter media used in CWs, such as the volcanic rock we use in Mexico, which probably contributed to the high removal of CBZ obtained in a previous study [9], as mentioned before. Therefore, the aim of this study was to evaluate the sorption capacity of ground *tezontle* for CBZ removal, as well as to evaluate the change in its capacity with the presence of biofilm developed through its exposition to wastewater.

2. Materials and Methods

2.1. Material Preparation

The material was purchased from a construction material store. In Mexico, *tezontle* is a porous material extensively used in construction. Approximately 5 kg of ground *tezontle* were washed with tap water in order to completely remove any dirt and dust, and then dried at 110 °C for 24 h. After drying, the material was left to reach room temperature and then it was sieved through appropriate AST sieves in order to classify it in two different particle sizes, i.e., 0.85–2 mm (PS1) and 4–4.75 mm (PS2). Once classified and separated according to the particle size, the material was stored in a desiccator at room temperature until its use to perform the adsorption experiments.

2.2. Characterization of the Ground Tezontle

The particle size distribution was determined through the dry-sieving technique [2,18]. By means of a grain-size distribution plot, d_{10} and d_{60} were estimated, while the uniformity coefficient was obtained as the ratio between d_{60} and d_{10} [2]. Additionally, the material porosity and bulk density was calculated according to Brix et al. [19]. On the other hand, *tezontle* structural properties were analyzed by X-ray diffraction (XRD) using a STOE diffractometer (Stoe, Darmstadt, Germany) with Cu Kα anode (λ = 1.5406 nm). In addition, textural properties of *tezontle*, such as the average pore diameter (APD), total pore volume (VP_T), and N_2 adsorption/desorption isotherms were determined by N_2-physisorption at the saturation temperature of liquid nitrogen (-196.5 °C) through the use of an

Autosorb gas sorption system (IQ model from Quantachrome, Boynton Beach, FL, USA). The specific surface area (S_{BET}) was calculated according to the Brunauer-Emmet-Teller (BET) equation.

2.3. Adsorption Kinetic Assays of CBZ onto Tezontle without Biofilm

In order to find out the best conditions for CBZ adsorption (particle size, temperature, and pH), adsorption experiments were performed by adding 1 g of *tezontle* to glass flasks containing 100 mL of carbamazepine (99% purity, Sigma-Aldrich, Saint Louis, MO, USA) solution at a concentration of 250 µg/L, buffered with 0.01 M of phosphate [20] in order to maintain the pH at the desired level, which was adjusted by using H_2SO_4 or NaOH 0.1 N. The flasks were stirred magnetically at 200 rpm for the evaluation of CBZ adsorption onto *tezontle* for 0.5, 1, 1.5, 2, 3, and 4 h. These assays were performed by triplicate with the two different particle size of *tezontle*, PS1 and PS2, at two different temperatures, 16 °C and 25 °C, and at two levels of pH, 6 and 8 (Figure 1).

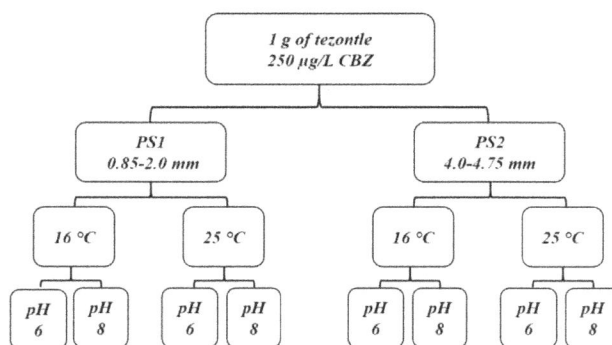

Figure 1. Experiment design for the adsorption kinetic assays of carbamazepine (CBZ) on *tezontle* without biofilm.

2.4. Adsorption Kinetic Assays of CBZ onto Tezontle with Biofilm

The growth of biofilm on the *tezontle* surface was carried out in 15 glass columns (dimensions, diameter = 2 cm; large = 25 cm) that were packed with the filter medium. The particle size used was that with which the higher adsorption capacity was obtained from the assays, as described in the previous section. All of the columns were fed with sedimented wastewater generated in the campus of Centro Universitario de la Ciénega with a flow rate of 0.19 ± 0.02 L/min (controlled by a valve for each column) and a hydraulic retention time of 13.79 ± 1.38 s. The wastewater was stored in a 10 L-plastic container and recirculated for 15 days by means of an 18 W-fountain pump and then replaced. An upward flow was maintained by feeding the wastewater through a tubing system joined to the bottom of each column; the wastewater was returned to the plastic container by tubings located at the top of the columns. This type of flow was used in order to keep flooding conditions inside the columns, as well as to maintain a constant flow rate.

Wastewater characterization included measurements of chemical oxygen demand (COD), biochemical oxygen demand (BOD), total nitrogen, phosphorus, total suspended solids (TSS), conductivity, and pH, which were determined as described in the Standard Methods for the examination of Water and Wastewater [21]. A column was removed from the whole experiment (15 glass columns) every week during 12 weeks and then, every month. The *tezontle* inside the columns was removed and deactivated by UV-radiation for 30 min from a distance of 12.0 cm [22] using a SPECTROLINE lamp, EA-160, (Spectronics Corporation, Westbury, NY, USA); immediately after, the *tezontle* was dried at 50 °C by 3 h. Then, the material was kept in a desiccator until its use to carry out adsorption kinetic experiments, as described in Section 2.3, under the optimum conditions determined in the essays described in that same section. In addition, a sample of this material was analyzed by Fourier Transform Infrared Spectroscopy (FTIR) and by Field Emission Scanning Electron Microscopy (FE-SEM) to confirm the presence of biofilm. Surface active site concentration (SASC) was also determined in each sample. For FTIR analysis, a Thermo Nicolet Nexus 470 spectrophotometer (Artisan Technology Group, Champaing, IL, USA) was used as well as the KBr method, while the FE-SEM micrographs were obtained by a MIRA 3 LMU Tescan microscope (TESCAN, Brno, South Moravian, Czech Republic) using a thin conductive film of gold on the sample surface and a current of 10 kV [2]. Finally, the SASC was quantified by the Boehm titration conventional method [23,24]. In all of these analyses a sample of *tezontle* without biofilm was included as a reference.

2.5. Aqueous Samples Preparation and CBZ Detection

From each essay, 100 mL of the supernatant was taken and filtered through a 20 µm filter paper (Whatman 41, Whatman Inc., Piscataway, NJ, USA)) and then through a 1.6 µm GF/A Whatman fiber glass filter. After that, CBZ was extracted from the aqueous sample by solid phase extraction (SPE) method using Phenomenex Strata-X (200 mg/6 cc) cartridges (Phenomenex, Torrance, CA, USA). The cartridges were conditioned with 5 mL of methanol and 10 mL of deionized water at a flow rate of 1.5–2 mL/min. Then, the aqueous sample was passed through the cartridge at the same flow rate. In order to remove any impurity, the cartridges were washed with 5 mL of deionized water and dried under vacuum for one hour. Thereafter, CBZ was eluted with 10 mL of methanol. Finally, the samples were filtered through a 0.2 µm PTFE filter. CBZ detection were performed by a Waters HPLC as described by [9].

2.6. Statistical Analysis

A factorial experimental design, specifically 2^3, was used to evaluate the adsorption kinetic assays of CBZ in aqueous solution by *tezontle*. The three analyzed factors and their corresponding levels were: particle size (PS1 and PS2), pH (6 and 8) and temperature (16 and 25 °C). The analysis of variance (ANOVA) was carried out using the STATGRAPHICS CENTURION XVII software (XVII, StatPoint Technologies, Inc., Warrenton, VA, USA). A significance level of $p = 0.05$ was used for all statistical tests and values reported are the average ± standard error of the mean.

3. Results and Discussion

3.1. Structural and Textural Properties of the Ground Tezontle

The structural and textural properties of the material used in this study are shown in Table 1. According to the particle size distribution, the ground *tezontle* has a wide particle size distribution, with 92.73% of the material having diameters between 0.425 and 4 mm. With regard to d_{10}, d_{60}, and the uniformity coefficient, they fall into the range of recommendable values for filter media used in CWs [25]. In addition, the apparent porosity was more than 50%, which is higher than those values reported for gravel; in this way, ground *tezontle* exhibits advantages as a support matrix in CWs, because, the higher amount of void space, the better the hydraulic conductivity [2].

Table 1. Structural and textural characteristics of ground *tezontle*.

d_{10} (mm)	0.48
d_{60} (mm)	1.9
Uniformity coefficient, U	3.95
Apparent porosity/void space (%)	56.2
Bulk density (kg/m^3)	1047
BET specific surface area, S_{BET} (m^2/g)	1.36
Total pore volume, VP_T (cm^3/g)	0.008
Average pore diameter, APD (nm)	26.64

On the other hand, as expected, the BET specific surface area was very small in comparison to adsorbents prepared or synthetized for drug removal but alike to the value reported by Alemayehu & Lennartz [26] for a similar volcanic scoria. In addition, the N_2 adsorption–desorption isotherm of *tezontle* (Figure 2) was of the type IV with a hysteresis loop of H3-type, according to the IUPAC classification [27]. The hysteresis loop of H3-type indicates the presence of non-rigid aggregates of plate-like particles giving rise to slit-type shaped pores [28]. The pore sizes calculated from the desorption branch of N_2 sorption isotherm were in the mesopore range with values between 2 and 50 nm, which is in line with the values reported by Vilchis-Granados et al. [29]. Finally, the X-ray diffraction pattern of *tezontle* evaluated in this study (Figure 3) revealed the crystalline structure of the material with crystalline peaks located at 2θ angles of $21.6°$, $27.5°$, and $35.3°$. The diffraction peak at $2\theta = 21.6°$ corresponds to the plane (110) of goethite ($FeO(OH)$), while the peak at $27.5°$ belongs to the plane (101) of quartz (SiO_2) according to Brooks et al. [30]. On the other hand, the peak at $35.3°$ corresponds to the crystalline plane (110) of hematite (Fe_2O_3) according to Farahmandjou & Soflaee [31]. These results, are in line to those crystalline planes reported for *tezontle* by Ponce et al. [32] who proposed a preliminary composition consisted mainly of quartz (SiO_2) and ferric oxides like hematite (Fe_2O_3).

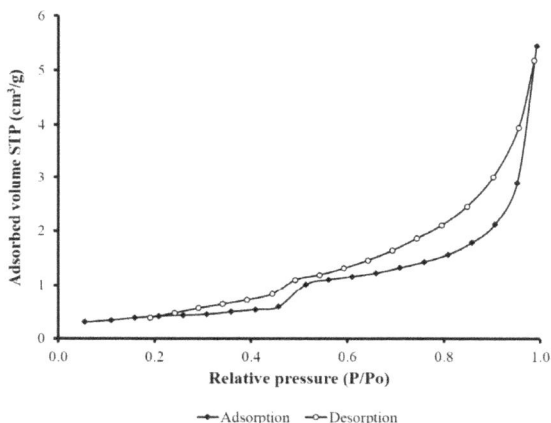

Figure 2. N_2 adsorption-desorption isotherm of *tezontle*.

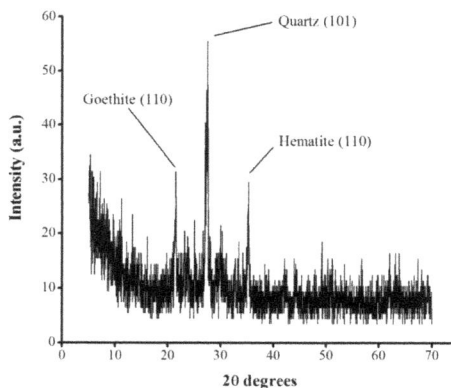

Figure 3. X-ray diffraction (XRD) pattern of *tezontle*.

3.2. Adsorption Kinetics of CBZ onto Tezontle without Biofilm

The kinetics of CBZ adsorption by *tezontle* under the different assessed conditions showed a similar behavior with respect to the time that the equilibrium was accomplished (Figure 4). With regard to the reached equilibrium concentrations, they showed in general, a low adsorption capacity of *tezontle*, with small variations at 16 °C irrespective of pH or particle size (Table 2); higher variations were observed at 25 °C. However, for both temperatures, the adsorption capacity of *tezontle* was higher with PS1; such results were expected, since it is well known that smaller particle sizes have larger surface areas available for adsorbate-adsorbent interactions.

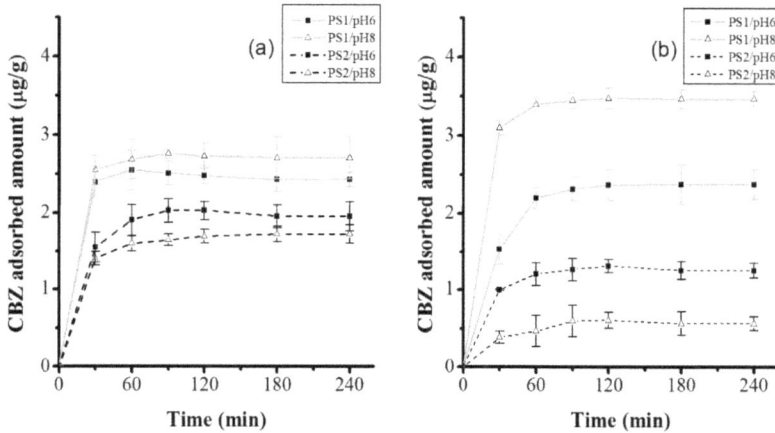

Figure 4. Adsorption kinetics of CBZ at (**a**) 16 °C and (**b**) 25 °C onto two particle sizes of *tezontle* (PS1, PS2) at pH 6 and pH 8.

Table 2. Equilibrium adsorption capacity (q_e) and its equivalent removal percentage at the different levels of particle size, pH and temperature.

Temperature	Parameter	PS1/pH 6	PS1/pH 8	PS2/pH6	PS2/pH8
16 °C	q_e (µg/g)	2.48	2.73	2.03	1.7
	Removal %	9.7	10.9	8.1	6.8
25 °C	q_e (µg/g)	2.3	3.48	1.3	0.61
	Removal %	9.3	14.67	5.2	2.5

The mathematical model of the factorial design [33] used to find the optimal conditions for CBZ removal by *tezontle* is shown in Equation (1). In addition, the experimental design along with the data obtained in the CBZ adsorption experiments are shown in Table 3.

$$y_{ijkl} = \mu + \tau_i + \beta_j + \gamma_k + (\tau\beta)_{ij} + (\tau\gamma)_{ik} + (\beta\gamma)_{ik} + (\tau\beta\gamma)_{ijk} + \varepsilon_{ijkl} \tag{1}$$

i = 1 and 2. j = 1 and 2. k = 1 and 2 for this particular case. μ, is the overall mean effect. τ_i, β_j, γ_k , are the effects of the i *th* level of factor A (particle size), $(\sum_i \tau_i = 0)$; of the j *th* level of factor B (temperature), $(\sum_j \beta_j = 0)$; and, of the k *th* level of factor C (pH), $(\sum_k \gamma_k) = 0$. $(\tau\beta)_{ij}$, $(\tau\gamma)_{ik}$, $(\beta\gamma)_{ik}$ and $(\tau\beta\gamma)_{ijk}$, are the effects of the interactions between A × B, A × C, B × C and A × B × C, respectively. ε_{ijkl}, is the random error in the combination *ijkl* and *l* are the replicates.

Table 3. CBZ adsorption experiments. A, particle size; B, temperature; C, pH. Factor levels, low (−1), high (+1).

Run	Coded Factors			Percent of CBZ Removal		
	A	B	C	Replicate 1	Replicate 2	Replicate 3
1	−1	−1	−1	11	10.6	12.1
2	1	−1	−1	12	8.3	6.5
3	−1	1	−1	9.8	12.2	10
4	1	1	−1	5.2	5.2	5.2
5	−1	−1	1	9.4	9	10.1
6	1	−1	1	6.1	6.1	6.1
7	−1	1	1	12.8	14.7	14.2
8	1	1	1	2.8	2.8	2.8

According to the ANOVA (Table 4), out of the three factors, only the particle size was significant ($p < 0.05$) for CBZ adsorption, confirming that PS1 functions better that PS2 as was aforementioned with regard to Figure 4. However, even more important, the interaction between the evaluated three factors (particle size, temperature, and pH) was significant ($p < 0.05$), and the best experimental conditions for CBZ adsorption onto *tezontle* was the combination of PS1, pH 8, and 25 °C. Under such conditions, the lowest equilibrium concentration and the corresponding highest removal of CBZ was reached (Table 2). It is important to point out that although the adsorptive capacity of *tezontle* was found to be low, the contribution of this porous medium to CBZ removal remains important because of the quantity of filter medium required in CWs, as well as the concentration of this drug in wastewater.

Table 4. Results from the ANOVA for CBZ adsorption.

Factors	*p*-Value
A: Particle size	0.0000
B: Temperature	0.1220
C: pH	0.0749
AB	0.0000
AC	0.0034
BC	0.0141
ABC	0.0343

The increase in *tezontle* adsorption capacity with temperature is probably due to the fact that temperature enhances the mobility of organic compounds, which may lead to a higher adsorption [15]. Nevertheless, this behavior was only observed in this particular case (PS1, pH 8), while in the other cases the adsorption capacity decreased with the increase in temperature. In this way, apparently temperature does not have a unique effect on the adsorption of CBZ onto *tezontle*. On the other hand, a better result at pH 8 in comparison to pH 6 coincides with the findings of other authors who affirm that under acidic conditions, the positively-charged adsorbent surface do not favor pharmaceutical sorption [13], in particular for neutral-organic compounds, such as CBZ [2].

With respect to the kinetic of the adsorption of CBZ on *tezontle*, the two most common models were evaluated, i.e., Lagergren pseudo-first-order model (Equation (2)) and pseudo second-order model (Equation (3)).

$$In(q_e - q_t) = In(q_e) - K_1 t \tag{2}$$

$$\frac{t}{q_t} = \frac{1}{K_2 q_e^2} + \frac{t}{q_e} \tag{3}$$

where K_1 is Lagergren rate constant (min^{-1}); K_2 is pseudo second-order rate constant (g/µg·min); q_e and q_t are the amount of pollutant adsorbed at equilibrium (µg/g) and at time t (min), respectively.

The pseudo-first-order model assumes a proportional relation for the rate of adsorption, while the pseudo-second-order equation suggests that the rate of adsorption is proportional to the square of the number of unoccupied sites [34]. According to the correlation coefficient R^2, the pseudo-first order model does not fit well to the data obtained from the experiments (Table 5). In contrast, a better fit to pseudo-second order model was found for all assays (Table 5), which suggests a CBZ chemisorption [35,36] onto *tezontle*.

The previous results were confirmed through the quantification of actives sites (in PS1), which determines the extent of a chemisorption process [37]. Similar to activated carbons [38], both acid and basic sites were found, in this case in 0.087 and 0.147 meq/g, respectively. Due to the capability of the carboxamide group present in the molecule of CBZ, in particular the $-NH_2$ group of forming hydrogen bond [39], one probable mechanism for CBZ adsorption was the formation of hydrogen bonds with π electrons from BAS on the *tezontle* surface. From studies on carbon surfaces, it has been found that basic properties are associated with Lewis sites located at the π electron-rich regions [38], characteristic of oxygen-containing functionality capable of acting as a basic center [40]. In this case, the presence of oxygen in a *tezontle* surface could be due to its main components, i.e., quartz and ferric oxides [32].

Table 5. Kinetic parameters for adsorption of CBZ onto *tezontle* at different experimental conditions.

Experimental Conditions		Pseudo First-Order			Pseudo Second-Order		
		K_1 (min^{-1})	q_e (μg/g)	R^2	K_2 (g/μg·min)	q_e (μg/g)	R^2
PS1, 0.85–2.0 mm							
16 °C	pH 6	0.0076	2.48	0.09	0.157	2.51	0.97
	pH 8	0.0196	2.73	0.53	0.481	2.63	0.98
25 °C	pH 6	0.0106	2.3	0.11	0.036	2.45	0.96
	pH 8	0.0099	3.48	0.43	0.088	3.2	0.97
PS2, 4.0–4.75 mm							
16 °C	pH 6	0.0268	2.03	0.4	0.241	2.03	1
	pH 8	0.0307	1.7	0.84	0.29	1.61	0.93
25 °C	pH 6	0.0182	1.3	0.65	0.63	1.25	0.96
	pH 8	0.0079	0.61	0.16	0.21	0.62	0.96

3.3. Adsorption Kinetic Assays of CBZ onto Tezontle with Biofilm

The characteristics of the sedimented wastewater used to promote biofilm formation on the *tezontle* with PS1 inside the columns are shown in Table 6. The characteristics were similar to those of the wastewater used in pilot-scale hybrid wetlands for CBZ removal [9].

Table 6. Characteristics of the wastewater fed to glass columns for biofilm growth on the *tezontle* (Average ± SD, n = 15).

Chemical Oxygen Demand, mg/L	107.7 ± 66.37
Biochemical Oxygen Demand, mg/L	45.9 ± 20
Total Nitrogen, mg/L	77.4 ± 44.23
Phosphorous, mg/L	6.8 ± 3.42
Total Suspended Solids, mg/L	35.1 ± 18.15
Electrical Conductivity, μS/cm	873.9 ± 315.92
pH	7.8 ± 0.24

Similar to the essays with *tezontle* without biofilm growth, the time for the equilibrium to be reached was 120 min for all of the assays. However, the CBZ equilibrium concentrations varied according to the exposition period of *tezontle* to wastewater. In general, the higher the time of exposition, the larger the adsorption capacity (Figure 5). These results indicate a modification on the basal state of contact surface of *tezontle* when being exposed to wastewater and suggest the presence of biofilm. It is recognized that the presence of biofilm implies the attachment and deposition of extracellular polymeric substances (EPS) along with bacterial cells and this complex matrix modify the physicochemical characteristics of carrier surfaces [41]. Moreover, the evolution in the sorption capacity of *tezontle* is probably related to the time required for the biofilm growth. Although the biofilm formation begins within a few minutes, the complete process to reach a mature biofilm capable of produce EPS, responsible of the sorption process, might require days [42].

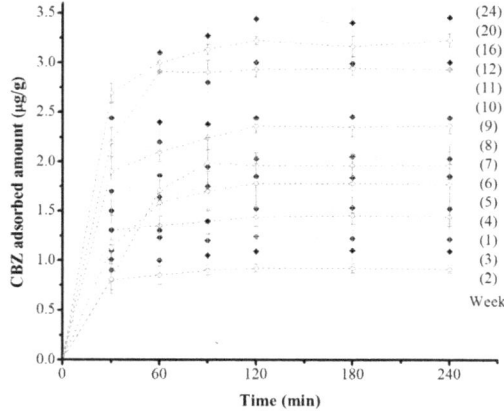

Figure 5. Adsorption kinetics of CBZ onto *tezontle* with different periods of biofilm formation. Error bars represent standard error of triplicates.

On the other hand, despite the increase in the adsorption capacity of *tezontle* along the time, with a noticeable increase in the removal percent of CBZ, the maximum value reached after 24 weeks was smaller than the value obtained with *tezontle* without biofilm. Figure 6 shows increments in the percent of CBZ removal, almost with a linear tendency ($R^2 = 0.95$) over time, starting in the second week until the end of the experiment. These results suggest that the removal percentage could probably reach and possibly surpass that obtained with free-biofilm *tezontle* with larger periods for biofilm growth.

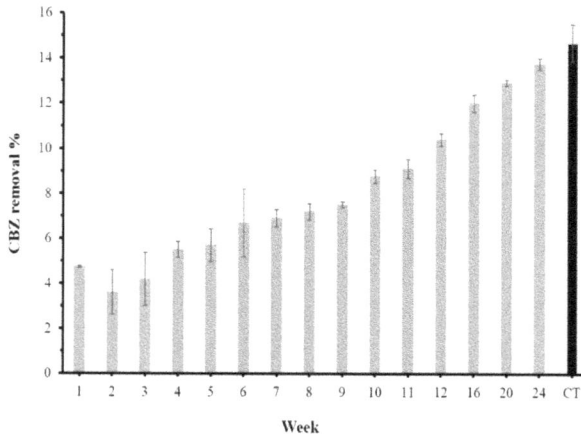

Figure 6. CBZ removal percentage (mean ± standard error) by *tezontle* with different periods of biofilm formation and clean *tezontle* (CT).

Furthermore, as expected, the experimental data obtained from each adsorption kinetic showed a better fit to pseudo-second order kinetic model (Table 7) alike to the adsorption kinetic with free-biofilm *tezontle*, indicating a chemisorption process. Once more, these results were confirmed through the quantification of active sites in each sample of *tezontle* with biofilm growth, throughout the study. After the first week, the concentration of BAS as well as the concentration of AAS showed a visible decrease in comparison to the concentration in free-biofilm *tezontle* and then, even more after two weeks (Figure 7); possibly as a consequence of the coating of basal active sites in the *tezontle* surface by bacterial attachment which changed its physicochemical characteristics as was aforementioned.

After this general reduction in the active sites concentration, the AAS showed a slight increase during the next two weeks and then, a clear tendency of decreasing until almost its disappearance after ~10 weeks, suggesting their minimal contribution to the adsorption process. In contrast, a noticeable increase was found in the BAS concentration along the time, suggesting that the BAS presence after the second week was due to the biofilm growth onto the *tezontle* exposed to wastewater and specifically, due to the release of EPS by forming-biofilm microbes. EPS are high-molecular-weight molecules consisting mainly of polysaccharides (40%), DNA, proteins, lipids, and humic acids [41,43]. Charged or hydrophobic polysaccharides and proteins are particularly responsible of organic compound sorption [44]. Some specific polysaccharide monomers detected by Andersson et al. [45] in EPS released by microbial consortia developed in wastewater are rhamnose, arabinose, galactose, glucose, mannose, ribose among others. In this way, this chemical structures with large π electron-rich region could have participated in the CBZ chemisorption process through hydrogen bonds [23]. In addition, a clear relationship was observed between BAS concentration and CBZ removal percent, highlighting this pathway as the main mechanism of adsorption.

The presence of biofilm on *tezontle* was confirmed through FTIR analysis by which the presence of characteristic biofilms peaks was observed (Figure 8). The small bands between 2900 and 3000 cm^{-1} are related to C-H stretching vibration [41] associated with bacterial biomass [46], whereas the peak at around 2400 cm^{-1} is due to the vibration of C-O functional groups likely from carboxylic acids which has been reported as a sorption active site present in the cell wall of Gram-positive bacteria [46]. Other characteristic biofilm bands corresponding to proteins (1637–1660 and 1272–1288 cm^{-1}) and polysaccharides (1000–1132 cm^{-1}) have been reported in the literature [41]; however, in this study they were not detected, apparently because of the wide and intense bands in the *tezontle* FTIR fingerprint between 400 and 1750 cm^{-1}, which interfered with the detection of these bands.

Table 7. Kinetic parameters for adsorption of CBZ onto *tezontle* with biofilm.

Weeks of Biofilm Formation	Pseudo First-Order			Pseudo Second-Order		
	K_1 (min^{-1})	q_e (µg/g)	R^2	K_2 (g/µg·min)	q_e (µg/g)	R^2
1	0.0212	1.22	0.5	3.5	1.25	0.99
2	0.0265	0.9	0.61	8.6	0.91	0.99
3	0.0048	1.09	0.53	0.035	0.9	0.90
4	0.0198	1.44	0.6	0.99	1.5	0.99
5	0.006	1.52	0.66	0.068	1.51	0.96
6	0.0048	1.78	0.054	0.31	1.8	0.97
7	0.0285	1.78	0.92	0.061	1.7	0.99
8	0.0049	1.96	0.34	0.058	1.86	0.96
9	0.0107	2.03	0.12	0.044	2.4	0.98
10	0.0069	2.36	0.07	2.8	1.9	0.97
11	0.006	2.44	0.18	1.26	2.39	0.99
12	0.0106	2.93	0.75	0.14	2.61	0.99
16	0.0181	3.0	0.84	0.0048	3.01	0.90
20	0.0049	3.22	0.096	0.072	3.27	0.99
24	0.0216	3.44	0.88	0.037	3.55	0.99

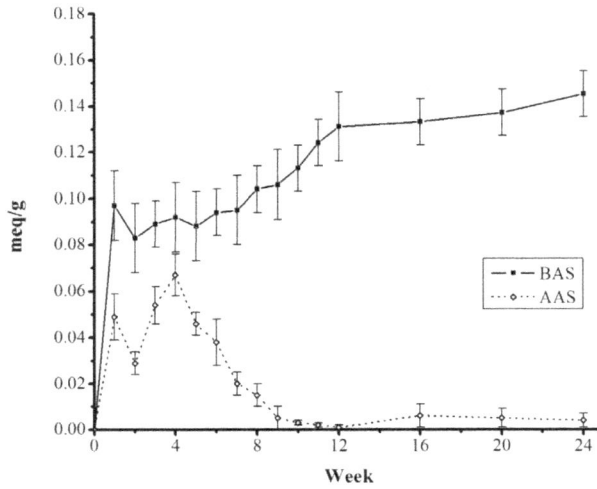

Figure 7. Concentration of active sites (mean ± standard error) by week of biofilm development, basic active sites (BAS) and acid active sites (AAS). Control concentration at 0.147 and 0.087 meq/g, respectively.

On the other hand, there is a clear difference between the spectra of free-biofilm *tezontle* and those of *tezontle* with biofilm growth, in the sense that the first one does not present the aforementioned bands that showed a noticeable evolution in the FTIR spectra along the time of experimentation.

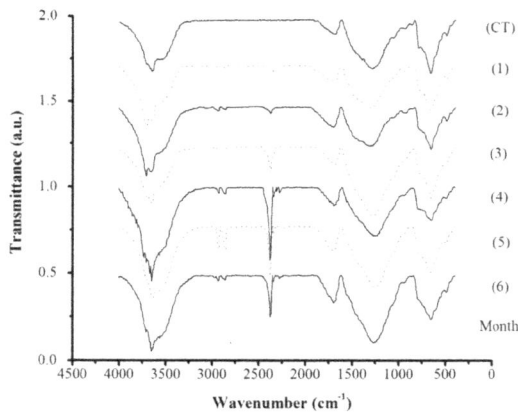

Figure 8. Infrared spectra of *tezontle* PS1 with biofilm development by month and clean *tezontle* (CT).

Finally, the FE-SEM micrographs revealed the presence of bridge-shape structures on the *tezontle* surface, which have been reported as a common physical structure of biofilms whose number usually increases through the time [47]. A comparison between free-biofilm *tezontle* and *tezontle* with two different periods of biofilm development (4 and 24 weeks) shows the highest density of biomass after 24 weeks (Figure 9).

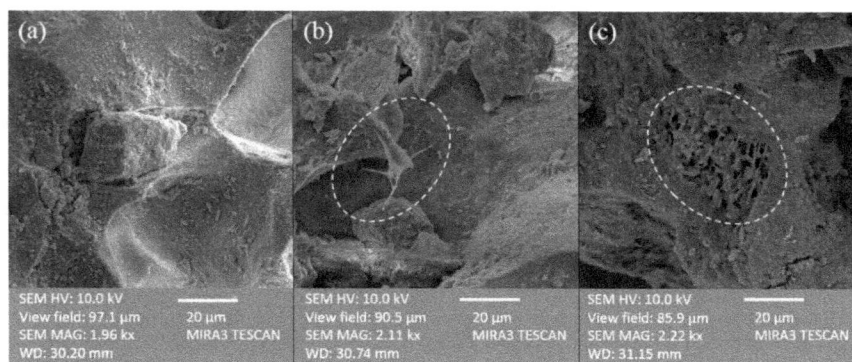

Figure 9. FE-SEM images of *tezontle*, (**a**) clean *tezontle*; (**b**) *tezontle* after 4 weeks of exposition to wastewater; and, (**c**) *tezontle* after 24 weeks of exposition to wastewater.

4. Conclusions

Tezontle, a common porous-filter medium for CWs in Mexico, was found to have some capacity for CBZ chemisorption through the presence of both AAS and BAS. This capacity was modified with biofilm formation; after an initial decrease in *tezontle* capacity, chemisorption took place through the EPS released by microbial consortia, which generated BAS (Lewis sites located at the π electron-rich regions). An increase in the adsorption capacity of *tezontle* with biofilm was obtained along the period of experimentation, with a tendency to possibly reach and maybe surpass the capacity of clean *tezontle*. These results confirm the essential role that filter media used in CWs might play for pharmaceutical removal.

Acknowledgments: Graduate scholarship for Allan Tejeda was provided by National Council for Science and Technology of México.

Author Contributions: F. Zurita conceived and designed the experiments; A. Tejeda performed the experiments, analyzed the data and wrote the paper under the supervision of F. Zurita; A. Barrera contributed to the manuscript with the material characterization and the analysis of the corresponding data.

Conflicts of Interest: The authors declare no conflict of interest.

References

1. Zhang, Y.; Geißen, S.U.; Gal, C. Carbamazepine and diclofenac: Removal in wastewater treatment plants and occurrence in water bodies. *Chemosphere* **2008**, *73*, 1151–1161. [CrossRef] [PubMed]
2. Dordio, A.V.; Estêvão Candeias, A.J.; Pinto, A.P.; Teixeira da Costa, C.; Palace Carvalho, A.J. Preliminary media screening for application in the removal of clofibric acid, carbamazepine and ibuprofen by ssf-constructed wetlands. *Ecol. Eng.* **2009**, *35*, 290–302. [CrossRef]
3. Fent, K.; Weston, A.A.; Caminada, D. Ecotoxicology of human pharmaceuticals. *Aquat. Toxicol.* **2006**, *76*, 122–159. [CrossRef] [PubMed]
4. Li, W.C. Occurrence, sources, and fate of pharmaceuticals in aquatic environment and soil. *Environ. Pollut.* **2014**, *187*, 193–201. [CrossRef] [PubMed]
5. Clara, M.; Strenn, B.; Ausserleitner, M.; Kreuzinger, N. Comparison of the behaviour of selected micropollutants in a membrane bioreactor and a conventional wastewater treatment plant. *Water Sci. Technol.* **2004**, *50*, 29–36. [PubMed]
6. Li, Y.; Zhu, G.; Ng, W.J.; Tan, S.K. A review on removing pharmaceutical contaminants from wastewater by constructed wetlands: Design, performance and mechanism. *Sci. Total Environ.* **2014**, *468–469*, 908–932. [CrossRef] [PubMed]

7. Ternes, T.A. Occurrence of drugs in german sewage treatment plants and rivers. *Water Res.* **1998**, *32*, 3245–3260. [CrossRef]

8. Zhang, D.; Gersberg, R.M.; Ng, W.J.; Tan, S.K. Removal of pharmaceuticals and personal care products in aquatic plant-based systems: A review. *Environ. Pollut.* **2014**, *184*, 620–639. [CrossRef] [PubMed]

9. Tejeda, A.; Torres-Bojorges, Á.X.; Zurita, F. Carbamazepine removal in three pilot-scale hybrid wetlands planted with ornamental species. *Ecol. Eng.* **2017**, *98*, 410–417. [CrossRef]

10. Matamoros, V.; Caselles-Osorio, A.; García, J.; Bayona, J.M. Behaviour of pharmaceutical products and biodegradation intermediates in horizontal subsurface flow constructed wetland. A microcosm experiment. *Sci. Total Environ.* **2008**, *394*, 171–176. [CrossRef] [PubMed]

11. Mailler, R.; Gasperi, J.; Coquet, Y.; Derome, C.; Buleté, A.; Vulliet, E.; Bressy, A.; Varrault, G.; Chebbo, G.; Rocher, V. Removal of emerging micropollutants from wastewater by activated carbon adsorption: Experimental study of different activated carbons and factors influencing the adsorption of micropollutants in wastewater. *J. Environ. Chem. Eng.* **2016**, *4*, 1102–1109. [CrossRef]

12. Suriyanon, N.; Permrungruang, J.; Kaosaiphun, J.; Wongrueng, A.; Ngamcharussrivichai, C.; Punyapalakul, P. Selective adsorption mechanisms of antilipidemic and non-steroidal anti-inflammatory drug residues on functionalized silica-based porous materials in a mixed solute. *Chemosphere* **2015**, *136*, 222–231. [CrossRef] [PubMed]

13. Zhang, S.; Dong, Y.; Yang, Z.; Yang, W.; Wu, J.; Dong, C. Adsorption of pharmaceuticals on chitosan-based magnetic composite particles with core-brush topology. *Chem. Eng. J.* **2016**, *304*, 325–334. [CrossRef]

14. Jiang, M.; Yang, W.; Zhang, Z.; Yang, Z.; Wang, Y. Adsorption of three pharmaceuticals on two magnetic ion-exchange resins. *J. Environ. Sci.* **2015**, *31*, 226–234. [CrossRef] [PubMed]

15. Thiebault, T.; Guégan, R.; Boussafir, M. Adsorption mechanisms of emerging micro-pollutants with a clay mineral: Case of tramadol and doxepine pharmaceutical products. *J. Colloid Interface Sci.* **2015**, *453*, 1–8. [CrossRef] [PubMed]

16. Imfeld, G.; Braeckevelt, M.; Kuschk, P.; Richnow, H. Monitoring and assessing processes of organic chemicals removal in constructed wetlands. *Chemosphere* **2009**, *74*, 349–362. [CrossRef] [PubMed]

17. Matamoros, V.; García, J.; Bayona, J.M. Behavior of selected pharmaceuticals in subsurface flow constructed wetlands: A pilot-scale study. *Environ. Sci. Technol.* **2005**, *39*, 5449–5454. [CrossRef] [PubMed]

18. Day, P.R. Particle fractionation and particle-size analysis. In *Methods of Soil Analysis*; Black, C.A., Evans, D.D., Ensminger, L.E., White, J.L., Clark, F.E., Eds.; American Society of Agronomy: Madison, WI, USA, 1965; pp. 545–567.

19. Brix, H.; Arias, C.A.; Del Bubba, M. Media selection for sustainable phosphorous removal in subsurface flow constructed wetlands. *Water Sci. Technol.* **2001**, *44*, 47–54. [PubMed]

20. Qin, X.; Liu, F.; Wang, G.; Li, L.; Wang, Y.; Weng, L. Modeling of levofloxacin adsorption to goethite and the competition with phosphate. *Chemosphere* **2014**, *111*, 283–290. [CrossRef] [PubMed]

21. American Public Health Association; American Water Works Association; Water Environment Federation. *Standard Methods for the Examination of Water and Wastewater*, 21st ed.; American Public Health Association: Washington, DC, USA, 2005.

22. Haughton, P.N.; Grau, E.G.; Lyng, J.; Cronin, D.; Fanning, S.; Whyte, P. Susceptibility of campylobacter to high intensity near ultraviolet/visible 395+/−5 nm light and its effectiveness for the decontamination of raw chicken and contact surfaces. *Int. J. Food Microbiol.* **2012**, *159*, 267–273. [CrossRef] [PubMed]

23. Boehm, H.P. Some aspects of the surface chemistry of carbon blacks and other carbons. *Carbon* **1994**, *35*, 759–769. [CrossRef]

24. Leyva, R.; Díaz, P.E.; Guerrero, R.M.; Mendoza, J.; Aragón, A. Adsorción de cd(ii) en solución acuosa sobre diferentes tipos de fibras de carbón activado. *J. Mex. Chem. Soc.* **2004**, *48*, 169–202.

25. Vymazal, J.; Kröpfelová, L. *Wastewater Treatment in Constructed Wetlands with Horizontal Sub-Surface Flow*; Springer: Dordrecht, The Netherlands, 2008; Volume 14.

26. Alemayehu, E.; Lennartz, B. Adsorptive removal of nickel from water using volcanic rocks. *Appl. Geochem.* **2010**, *25*, 1596–1602. [CrossRef]

27. Sing, K.S.W.; Everett, D.H.; Haul, R.A.W.; Moscou, L.; Pierotti, R.A.; Rouquérol, J.; Siemieniewska, T. Reporting physisorption data for gas/solid systems with special reference to the determination of surface area and porosity. *Pure Appl. Chem.* **1985**, *57*, 603–619. [CrossRef]

28. Lowell, S.; Shields, J.E.; Thomas, M.A.; Thommes, M. *Characterization of Porous Solids and Powders: Surface Area, Pore Size and Density*; Springer: Dordrecht, The Netherlands, 2004; Volume 16.
29. Vilchis-Granados, J.; Granados-Correa, F.; Barrera-Díaz, C.E. Surface fractal dimensions and textural properties of mesoporous alkaline-earth hydroxyapatites. *Appl. Surf. Sci.* **2013**, *279*, 97–102. [CrossRef]
30. Brooks, W.E.; Piminchumo, V.; Suárez, H.; Jackson, J.C.; McGeehin, J.P. Mineral pigments at huaca tacaynamo (Chan Chan, Peru). *Bull. Inst. Fr. D'études Andin.* **2008**, *37*, 441–450. [CrossRef]
31. Farahmandjou, M.; Soflaee, F. Synthesis and characterization of α-Fe$_2$O$_3$ nanoparticles by simple co-precipitation method. *Phys. Chem. Res.* **2015**, *3*, 191–196.
32. Ponce, B.; Ortiz, A.; Otazo, E.M.; Reguera, E.; Acevedo, O.A.; Prieto, F.; González, C.A. Physical characterization of an extensive volcanic rockin méxico: "Red tezontle" from cerro de la cruz, in tlahuelilpan, hidalgo. *Acta Univ. Univ. Guanaj.* **2013**, *23*, 20–27.
33. Gutiérrez Pulido, H.; De la Vara Salazar, R. *Análisis y Diseño de Experimentos*, 3rd ed.; McGraw-Hill Educación: Mexico City, Mexico, 2012.
34. Ali, R.M.; Hamad, H.A.; Hussein, M.M.; Malash, G.F. Potential of using green adsorbent of heavy metal removal from aqueous solutions: Adsorption kinetics, isotherm, thermodynamic, mechanism and economic analysis. *Ecol. Eng.* **2016**, *91*, 317–332. [CrossRef]
35. Chiou, M.S.; Li, H.Y. Equilibrium and kinetic modeling of adsorption of reactive dye on cross-linked chitosan beads. *J. Hazard. Mater.* **2002**, *B93*, 233–248. [CrossRef]
36. Vadivelan, V.; Kumar, K.V. Equilibrium, kinetics, mechanism, and process design for the sorption of methylene blue onto rice husk. *J. Colloid Interface Sci.* **2005**, *286*, 90–100. [CrossRef] [PubMed]
37. Ho, Y.S.; McKay, G. Pseudo-second order model for sorption processes. *Process Biochem.* **1999**, *34*, 451–465. [CrossRef]
38. Lopez-Ramon, M.V.; Stoeckli, F.; Moreno-Castilla, C.; Carrasco-Marin, F. On the characterization of acidic and basic surface sites on carbons by various techniques. *Carbon* **1999**, *37*, 1215–1221. [CrossRef]
39. Liu, A.; Wang, J.; Lu, Z.; Yao, L.; Li, Y.; Yan, H. Hydrogen-bond detection, configuration assignment and rotamer correction of side-chain amides in large proteins by nmr spectroscopy through protium/deuterium isotope effects. *ChemBioChem* **2008**, *9*, 2860–2871. [CrossRef] [PubMed]
40. Suárez, D.; Menéndez, J.A.; Fuente, E.; Montes-Morán, M.A. Pyrone-like structures as novel oxygen-based organic superbases. *Angew. Chem. Int. Ed.* **2000**, *112*, 1376–1379. [CrossRef]
41. Chen, Y.P.; Zhang, P.; Guo, J.S.; Fang, F.; Gao, X.; Li, C. Functional groups characteristics of eps in biofilm growing on different carriers. *Chemosphere* **2013**, *92*, 633–638. [CrossRef] [PubMed]
42. Andersson, S. Characterization of Bacterial Biofilms for Wastewater Treatment. Ph.D. Thesis, Royal Institute of Technology, Stockholm, Sweden, 2009.
43. Fang, F.; Lu, W.T.; Shan, Q.; Cao, J.S. Characteristics of extracellular polymeric substances of phototrophic biofilms at different aquatic habitats. *Carbohydr. Polym.* **2014**, *106*, 1–6. [CrossRef] [PubMed]
44. Flemming, H.C.; Wingender, J. The biofilm matrix. *Nat. Rev. Microbiol.* **2010**, *8*, 623–633. [CrossRef] [PubMed]
45. Andersson, S.; Dalhammar, G.; Kuttuva Rajarao, G. Influence of microbial interactions and eps/polysaccharide composition on nutrient removal activity in biofilms formed by strains found in wastewater treatment systems. *Microbiol. Res.* **2011**, *166*, 449–457. [CrossRef] [PubMed]
46. Eduok, U.; Khaled, M.; Khalil, A.; Suleiman, R.; El Ali, B. Probing the corrosion inhibiting role of a thermophilic bacillus licheniformis biofilm on steel in a saline axenic culture. *RSC Adv.* **2016**, *6*, 18246–18256. [CrossRef]
47. Melo, L.F. Biofilm formation and its role in fixed film processes. In *Handbook of Water and Wastewater Microbiology*; Mara, D., Horan, N., Eds.; Academic Press: London, UK, 2003; pp. 337–349.

water

MDPI

Article

Removing Organic Matter and Nutrients from Swine Wastewater after Anaerobic–Aerobic Treatment

Rubén Alfonso Saucedo Terán [1,*,†], **Celia de la Mora Orozco** [2], **Irma Julieta González Acuña** [3], **Sergio Gómez Rosales** [4], **Gerardo Domínguez Araujo** [2] and **Héctor Osbaldo Rubio Arias** [5]

[1]   Sitio Experimental La Campana-INIFAP, Aldama, Chihuahua 32910, Mexico
[2]   Campo Experimental Central Altos de Jalisco-INIFAP, Tepatitlán de Morelos, Jalisco 47600, Mexico; delamora.celia@inifap.gob.mx (C.d.l.M.O.); dominguez.gerardo@inifap.gob.mx (G.D.A.)
[3]   Campo Experimental Santiago Ixcuintla-INIFAP, Santiago Ixcuintla, Nayarit 63300, Mexico; gonzalez.irmajulieta@inifap.gob.mx
[4]   CENID Fisiología-INIFAP, Ajuchitlán, Querétaro 76280, Mexico; gomez.sergio@inifap.gob.mx
[5]   Facultad de Zootecnia y Ecología-Universidad Autónoma de Chihuahua, Chihuahua 31031, Mexico; rubioa1105@hotmail.com
*   Correspondence: rasteran@yahoo.com.mx; Tel.: +52-614-184-8582
†   Retired, Former Researcher.

Received: 2 August 2017; Accepted: 16 September 2017; Published: 25 September 2017

Abstract: Anaerobic digesters generate effluent containing about 3000 mg L^{-1} of organic matter in terms of chemical oxygen demand (COD). This effluent must be treated before being reused or discharged into the environment. The objective of this study was to evaluate the efficiency of a trickling filter packed with red volcanic rock for the treatment of anaerobic digester effluent with COD concentrations of around 3000 mg L^{-1}. The trickling filter consisted of an aluminum cylinder, 2 mm thick, 3 m high, and 1 m in diameter. To evaluate the efficiency of the treatment system, there were three experimental runs, each lasting 20 days (d). The predictor variable was the initial COD concentration, which ranged from 2002 to 3074 mg L^{-1}. The hydraulic retention time was 9 h. The influent flow was 2.2 L min^{-1}, which amounts to a hydraulic load of 4033 m^3 m^{-2} day^{-1} and an organic load of 0.006342 to 0.009738 kg m^{-3} day^{-1} of COD. Independent of the initial concentration, COD removal efficiency was very high, varying from 90 to 96%. Final effluents met all the maximum permissible limits to be used as irrigation water, as well as for its release into natural or artificial water reservoirs, stored for agricultural crop irrigation.

Keywords: trickling filter; anaerobic digester; swine wastewater; organic matter; COD

1. Introduction

Pig farming represents the third most important livestock activity in Mexico. According to official statistics, the national inventory of pigs is estimated at more than 15.2 million heads, ranking as the third most important livestock animal in Mexico. Pig farming is concentrated in central and northern Mexico, mainly in the states of Jalisco and Sonora, which accounts for almost 49% of total production [1,2]. In these regions, pig-farming stands out not only because of its economic importance, but also its significant impact on the environment owing to the large volumes of solid and liquid wastes generated, altering the physical, chemical, and microbiological composition of soils and water bodies. In the case of liquid waste, a medium-sized farm generates between 30 and 35 m^{-3} day^{-1} of sewage, which contains high concentrations of solids, organic matter, nitrogen, and phosphorous, among other contaminants. Even with technologically advanced farms, which account for 56.9% of the total, treatment of wastes is a low priority. The vast majority of waste matter is discharged into the environment without any treatment and, evidently, without complying with official requirements.

The few pig producers that treat animal wastewater use anaerobic digesters. Nationally, there are 479 digesters registered in the states of Coahuila, Chihuahua Guanajuato, Durango, Guanajuato, Jalisco, Michoacan, Nuevo Leon, Puebla, Queretaro, Sonora, Veracruz, and Yucatan. Of these, only 82% are in operation and most of these are characterized by problems like oversizing, failures in the agitation systems and burners, irregular maintenance, and lack of knowledge of the operating systems among farmers [3]. Under normal operating conditions, anaerobic systems generate effluents containing organic matter of about 3000 mg L^{-1} in terms of chemical oxygen demand (COD), which is equivalent to five times the organic matter content of domestic wastewater, highlighting the level of contamination. Therefore, the effluent of anaerobic digesters must be treated before being reused or discharged into the environment. In this sense, aerobic systems present an important alternative, because they require short hydraulic retention times and do not generate bad odors, which is of particular importance because the majority of pig farms in Mexico are located in suburban areas.

Among the aerobic systems, trickling filters stand out. This is a widely-used technology for the treatment of industrial wastewater, which was recently adapted for the treatment of bio-waste. In the treatment of household wastewater, efficiencies of above 90% in the reduction of organic matter are reported, generating effluents with maximum COD concentrations of 30 mg L^{-1}, which complies with the quality standards for wastewater disposal. It is also reported that trickling filters can reduce dissolved organic nitrogen by up to 72%, resulting in effluents with less than 1.8 mg L^{-1} of biodegradable dissolved nitrogen [4]. In the case of a dairy processing plant, an efficiency rate of 96% in COD removal was obtained, with a hydraulic retention time of 7 h and an organic matter concentration in the influent of about 1700 mg L^{-1} COD [5]. In the same study, an efficiency of over 70% was reported in the removal of total nitrogen. Dairy wastewater has been successfully treated with organic loads up to 2700 mg L^{-1} COD and hydraulic retention times of 5 to 7 h. The main factors that limit the ability of trickling filter denitrification are excessive organic loads and the emergence of large populations of aquatic snails [6]. The key factors in the functioning of trickling filters are the hydraulic retention time, the concentration and type of organic matter in the influent, and the porosity and size of the particles that constitute the support material in which the degrading microorganisms of organic matter contaminants develop [7,8].

There is little information about trickling filters for treating wastewater from pig farms. In a work similar to ours, Szogi et al. [9] obtained a 54% reduction in the COD content in an anaerobic lagoon in which the initial concentration was 869 mg L^{-1}. Morton and Auvermann [10] also assessed the treatment of effluent from a lagoon storing wastewater from a pig farm, and reported very low removal efficiencies, including in some cases an increase in concentrations of COD, NH_3-N, and NO_3-N. Garzon-Zuñiga et al. [11] assessed the performance of a trickling filter with initial COD concentrations of 8668 to 19,320 mg L^{-1}, which were reduced to 1200 to 2400 mg L^{-1} after 100 days of operation. These authors indicate that the aeration rate is an important factor in the efficiency of COD removal of trickling filters packed with organic matter. Duda and Alves de Oliveira [12] obtained COD removal efficiencies of up to 96% using a treatment series system composed of a UASB reactor, an anaerobic filter, and a trickling filter. In addition to the efficiency of the trickling filters, a theme that has been amply studied is the search for the best support materials. In addition to PVC, other materials have been assessed such as gravel [9], plastic Bioballs™ (Meyer Aquascapes, Inc., Harrison, OH, USA), recycled soda six-pack rings [10], peat [11], bamboo rings [12], rubber, polystyrene, stone [13], sponge [14], and cotton sticks [15].

In general terms, the main advantages of trickling filters are the simplicity of operation, low environmental impact, low energy requirements for operation, and a very favorable cost–benefit ratio [16,17]. However, the effectiveness of trickling filters on the treatment of anaerobic digester effluents from pig farms is unknown. Similarly, we found no information on the use of volcanic rock as a substitute for PVC particles (Engineering360, Tulsa, OK, USA), which is the traditional support material used in trickling filters. The aim of this study was to evaluate the efficiency of a trickling filter packed with red volcanic rock for the treatment of effluents with COD concentrations of about

3000 mg L^{-1} from anaerobic digesters installed on pig farms. Volcanic rock is characterized by high degrees of porosity and absorption, it is widely available and inexpensive. These characteristics result in volcanic rock having great potential for use in trickling filters. Potential users of the information reported are pig producers, professional service providers of technical assistance, and government agencies related to this subsector of production.

2. Materials and Methods

This study was carried out on the Santa Maria pig farm, located at km 24 of San Miguel El Alto—Atotonilco highway, in the municipality of Arandas, Jalisco. The pigs are produced for slaughter and the farm has an inventory of 12,000 pigs. The anaerobic digester has a capacity of 9518 m^3 and generates approximately 2000 m^3 of bio-gas per day. The effluent of the anaerobic digester is sent to an artificial lagoon where it is stored and used as pasture irrigation water. The wastewater stored in the lagoon was the influent in this study (Figure 1).

Since the effluent from the anaerobic digester had a COD concentration of approximately 7160 mg L^{-1}, the lagoon water was diluted with well water to obtain the desired maximum COD concentration of 3000 mg L^{-1}, which is the average concentration of effluents from anaerobic digesters operating under normal conditions. The wastewater from the lagoon was pumped into a 10,000-liter tank with a submersible pump. The tank was equipped with a mechanical stirrer that was activated in accordance with the on–off cycles of the pump, which in turn were regulated with a float. The water tank was placed on the edge of the lagoon, 4 m above the trickling filter to ensure that residual diluted water flowed by gravity from the water tank to the trickling filter. The flow was controlled by a rotameter with a ball valve and an operating range from 0 to 7.5 liters per minute (L min^{-1}).

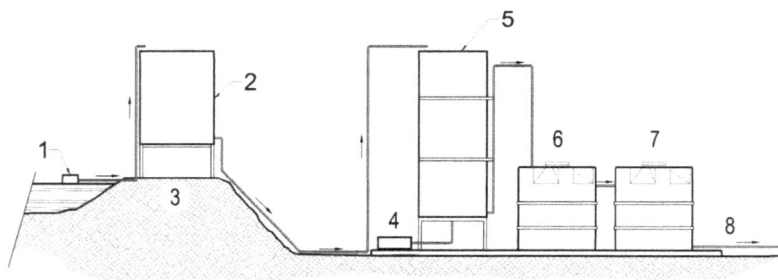

Figure 1. Schematic representation of the wastewater treatment system. (1) Pump; (2) Dilution tank; (3) Lagoon board; (4) Compressor; (5) Trickling filter; (6) Pre-clarifier; (7) Clarifier; (8) Final effluent.

The trickling filter consisted of an aluminum cylinder, 2 mm thick, 3 m high, and 1 m in diameter, with a cylindrical aluminum lid on top held up by four metal supports on the inner side of the trickling filter. The lid, which was placed at a height of 20 cm below the upper edge of the cylinder, functioned as a radial distribution system of the influent. For this purpose, the lid had multiple radial perforations, 1 cm in diameter each. The exterior edge of the lid was coated with a rubber gasket that prevented the flow through the inner wall of the cylinder. The wastewater was directed to the top part of the trickling filter and poured into the center of the lid. The trickling filter was filled with spherically-shaped red volcanic rock approximately 2–4 cm in diameter, which served as support material to the bacteria that degraded the organic matter in the wastewater under treatment (Table 1). The working group did not assess the physical characteristics of the red volcanic rock. However, there are several reports in this regard. Rodriguez Diaz et al. [18] reported that the volcanic rock from the site from which the rock used in this investigation was obtained has a total porosity of 55.5% and an aeration porosity of 40.7%. Total porosity reported in other studies range from 67 to 74.7%, with aeration porosity levels of 39.2 to 44.4%, and a real density of 2.45 g cm^{-3} [19,20]. The packing depth was 2.80 m. The average temperature of the influent was 20.84 °C and the pH level was in the range 7.72 to 8.53, with an average of 8.17. A radial aeration system was installed at the bottom of the trickling filter and connected to a compressor. Air was injected through a 1-npt spigot nozzle at a flow rate of 10 L min^{-1}. The dissolved oxygen (DO) content in the influent was in the range of 0.1 to 0.4 mg L^{-1}, with an average of 0.17 mg L^{-1}. Two 5000-liter water tanks were installed to separate the outgoing solids, each with a sedimentation system in series consisting of a pre-clarifier and clarifier. Both the pre-clarifier and clarifier had a purging and sewage collection system controlled by a ball valve. The sewage was purged weekly, collecting the sediments at a rate of 20 liters per day (L day^{-1}) from each sedimentation tank.

Table 1. General operating conditions.

Support Material	Red Volcanic Rock
Packing depth (m)	2.8
Inflow (L min^{-1})	2.2
Hydraulic retention time (h)	9
Hydraulic load (m^{-3} m^{-2} day^{-1})	4033
Air flow rate (L min^{-1})	10
Influent temperature (°C)	20.84 ± 2.07 (Mean ± standard deviation)
Influent COD concentration (mg L^{-1})	2002–3074
Organic load (kg m^{-3} day^{-1} of COD)	0.006342–0.009738
Influent total N concentration (mg L^{-1})	138.75–151.33
Influent NH$_3$.N concentration (mg L^{-1})	65.70–71.22
Influent total P concentration (mg L^{-1})	65.00–78.0
Influent EC concentration (mS cm^{-1})	1.24–1.75
Influent pH (dimensionless)	7.72–8.53
Influent DO concentration (mg L^{-1})	0.1–0.4

To evaluate the efficiency of the treatment system, an experiment was run for 60 days. A 20-d experimental adaptation was carried out previously, in which the trickling filter was inoculated with activated sludge from a suspended growth process wastewater treatment plant. The predictor variable was the COD concentration in the influent, which ranged from 2002 to 3074 mg L^{-1}. The hydraulic retention time for all experimental runs was 9 h. The influent flow was at 2.2 L min^{-1}, which resulted in a hydraulic load of 4033 m^{-3} m^{-2} day^{-1} and an organic load of 0.006342 to 0.009738 kg m^{-3} day^{-1} of COD, representing concentrations of 2002 and 3074 mg L^{-1}, respectively. The 12 samples of influent and effluent collected during the experimental run were analyzed measuring the following variables: COD, total nitrogen, total ammonia nitrogen, total phosphorus, electrical conductivity, and dissolved oxygen. COD was quantified by an oxidation potassium dichromate technique, using a digester and a colorimeter Hach, model 800. Total nitrogen was analyzed by the persulfate digestion method. To determine the concentration of ammonia nitrogen, a salicylate method was applied. Total phosphorous was determined using a molybdovanadate method with acid persulfate digestion. Electrical conductivity was measured with a MW 801 Milwaukee sensor. Dissolved oxygen was determined in the field, using a potentiometer JPB, model 607A. The paired difference test was applied to compare the means of influents and effluents, using a significance level of 0.01 ($\alpha = 0.01$).

3. Results

There were significant differences between the means of influents and effluents for COD and the other parameters ($p < 0.01$). Table 2 shows the results of COD removal from the trickling filter. As can be seen, the COD concentration in the influent ranged from 2002 to 3074 mg L^{-1}. Independent of the initial concentration, the removal efficiency was very high, varying from 90 to more than 96%, with an average of 93%. The average COD in the effluent was 172 mg L^{-1}, which is considered an acceptable quality level for wastewater used for the irrigation of pastures and for its disposal in water bodies. In this context, the legal standard indicates that the organic matter content, expressed in terms of biochemical oxygen demand (BOD_5), should not exceed the maximum allowed limit of 200 mg L^{-1}, applicable to wastewater released into rivers whose water is used for agricultural irrigation [21]. There is no maximum permissible limit for COD in Mexico. However, given that BOD_5 is equivalent to 1.6 times the organic matter content expressed by COD [22], the organic matter concentration of the effluent complied with the Mexican standards. International standards are stricter than those in Mexico. For example, the maximum permissible limit for using wastewater in agriculture is 60 mg L^{-1} in France and 100 mg L^{-1} in Italy [23]. Likewise, the COD levels obtained in this research were above the standard for wastewater reuse in the Middle East, with 100 mg L^{-1} in Jordan and Kuwait, and 150 mg L^{-1} in Oman [24].

Table 2. Efficiency of a trickling filter in removing COD from the effluent of an anaerobic digester.

Date	COD in the Influent (mg L^{-1})	COD in the Effluent (mg L^{-1})	Removal Efficiency (%)
2 June 2014	2100	198	90.6
7 June 2014	2002	200	90.0
12 June 2014	2678	183	93.2
17 June 2014	2560	163	93.6
22 June 2014	2484	165	93.4
27 June 2014	2522	140	94.4
2 July 2014	2410	140	94.2
7 July 2014	2216	162	92.7
12 July 2014	3054	117	96.2
17 July 2014	3006	196	93.5
22 July 2014	3010	201	93.3
27 July 2014	3074	203	93.4
Means	2593 [a]	172 [b]	93

Notes: [a,b]: different letters indicate significant differences ($p < 0.01$) between influent and effluent means.

Table 3 shows the results for total nitrogen removal. The average nitrogen concentration was 145 mg L^{-1} in the influent and 75 mg L^{-1} in the effluent, so the average removal rate was 48%. The final concentration of nitrogen exceeded the maximum permissible limit (60 mg L^{-1}) for release into rivers [21]. Thus, using this water for irrigation helps to reduce the nitrogen concentration before the water reaches rivers and natural or artificial water reservoirs. With additional treatment, the effluent from the system could meet more stringent standards for the water to be used for washing pens on farms.

Table 3. Efficiency of a trickling filter in removing total nitrogen from an anaerobic digester effluent.

Date	Total-N in the Influent (mg L^{-1})	Total-N in the Effluent (mg L^{-1})	Removal Efficiency (%)
2 June 2014	145.22	78.00	46.3
7 June 2014	142.01	79.00	44.4
12 June 2014	138.75	69.00	50.3
17 June 2014	140.22	65.00	53.6
22 June 2014	151.33	88.00	41.8
27 June 2014	149.20	76.50	48.7
2 July 2014	147.10	82.30	44.1
7 July 2014	145.30	81.78	43.7
12 July 2014	146.77	68.02	53.7
17 July 2014	142.12	65.12	54.2
22 July 2014	142.20	71.02	50.1
27 July 2014	151.10	81.22	46.2
Means	145 [a]	75 [b]	48

Notes: [a,b]: different letters indicate significant differences ($p < 0.01$) between influent and effluent means.

Table 4 shows the results for ammonia nitrogen removal. The concentration of ammonia nitrogen in the influent ranged between 66 and 71 mg L^{-1}. After treatment in the trickling filter, the concentration of ammonia nitrogen decreased by almost 99%, leaving an average residual concentration of 2.4 mg L^{-1}, which was low enough to even meet drinking water standards.

Table 4. Efficiency of a trickling filter in removing ammonia nitrogen from an anaerobic digester effluent.

Date	NH$_3$.N in the Influent (mg L^{-1})	NH$_3$.N in the Effluent (mg L^{-1})	Removal Efficiency (%)
2 June 2014	66.32	4.50	93.2
7 June 2014	65.70	2.00	97.0
12 June 2014	72.80	5.10	93.0
17 June 2014	70.12	1.30	98.1
22 June 2014	71.00	4.00	94.4
27 June 2014	68.75	3.50	94.9
2 July 2014	69.00	1.25	98.2
7 July 2014	71.22	1.00	98.6
12 July 2014	68.50	2.30	96.6
17 July 2014	71.00	2.20	96.9
22 July 2014	69.25	1.00	98.6
27 July 2014	70.00	0.98	98.6
Means.	69 [a]	2.4 [b]	98

Notes: [a,b]: different letters indicate significant differences ($p < 0.01$) between influent and effluent means.

Table 5 shows the results of the removal of total phosphorous. As can be seen, the efficiency of phosphorous removal was between 43 and 68%, starting from around 70 mg L^{-1} and resulting in an average of 29 mg L^{-1} in the effluent. This concentration was below the 30 mg L^{-1} maximum limit for releasing wastewater into rivers and water reservoirs to be used in agricultural irrigation [21]. However, because the concentration of phosphorous in the effluent varied (29 ± 5.6), batch-testing for phosphorous is needed to avoid non-compliance with regulations. Electrical conductivity decreased through the treatment by approximately 35%, going from an initial concentration of 1.57 to a final concentration of 1.02 mS cm^{-1} (Figure 2). Although there are no regulatory limits for this variable, it is an indicator of dissolved salt content in water. According to the final concentration of electrical conductivity, the treated water should be moderately restricted for agricultural irrigation, depending on the tolerance of the specific crop.

Table 5. Efficiency of a trickling filter in removing total phosphorous from an anaerobic digester effluent.

Date	Total-P in the Influent (mg L^{-1})	Total-P in the Effluent (mg L^{-1})	Removal Efficiency (%)
2 June 2014	69.57	34.28	50.7
7 June 2014	69.53	39.13	43.7
12 June 2014	78.00	38.45	50.7
17 June 2014	67.90	26.90	60.4
22 June 2014	70.22	25.50	63.7
27 June 2014	68.90	29.50	57.2
2 July 2014	66.50	28.20	57.6
7 July 2014	65.00	21.00	67.7
12 July 2014	70.22	23.40	66.7
17 July 2014	69.45	28.90	58.4
22 July 2014	70.31	31.60	55.1
27 July 2014	71.70	26.50	63.0
Means	70 [a]	29 [b]	58

Notes: [a,b]: different letters indicate significant differences ($p < 0.01$) between influent and effluent means.

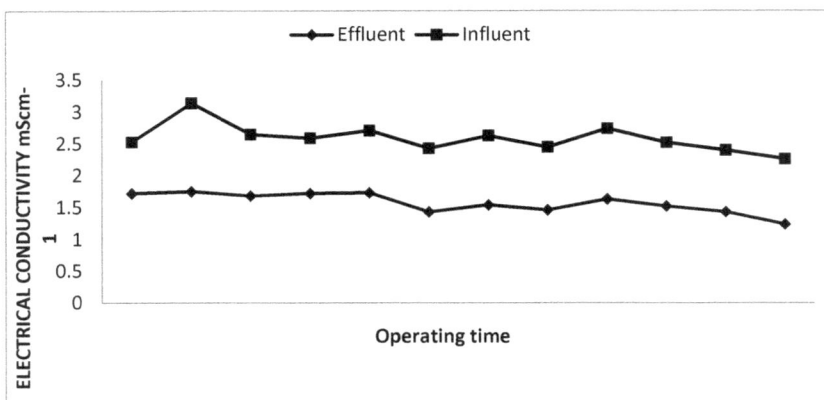

Figure 2. Efficiency of a trickling filter in removing electrical conductivity from an anaerobic digester effluent.

Figure 3 shows the changes in the dissolved oxygen concentration, which increased from 0.2 mg L^{-1} to an average of 3.1 mg L^{-1}. This variable is an indicator of water quality for the wellbeing of different aquatic organisms. Concentrations below 3 mg L^{-1} reduce the chances of survival of biotic communities and represent an imminent threat to the conservation of biodiversity in aquatic ecosystems [25]. Dissolved oxygen is also important for the efficient operation of the trickling filter, in which the bacteria aerobically degrades the organic matter in the wastewater under treatment [8]. The lower limit for the development of aerobic biofilms is around 0.57 mg L^{-1}, so that the oxygen concentration obtained in this study guarantees the adequate functioning of the treatment system.

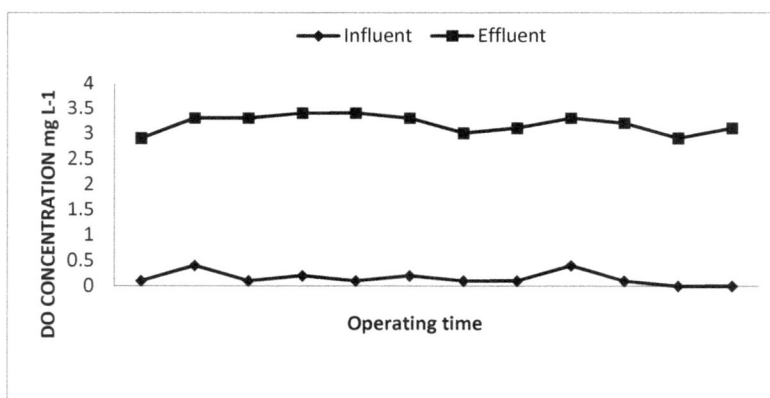

Figure 3. Efficiency of a trickling filter in increasing dissolved oxygen in wastewater from an anaerobic effluent.

4. Discussion

The efficiency of COD removal in this study was 90 to 96%, which exceeds the levels obtained in other studies. Reyes-Lara and Reyes-Mazzoco [26] evaluated a trickling filter with organic feed concentrations from 2114.8 to 3814.4 mg L^{-1} COD and reported removal rates of 54 to 66%. Braulio-Villalobos et al. [27] obtained a 72% organic matter removal rate from an influent with 300 mg L^{-1} COD. Gilbert et al. [28] also evaluated a trickling filter for treating wastewater from pig farms and observed a higher level of efficiency than that obtained in the present study, reducing COD content from 15,300 mg L^{-1} in the influent to 330 mg L^{-1} in the effluent. However, their trickling filter operated with a hydraulic load of 0.017 m^3 m^{-2} day^{-1}, which was much lower than the 4033 m^3 m^{-2} day^{-1} in our study. Similarly, Buelna et al. [29] obtained efficiencies in the removal of organic matter of up to 95% in a trickling filter designed for the treatment of 12 m^3 day^{-1} of pig wastewater with 10,000 to 20,000 mg L^{-1} BOD_5. According to the volume of the filter and the influent flow, the retention time was much longer in that study than in ours. Beyenal and Lewandowski [30] stated that the capacity of trickling filters to remove organic matter depends on the diffusion in the biofilm, which is directly proportional to the organic load, if there is no another limiting factor, such as oxygen availability. In this respect, Reyes-Lara and Reyes-Mazzoco [26] found that with low organic matter concentrations, similar to the levels evaluated in our study, the substrate may be the limiting reactant and not the available oxygen. The porosity of materials like red volcanic rock provides a larger surface area for adhesion of biofilms than commonly used particles [31]. This allows for the majority of particles to be coated with the biofilm in a maximum of three weeks, which results in a stable operational state within this period [32]. This was evident in our work in which, from the first day of operation, COD was reduced by 90.6%, going from an initial concentration of 2100 to a final concentration of 198 mg L^{-1}. According to Metcalf and Eddy [22], a steady operating state in trickling filters is usually reached in about four weeks of continuous operation, although other authors have observed that the steady state for COD removal can occur anywhere from three days to seven weeks [33,34]. Naz et al. evaluated different media in a trickling filter, and observed that the highest COD removal efficiency was obtained by stone (93.4%), outperforming plastic (89.4%), polystyrene (86.3%), and rubber (81.9%) [13].

It is difficult to compare our results to those of other studies in terms of nitrogen removal efficiency because of the differences in operating conditions. Gilbert et al. [28] obtained a removal efficiency of 75% for total nitrogen when treating pig wastewater with concentrations of 3200 mg L^{-1}, but with a very low hydraulic load. This study involved lower retention times and a lower nitrogen load (0.067 kg m^{-2} day^{-1}) than those of our study (0.559 kg m^{-2} day^{-1}). Buelna et al. [29] obtained a nitrogen removal rate of 26% in the treatment of swine wastewater, with a total nitrogen load of 2300 mg L^{-1}, and a very low hydraulic load (0.017 m^{-3} m^{-2} day^{-1}). Garzon Zuñiga et al. [11] reported an efficiency of 50% in the removal of total nitrogen content in swine wastewater with 2080 mg L^{-1} of nitrogen and a very low hydraulic load (0.5 m^{-3} m^{-2} day^{-1}). The main factors that limit the capacity of trickling filters for denitrification are excessive organic loads and the development of high populations of aquatic snails [6]. The efficiency of ammonia nitrogen removal in our study was very high (98%) but its concentration in the influent was very low (69 mg L^{-1}), which is equivalent to a mass load of 0.278 kg m^{-3} day^{-1}. Sabbah et al. [35] obtained similar results to those of the present study, reducing the content of ammonia nitrogen by up to 95%, from an initial concentration of 77.9 mg L^{-1}, under a much lower hydraulic load (0.093 m^{-3} m^{-2} day^{-1}). Hort et al. [36] were able to reduce ammonia nitrogen content by 94%, with a mass load of NH_4 of 0.0604 kg m^{-3} day^{-1}. Ying-Xu et al. [37] found a removal efficiency of 95–99% of ammonia nitrogen, with a very high concentration of ammonia in the influent (110 mg m^{-3}), but with a load rate similar to that in this research (0.243 kg m^{-3} day^{-1}).

We found few studies on the performance of trickling filters with respect to the removal of total phosphorus and the changes in electrical conductivity and dissolved oxygen. Buelna et al. [29] reported an efficiency of phosphorus removal of 71% from an influent with a concentration of 180 mg L^{-1} and a very low hydraulic load (0.017 m^{-3} m^{-2} day^{-1}). Garzon-Zùñiga et al. [11] evaluated various aeration rates, and observed effluents with dissolved oxygen contents from 7 to 8.5 mg L^{-1}, from an influent with 0.05 mg L^{-1}. No reports were found regarding changes of electrical conductivity in treated swine wastewater with trickling filters. Katukiza et al. [38] attributed the removal of dissolved nitrogen and phosphorous mainly to precipitation. In addition, adsorption and ionic exchange have been found to contribute to the removal of phosphates from wastewater [39]. However, the removal of dissolved nitrogen and phosphorus by adsorption is limited when the pH of the influent is above 7 [40].

In the present study, the pH of the influent ranged from 7.72 to 8.53. In addition, particulate nitrogen can be removed by straining during the trickling filter operation [38]. The physical and chemical adsorption of NH_4 in organic matter, and hence, its microbial assimilation, could be responsible for the removal of significant amounts of N from wastewater [40]. However, this could be supported by significant levels of nitrifying and denitrifying activity. Interestingly, Patel et al. [31] observed nitrifiers and denitrifiers in both anoxic and aerobic biofilms, which suggests a highly complex structure of multispecies biofilm. Although the existence of denitrifiers in the aerobic layer can be attributed to limited oxygen diffusion in the biofilm, the emergence of nitrifying bacteria in the anoxic bed was surprising. To make the results more complex, higher levels of phosphorus assimilating bacteria have been reported in the anoxic than in the aerobic biofilm [31]. Since there have been few studies on nitrogen and phosphorus removal mechanisms in volcanic rock biofilms, it is difficult to identify the factors that determined the high removal efficiency obtained in the present research. It was probably a combination of factors like precipitation, adsorption, straining, microbial denitrifying, and denitrifying transformation of nitrogen.

5. Conclusions

The trickling filter packed with red volcanic rock proved to be highly efficient in treating anaerobic digester effluent with COD concentrations of around 3000 mg L^{-1}. Final effluents met all the specifications in terms of maximum permissible limits for their use as irrigation water, as well as for their release into rivers and natural or artificial reservoirs that store water for agricultural crop irrigation. A nine-hour hydraulic retention time was used in the present research. It is highly recommended to evaluate shorter retention times in future studies.

Acknowledgments: The authors thank the support given by the Instituto Nacional de Investigaciones Forestales, Agrícolas y pecuarias-INIFAP that was responsable financially. The owners of Granja Santa María pig farm deserve special thanks for the facilities offered to this research.

Author Contributions: R.A.S.T. conceived and designed the study, performed sample collection, conducted field analysis, and drafted the manuscript. C.d.l.M.O. and I.J.G.A. carried out the physical-chemical analysis of all samples and helped in the data´s interpretation. S.G.R. and G.D.A. interpreted the statistical analysis and participated in drafted the manuscript. H.O.R.A. suggested and performed the statistical analysis and helped in the interpretation of the levels of all parameters.

Conflicts of Interest: The authors declare no conflict of interest.

References

1. Servicio de Información Agroalimentaria y Pesquera (SIAP). *Atlas Agroalimentario 2014*; Primera Edición: México, D.F., México, 2014; p. 193. Available online: https://www.gob.mx/siap/acciones-y-programas/atlas-2014?idiom=es (accessed on 18 July 2017). (In Spanish)
2. Bobadilla Soto, E.E.; Espinoza Ortega, A.; Martínez, C.; Castañeda, F.E. Dinámica de la producción porcina en México de 1980 a 2008. *Rev. Mex. Cienc. Pecu.* **2010**, *1*, 251–268. (In Spanish)
3. FIRCO-SAGARPA. Diagnóstico General de la Situación Actual de los Sistemas de Biodigestión en México. México, D.F., 2009; p. 32. Available online: http://www.ecotec.unam.mx/Ecotec//wpcontent/uploads/Diagnostico-Nacional-de-los-Sistemas-de-Biodigestion.pdf (accessed on 18 July 2017). (In Spanish and English).
4. Simsek, H.; Kasi, M.; Wadhawan, T.; Bye, C.; Blonigen, M. Fate of dissolved organic nitrogen in two stage trickling filter process. *Water Res.* **2012**, *46*, 5115–5126. [CrossRef] [PubMed]
5. Mehrdadi, N.; Nabi Bidhendi, G.R.; Shokouhi, M. Determination of dairy wastewater treatability by bio-trickling filter packed with lava rocks—Case study PEGAH dairy factory. *Water Sci. Technol.* **2012**, *65*, 1441–1447. [CrossRef] [PubMed]
6. Van den Akker, B.; Holmes, M.; Short, M.D.; Cromar, N.J.; Fallowfield, H.J. Application of high rate nitrifying trickling filters to remove low concentrations of ammonia from reclaimed municipal wastewater. *Water Sci. Technol.* **2010**, *61*, 2425–2432. [CrossRef] [PubMed]

7.   Habte, H.L.; Eckstadt, H. Performance of a trickling filter for nitrogen and phosphorous removal with synthetic brewery wastewater in trickling filter biofilm. *Int. J. Appl. Microbiol. Biotechnol. Res.* **2014**, *2*, 30–42.

8.   Norsker, N.H.; Nielsen, P.H.; Hvitved-Jacobsen, T. Influence of oxygen on biofilm growth and potential sulfate reduction in gravity sewer biofilm. *Water Sci. Technol.* **1995**, *31*, 159–167.

9.   Szogi, A.; Humenik, F.; Rice, J.; Hunt, P. Swine wastewater treatment by media filtration. *J. Environ. Sci. Health* **1997**, *832*, 831–843. [CrossRef] [PubMed]

10.   Morton, A.; Auvermann, B. Comparison of plastic trickling filter media for the treatment of swine lagoon effluent. In Proceedings of the 2001 ASAE Annual International Meeting, Sacramento, CA, USA, 29 July–1 August 2001; Paper Number: 01-2286. Available online: http://amarillo.tamu.edu/files/2011/01/comparisonof_21.pdf (accessed on 29 August 2017).

11.   Garzón-Zúñiga, M.A.; Lessard, P.; Aubry, G.; Buelns, G. Aeration effect on the efficiency of swine manure treatment in a trickling filter packed with organic materials. *Water Sci. Technol.* **2007**, *55*, 135–143. [CrossRef] [PubMed]

12.   Duda, R.; Alves de Oliveira, R. Treatment of swine wastewater in UASB reactor and anaerobic filter in series followed of trickling filter. *Eng. Sanit. Ambient.* **2011**, *16*, 91–100. [CrossRef]

13.   Naz, I.; Sarojb, D.; Mumtaza, S.; Alia, N.; Ahmeda, S. Assessment of biological trickling filter systems with various packing materials for improved wastewater treatment. *Environ. Technol.* **2014**, *35*, 1–11.

14.   Sánchez Guillén, J.; Jayawardana, L.; Lopez Vazquez, C.; de Oliveira Cruz, L.; Brdjanovic, D.; van Lier, J. Autotrophic nitrogen removal over nitrite in a sponge-bed trickling filter. *Bioresour. Technol.* **2015**, *187*, 314–325. [CrossRef] [PubMed]

15.   Ahson Aslam, M.; Khan, Z.; Sultan, M.; Niaz, Y.; Mahmood, M.; Shoaib, M.; Shakoor, A.; Ahmad, M. Performance Evaluation of Trickling Filter based Wastewater Treatment System utilizing Cotton Sticks as Filter Media. *Pol. J. Environ. Stud.* **2017**, *10*, 1–17. [CrossRef]

16.   Daigger, G.T.; Boltz, J.P. Trickling filter and trickling filter-suspended growth process design and operation: A state-of-the-art review. *Water Environ. Res.* **2011**, *83*, 388–404. [CrossRef] [PubMed]

17.   Zhao, Q.L.; Zhong, H.Y.; Liu, J.L.; Liu, Y. Integrated coagulation-trickling filter–ultrafiltration processes for domestic wastewater treatment and reclamation. *Water Sci. Technol.* **2012**, *65*, 1599–1605. [CrossRef] [PubMed]

18.   Rodríguez Díaz, E.; Salcedo Pérez, E.; Rodríguez Macias, R.; González Eguiarte, D.; Mena Munguía, S. Reúso del tezontle: Efecto en sus características físicas y en la producción de tomate (Lycopersicon esculentum Mill). *Terra Latinoam.* **2013**, *31*, 275–284. (In Spanish)

19.   Ojodeagua Arredondo, J.; Castellanos Ramos, J.; Muñoz Ramos, J.; Alcántar González, G.; Tijerina Chávez, L.; Vargas Tapia, P.; Enríquez Reyes, S. Eficiencia de suelo y tezontle en sistemas de producción de tomate en invernadero. *Rev. Fitotec. Mex.* **2008**, *31*, 367–374. (In Spanish)

20.   Trejo-Téllez, L.; Ramírez-Martínez, M.; Gómez-Merino, F.; García-Albarado, J.; Baca-Castillo, G.; Tejeda-Sartorius, O. Evaluación física y química de tezontle y su uso en la producción de tulipán. *Rev. Mex. Cienc. Agríc.* **2013**, *4*, 863–876, (In Spanish and English).

21.   NOM-001-SEMARNAT-1996. Norma Oficial Mexicana que Establece los Límites Máximos Permisibles de Contaminantes en las Descargas de Aguas Residuales en Aguas y Bienes Nacionales. Available online: http://biblioteca.semarnat.gob.mx/janium/Documentos/Ciga/agenda/DOFsr/DO2470.pdf (accessed on 18 July 2017). (In Spanish)

22.   George, T.; Metcalf, E. *Wastewater Engineering: Treatment, Disposal and Reuse*, 3rd ed.; Tchobanoglous, G., Burton, F.L., Eds.; McGraw-Hill, Inc.: New York, NY, USA, 1991.

23.   Wriedt, G.; Van der Velde, M.; Aloe, A.; Bouraoui, F. Water Requirements for Irrigation in the European Union. JCR Scientific and Technical Reports 2008, p. 64. Available online: http://www.enorasis.eu/uploads/files/Water%20Governance/5.JRC46748_Report_Irrigation_EUR_23453_EN.pdf (accessed on 29 August 2017).

24.   World Health Organization (WHO). A Compendium of Standards for Wastewater Reuse in the Eastern Mediterranean Region. WHO-EM/CEH/142/E, p. 19. Available online: http://applications.emro.who.int/dsaf/dsa1184.pdf (accessed on 29 August 2017).

25.   Pérez-Castillo, A.G.; Rodríguez, A. Índice fisicoquímico de la calidad de agua para el manejo de lagunas tropicales de inundación. *Rev. Biol. Trop.* **2008**, *56*, 1905–1918. (In Spanish) [PubMed]

26.   Reyes-Lara, S.; Reyes-Mazzoco, R. Efecto de las cargas hidráulica y orgánica sobre la remoción másica de un empaque estructurado en un filtro percolador. *Rev. Mex. Ing. Quim.* **2009**, *8*, 101–109. (In Spanish)

27. Braulio-Villalobos, M.A.; Sandoval-Silva, E.A.; Aréchiga-Viramontes, J.U. Operación y rediseño de una tecnología para el tratamiento de aguas residuales en Cuemanco. *Rev. Mex. Ing. Quim.* **2006**, *5*, 5–9. (In Spanish)

28. Gilbert, Y.; Le Bihan, Y.; Aubry, G.; Veillette, M.; Duchaine, C. Microbiological and molecular characterization of denitrification in biofilters treating pig manure. *Bioresour. Technol.* **2008**, *99*, 4495–4502. [CrossRef] [PubMed]

29. Buelna, G.; Dubé, R.; Turgeon, N. Pig manure treatment by organic bed biofiltration. *Desalination* **2008**, *231*, 297–304. [CrossRef]

30. Beyenal, H.; Lewandowski, Z. Combined Effect of Substrate Concentration and Flow Velocity on Effective Diffusivity in Biofilms. *Water Res.* **2000**, *34*, 528–538. [CrossRef]

31. Patel, A.; Nakhla, G.; Zhu, J. Detachment of multispecies biofilm in circulating fluidized bed reactor. *Biotechnol. Bioeng.* **2005**, *92*, 427–437. [CrossRef] [PubMed]

32. Chowdurry, N.; Nakhla, G.; Zhu, J. Load maximization of a liquid-solid circulating fluidized bed reactor for nitrogen removal from synthetic municipal wastewater. *Chemosphere* **2008**, *71*, 807–815. [CrossRef] [PubMed]

33. Moore, R.; Quarmby, J.; Stephenson, T. The effects of media size on the performance of biological aerated filters. *Water Res.* **2001**, *35*, 2514–2522. [CrossRef]

34. Yu, Y.; Feng, Y.; Qiu, L.; Han, W.; Guan, L. Effect of grain-slag media for the treatment of wastewater in a biological aerated filter. *Bioresour. Technol.* **2008**, *99*, 4120–4123. [CrossRef] [PubMed]

35. Sabbah, I.; Baransia, K.; Massalhaa, N.; Dawasa, A.; Saadic, I.; Nejidat, A. Efficient ammonia removal from wastewater by a microbial biofilm in tuff-based intermittent biofilters. *Ecol. Eng.* **2013**, *53*, 354–360. [CrossRef]

36. Hort, C.; Gracy, S.; Platel, V.; Moynault, L. Evaluation of sewage sludge and yard waste compost as a biofilter media for the removal of ammonia and volatile organic sulfur compounds (VOSCs). *Chem. Eng. J.* **2009**, *152*, 44–53. [CrossRef]

37. Ying-Xu, C.; Jun, Y.; Kai-Xiong, W. Long-term operation of biofilters for biological removal of ammonia. *Chemosphere* **2005**, *58*, 1023–1030.

38. Katukiza, A.; Ronteltap, M.; Niwagaba, C.; Kansiime, F.; Lens, P. A two-step crushed lava rock filter unit for grey water treatment at household level in an urban slum. *J. Environ. Manag.* **2014**, *133*, 258–267. [CrossRef] [PubMed]

39. Mann, R.; Bavor, H. Phosphorus removal in constructed wetlands using gravel and industrial waste substrata. *Water Sci. Technol.* **1993**, *27*, 107–113.

40. Achak, M.; Mandi, L.; Ouazzani, N. Removal of organic pollutants and nutrients from olive mill wastewater by a sand filter. *J. Environ. Manag.* **2009**, *90*, 2771–2779. [CrossRef] [PubMed]

water

Article

Effect of Substrate, Feeding Mode and Number of Stages on the Performance of Hybrid Constructed Wetland Systems

José Alberto Herrera-Melián *, Alejandro Borreguero-Fabelo, Javier Araña,
Néstor Peñate-Castellano and José Alejandro Ortega-Méndez

Institute of Environmental Studies and Natural Resources (i-UNAT), University of Las Palmas de Gran Canaria,
35017 Las Palmas, Spain; a.borreguero@hotmail.com (A.B.-F.); javier.arana@ulpgc.es (J.A.);
nestorp1990@gmail.com (N.P.-C.); alejandro.ortega@ulpgc.es (J.A.O.-M.)
* Correspondence: josealberto.herrera@ulpgc.es

Received: 30 October 2017; Accepted: 2 January 2018; Published: 5 January 2018

Abstract: A hybrid constructed wetland mesocosm has been used for the treatment of raw urban wastewater. The first stage was a mulch-based, subsurface, horizontal flow constructed wetland (HF). The HF achieved good removals of COD (61%; 54 g/m^2·day) and Total Suspended Solids (84%; 29 g/m^2·day). The second stage was composed of vertical flow constructed wetlands (VF) that were employed to study the effect of substrate (gravel vs. mulch), feeding mode (continuous vs. intermittent) and the number of stages (1 vs. 2) on performance. High hydraulic and organic surface loadings (513–583 L/m^2·day and 103–118 g/m^2·day of COD) were applied to the reactors. The mulch was more efficient than gravel for all the parameters analyzed. The continuous feeding allowed a 3 to 6-fold reduction of the surface area required.

Keywords: forest waste; palm mulch; constructed wetlands; vertical flow

1. Introduction

The supply of water has always been a matter of great concern for the inhabitants of the Canary Islands (Spain), particularly in the second half of the 20th century when a remarkable increment of the population coincided with a strong decreasing trend in precipitation [1]. Additionally, the steep orography with altitudes up to 3700 m and the presence of many disseminated small communities, reinforce the idea of the adequacy of non-conventional or decentralized systems for wastewater treatment and reuse on the islands [2].

In the last decades constructed wetlands (CWs) have gained increasing popularity for wastewater treatment in small communities. CWs are easily designed and constructed, and maintenance is simple and economic as it does not require highly skilled personnel or expensive machinery. The cost of domestic wastewater treatment with CWs varies with land price but it can be about 2–3 times lower than that of conventional treatment processes [3]. Besides being highly efficient and robust, CWs also add aesthetic, ecological and cultural values [4]. Life cycle comparisons of CWs vs. activated sludge technology have shown that the former emit less greenhouse gases and cause less environmental impact [5]. However, two important disadvantages of CWs can limit their implantation: the large surface area required and in the case of subsurface flow CWs, the clogging of the substrate [6].

Vertical flow CWs (VFs) require less surface area than horizontal flow CWs (HFs) because of the higher substrate aeration efficiency of the former [7]. Consequently, the applicability of VFs or hybrid systems including VFs is expected to be higher in places where the land is costly or scarce, or if the reclaimed water is intended to be used in irrigation, since water loss by evapotranspiration and the consequent salinity increment will be lower. This is the situation in many regions with

Mediterranean-like weather like the Canary Islands [8–10]. An example of a remarkably efficient VF, capable of treating raw domestic wastewater without primary settling is the so-called "French system". The classical design consists of two stages of unsaturated VFs in series with feeding/rest periods of 1 or 2 weeks [11]. Besides, the environmental impact of VFs is smaller than that of HFs because the former emit fewer greenhouse gases during wastewater treatment and have lower construction requirements. The construction impacts could be significantly reduced by using local materials so that transportation of the wetland substrate would be minimized [5,12]. Nevertheless, VFs can become clogged more easily than HFs because of the use of substrates with smaller particle size [6,13]. Consequently, it has been suggested that in hybrid CWs, the HF should be the first stage and VFs the second one [8].

Gravel is the conventional substrate of CWs. This mineral material supports the attached-growth biomass and plants but has a low capacity for sorption and precipitation [14]. Hence, other materials such as rice husk [15] and peat/crushed pine bark [16] have been successfully tested. Organic substrates have also been used as electron donors for sulfate reducers in passive remediation systems for the treatment of acid mine drainage [17]. Another remarkable application of organic substrates is the treatment of low C/N wastewater as extra carbon is needed to enhance denitrification efficiency [18]. Agricultural and forest organic wastes can be good substrates of CWs. This practice can have several environmental and economic advantages compared with mineral substrates (gravel and sand) by: (i) providing a viable solution to reduce waste materials in a cheap and eco-friendly way, (ii) adding economic value to the waste, (iii) reducing the impact of CWs construction as a renewable, locally abundant material would be used. Additionally, these materials offer an interesting advantage, which is their capacity to work as low cost bio-sorbents [19,20]. Ribé et al. [21] observed that pine bark was able to efficiently remove heavy metals from landfill leachates. Gao et al. [22] studied a 600 m² VF and with a substrate that contained about 37% organic matter including wood turf, organic compost, activated sludge and pine bark. The authors claimed that the substrate had good porosity to prevent clogging which is a fatal threat for the subsurface-flow CW.

The Canarian palm tree (*Phoenix canariensis*) is native to the Canary Islands and has been introduced throughout the world as an ornamental plant. The plant shows good resistance to hot and dry environments and adapts well to drought [23]. Its stipe can reach 20 m in height and 30–40 cm in diameter. The pinnate leaves are 5–6 m in length. Thus, taking into account that palm mulch is an abundant, cheap, renewable material, the main goals of this research were:

- To check the performance of a mulch-based HF as the first stage of a hybrid CW after 3 years in operation.
- Regarding the second stage VF:

  ○ To compare gravel with mulch as substrates for VFs.
  ○ To compare the continuous feeding mode with intermittent feeding mode.
  ○ To determine the number of VFs in series to meet the European legal limits for effluent discharge regarding TSS (35 mg/L) and organic matter (BOD: 25 mg/L, COD: 125 mg/L) [24].

2. Materials and Methods

The influent, raw wastewater from the Campus of the University of Las Palmas de Gran Canaria (Canary Islands, Spain) was collected from a 17-m³ tank with a timer-controlled, triturating pump located at the bottom of the tank. The pump-timer was programmed to function every 2 h (12 times a day) for 1 min. Daily inflows to the primary CWs were determined by measuring the influent volume with graduated recipients.

2.1. Constructed Wetland (CW) Mesocosms

The first stage HF (Figure 1) was built with three 265-L polypropylene recipients (length: 125 cm, height: 57 cm, width: 56 cm, surface area: 0.7 m², Prograrden, Italy). The three recipients contained only palm mulch as substrate and were planted with common reed and papyrus. The HF has been in operation since September 2011 [25]. The surface area of this CW was 2.1 m^2.

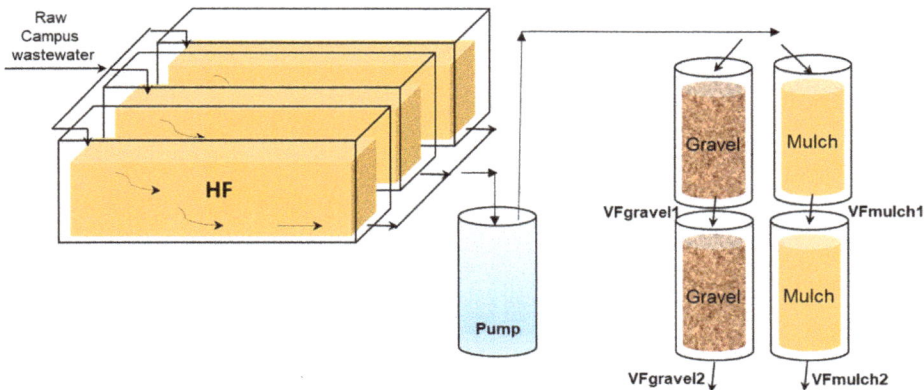

Figure 1. Layout of the hybrid CW: the mulch-based, horizontal flow (HF) and VFs with gravel (VFgravel) and mulch (VFmulch).

The HF1 effluent was collected in a recipient from which it was pumped into two lab-scale VFs containing only gravel (VFgravel) and only palm mulch (VFmulch). The gravel was basaltic with 49% porosity and average diameter of 6.5 mm. Both reactors were composed of two plastic, cylindrical recipients in series, each with a height of 80 cm and a surface area of 0.1 m^2. The mulch was a heterogeneous material obtained from the trituration of dry branches of the Canarian palm tree (*Phoenix canariensis*). The mulch had a porosity of 54% and a hygroscopicity of 10%. These VFs were designed to determine the effect of the influent feeding mode (continuous vs. pulse) and to determine the number of VFs in series to meet the European legal limits for the discharge of treated wastewater into the environment. These reactors were in operation between July and December 2013 and between February and July 2014.

All the reactors were placed outdoors at the Campus of Tafira, Gran Canaria, Canary Islands, Spain (latitude: 28°4' North, longitude: 15°27' West). The height above the sea level is 305.5 m. The climate is spring-like the year round because of the influence of the trade winds. The average summer temperatures are mild (22 °C) and not very different from those of the winter (13 °C minimum). Rainfall is extremely scarce with annual averages ranging between 150 and 200 mm [26]. Evapo-transpiration is about 65% of the average annual rainfall [10].

2.2. Water Analysis

Water quality parameters were measured in unfiltered, homogenized samples as described by standard methods [27]. Hence, total BOD$_5$ and COD were measured. BOD$_5$ (henceforth BOD) can include nitrification as no inhibitor was added. NH$_4^+$, and Na$^+$ ions were determined with selective electrodes from Crison (Barcelona, Spain). PO$_4^{3-}$ ions were dissolved, molybdate-reactive phosphates. Permanent hardness (Ca^{2+} + Mg^{2+}) was determined by the EDTA titrimetric method. The concentration of fecal coliforms (FC) was determined by the membrane filter method and incubation at 44 °C for 24 h with the Chapmann-TTC agar medium.

2.3. Statistics

The statistics applied in this study have been described in detail in [25]. In brief, average values of concentrations, surface loadings and removals were compared by means of the Anova if the data were homocedastic (Bartlett test) and normally distributed (Shapiro-Wilk test). If these conditions were not met the Kruskal-Wallis non parametric test was used. In all cases a significance level of 95% (*p*-value > 0.05) was used. Correlation between variables was tested with Pearson and Spearman tests with the same significance level.

3. Results and Discussion

3.1. Characteristics of the Influent

The influent was raw wastewater from the Campus and included those from cafeterias, laboratories and toilets. The influent was collected from a 17-m^3 tank with a timer-controlled triturating pump placed at the bottom of the tank. The pump timer was programmed to function every day for 1 min every 2 h during all the experimental period. Table 1 shows the features of the influent during the experimental time. According to the concentrations of organic matter, solids and ammonia the influent can be considered a medium to strong urban wastewater. Additionally, the high variability shown by the standard deviation values can be influenced by the daily operations within the Campus.

Table 1. Characteristics of the wastewater used in this work. Average concentrations ± standard deviation and number of data (*n*) between January 2013 and July 2014.

Parameter	Value	Units
BOD$_5$	444 ± 131, *n*: 59	mg/L of O$_2$
COD	552± 162, *n*: 71	mg/L of O$_2$
TSS	252 ± 133, *n*: 64	mg/L
Turbidity	209 ± 97, *n*: 82	NTU
NH$_4$$^+$	68 ± 21, *n*: 60	mg/L
PO$_4$$^{3-}$	34 ± 8, *n*: 18	mg/L
FC	1.91 (±1.48) × 10^7, *n*: 15	CFU/100 mL
Na$^+$	155 ± 40, *n*: 10	mg/L
pH	6.93 ± 0.27, *n*: 29	pH units
Electrical conductivity	1665 ± 590, *n*: 29	mS/cm
Permanent hardness (Ca^{2+} + Mg^{2+})	1.91 ± 0.22, *n*: 10	meq/L

3.2. Performance of the First Stage HF

A mulch-based HF can provide remarkable results in the treatment of urban wastewaters with no evident clogging symptoms even at high surface loading rates (LRs) [25]. However, to our knowledge no research has been devoted to determine performance of mulch-based CWs in the long term. HF has been in operation with different configurations since September 2011. Hence, one of the goals of this study was to determine HF performance and clogging after 3 years. The results considered in this study comprise those obtained between January 2013 and July 2014. Table 2 shows the average LRs (±standard deviation) and removals obtained by the HF.

Table 2. Average LRs and removals (±standard deviation) of the HF.

Parameter	LR	Removal, %
HLR, L/m^2·day	146 ± 52	-
BOD, g/m^2·day	64 ± 23	68 ± 19
COD, g/m^2·day	88 ± 38	61 ± 14
TSS, g/m^2·day	35 ± 39	84 ± 8
Turbidity, NTUxL/m^2·day	65 ± 22	77 ± 12
NH$_4$$^+$, g/m^2·day	9 ± 3	−21 ± 25
PO$_4$$^{3-}$, g/m^2·day	4 ± 1	−11 ± 19
Fecal coliforms, CFU/m^2·day	2.6 (±2.3) × 10^{10}	75 ± 24

As can be observed on Table 2, removals of TSS (84%) and turbidity (78%) were quite good. TSS are the main cause of clogging in CWs [6]. Thus, it is important to achieve high TSS removal if the following treatment stage in the hybrid CW is a VF.

COD removal (61%, 54 g/m²·day) was better than those achieved by conventional HFs and VFs, that range between 10 [28] and 20 g/m²·day [29]. Greater performances have been obtained with non-conventional CWs such as intermittently aerated VFs (57 g/m²·day) [30] and tidal flow CWs (62 g/m²·day) [31] but these reactors require more energy input and/or device implementation.

Ammonia can be removed from CW water by different mechanisms that include volatilization, nitrification or plant uptake [14]. In the present study the average influent concentration of ammonia was 68 mg/L. The ion concentration was increased by 21%. Such increment can be caused by ammonification, i.e., the release of ammonia from organic N, in addition to the lack of enough dissolved oxygen for nitrification. Ammonification process is faster than nitrification [32]. Ammonification occurs in aerobic, facultative and anaerobic conditions but reaction becomes slower with reduced concentrations of dissolved oxygen [14].

No phosphate removal was achieved by HF (−11%). In fact, the increment observed could be caused by desorption from the mulch or by mineralization of organic phosphorus. In this case, phosphate desorption from the substrate is not likely as it has been in operation for 3 years. Figure 2 illustrates the concentrations of TSS, COD, turbidity and fecal coliforms in the effluent of HF vs. LR.

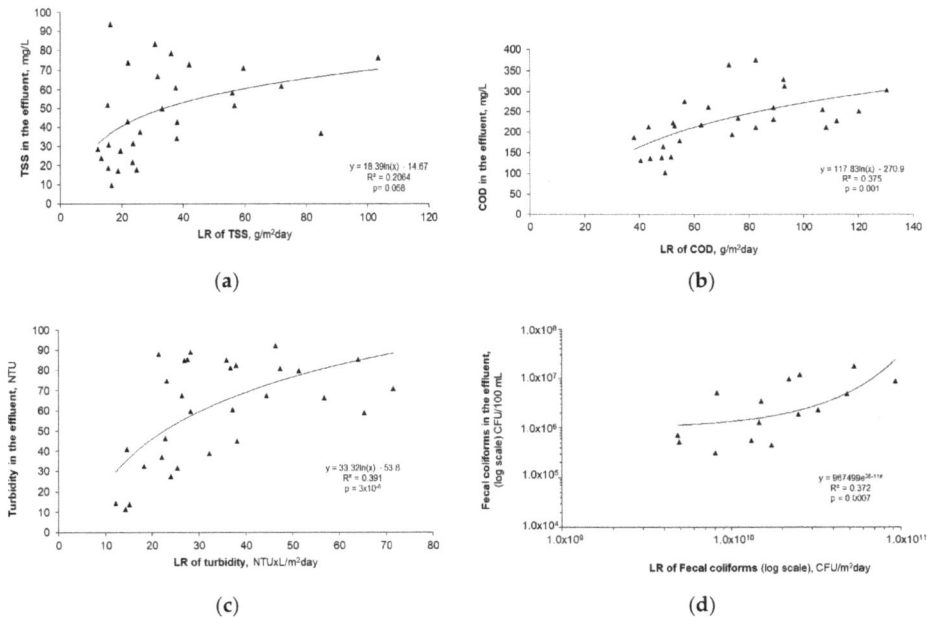

Figure 2. Concentration of (**a**) TSS, (**b**) COD, (**c**) turbidity and (**d**) fecal coliforms in the effluent of HF vs. LR. Values of R^2 and Spearman correlation coefficient are provided.

Although R^2 values were not particularly high, the best correlations between LRs and effluent concentrations were logarithmic for TSS, COD and turbidity (Figure 2). This results shows that the effluent concentrations were increased with LRs until an upper limit. At higher LR values, the effluent concentrations were independent of LRs. In the case of fecal coliforms, the best correlation was exponential, indicating the low robustness of HF regarding the removal of this parameter. In fact, the average removal of coliforms was relatively poor (75%, Table 2).

According to the results obtained in this study (Figure 2), the design guidelines for a mulch-based HF used as a first stage of a hybrid CW, would be 15–30 g/m^2·day for TSS and 40–60 g/m^2·day for COD. These guidelines can be considered conservative as correspond to the lowest LRs used in these experiments. During this 3-year study, mulch (about 10%) has been added to the reactors in several occasions because of degradation, however no clogging symptoms have been observed. This can be explained by the good porosity of the mulch, the presence of the plants, and the rest periods imposed by the low activity in the Campus during the students' holydays (Christmas, Easter and summer). During these rest periods, the fragmentation and degradation of the deposited organic solids should be accomplished. Paing et al. claimed that rest periods seem to be indispensable to achieve their remarkable performance and to delaying clogging in French VFs [11].

3.3. Second Hybrid CW Stage: the VFs

Mulch was compared with gravel as substrates of secondary VFs by determining the effect of the influent feeding mode (pulse versus continuous) and the effect of the number of VFs (1 or 2) on performance. The European legal limits for the discharge of treated wastewater into the environment regarding organic matter (COD: 125 mg/L, BOD: 25 mg/L) and TSS (35 mg/L) were taken as reference.

3.3.1. Effect of the Number of VFs

These experiments were performed between July and November 2013. During this period the reactors were fed continuously with a peristaltic pump with HF1 effluent (Figure 1). Samples were taken in the influent, effluent of the first VFs with gravel (VFgravel1) and mulch (VFmulch1) and in the effluents of the second VFs with gravel (VFgravel2) and mulch (VFmulch2). The HLR of the VFgravel (558 ± 213) L/m^2·day was similar to that of the VFmulch (556 ± 216) L/m^2·day. Note that these HLR are remarkably higher than those used in French VFs (median HLR: 60 L/m^2·day) and German VFs (median HLR: 300 L/m^2·day) [6]. Moreover, in this study all the LR of the VFgravel were similar to those of the VFmulch ($p > 0.2$). Table 3 summarizes the results obtained.

Table 3. Average concentrations (± standard deviation) of COD, BOD, TSS, turbidity and ammonia in the influent and the effluents of the first VFs (VFgravel1 and VFmulch1) and second VFs (VFgravel2 and VFmulch2). The number of data of each sample is 12.

Parameter	Influent	VFgravel1	VFgravel2	VFmulch1	VFmulch2
COD, mg/L	214 (±48)	153 (±46)	113 (±28)	128 (±30)	99 (±32)
BOD, mg/L	170 (±53)	80 (±22)	33 (±8)	59 (±25)	17 (±15)
TSS, mg/L	34 (±22)	9 (±3)	-	4 (±1)	-
Turbidity, NTU	23 (±8)	13 (±1)	4.5 (±2.7)	3.1 (±1.2)	1.3 (±0.5)
NH$_4^+$, mg/L	76 (±19)	40 (±11)	20 (±6)	35 (±20)	13 (±11)

The LR of COD for VFgravel (103 ± 45 g/m^2·day) and VFmulch (118 ± 45 g/m^2·day) were statistically similar. However, the COD concentration of VFgravel1 (153 mg/L) is relatively far from the reference given by the European legislation (125 mg/L) while that of VFmuclh1 was closer (128 mg/L). Nonetheless, there is no significant difference regarding COD effluent concentrations and removals between VFgravel1 and VFmulch1. The effluents of both the second VFs met the COD legal limit, with the VFmulch providing a slightly lower value (VFgravel2: 113 mg/L, VFmulch2: 99 mg/L, Table 3). The presence of the second VFs significantly improved COD removals in both reactors.

The LR of BOD (VFgravel: 82 g/m^2·day, VFmulch: 104 g/m^2·day) were also statistically similar. The average concentrations of BOD in the effluents of VFgravel1 (80 mg/L) and VFmulch1 (59 mg/L, Table 3) were also similar ($p = 0.0787$) and above the reference value of 25 mg/L. Nonetheless, BOD concentrations were significantly reduced to 33 mg/L in VFgravel2 ($p = 0.0005$) and 17 mg/L in VFmulch2 ($p = 0.00027$), being that of VFmulch2 significantly lower. These results indicate the positive effect of the presence of the second VFs and the better performance of the mulch.

Performance regarding turbidity was remarkable because the influent value (23 NTU) was reduced to 13 NTU in VFgravel1 while in VFmulch1 the average turbidity was remarkably lower (3.1 NTU, Table 3, $p = 1.036 \times 10^{-5}$). This might be caused by the notably better filtering and retention capacity of particles and colloidal matter of the mulch in the upper part of the VF. Turbidity in VFmulch2 effluent (1.3 NTU) was also clearly lower than in VFgravel2 (3.1 NTU, $p = 5.336 \times 10^{-7}$) and below the limit recommended by World Health Organization and the U.S. Environmental Protection Agency for water intended for irrigation (2 NTU) [33]. The second VFs significantly improved turbidity removals (VFgravel, $p = 0.0011$; VFmulch, $p = 7.74 \times 10^{-5}$).

The average concentration of TSS of the influent (34 mg/L) was reduced to 9 mg/L in VFgravel1 and to 4 mg/L in VFmulch1 (Table 3), with VFmulch1 being significantly more efficient.

The ammonia concentration of the influent (76 mg/L) was reduced to 40 mg/L in VFgravel1 and 35 mg/L in VFmulch1 (Table 3) with no significant differences between both values. However, the concentration of ammonia in VFgravel2 (20 mg/L) was significantly greater than that of VFmulch2 (13 mg/L). Additionally, ammonia removal was significantly improved with the addition of the second VFs (VFgravel, $p = 2.75 \times 10^{-6}$; VFmulch, $p = 0.0012$). Thus, it is possible to achieve a high enough efficiency to meet the European legal limits with high LR by using two continuously fed, mulch-based CW in series.

3.3.2. Effect of Influent Feeding Mode (Pulse vs. Continuous)

The conventional operation of VFs implies the pulse feeding of the influent into the reactors. This way, flooding-drainage periods are alternated and the substrate aeration and biomass contact with the influent are optimized. Consequently, high performance regarding organic matter, nitrification and bacteria can be obtained [34]. Nonetheless, the continuous feeding of VFs can also provide remarkable results [9]. Consequently, it was decided to determine the effect of the influent feeding mode on the performance of the VFs.

The continuous feeding mode was used from July to December 2013. From February to May 2014 the feeding was made pulse and continuous again from May to July 2014. In this way, the possible "memory effects" and that of temperature would be counteracted. The average HLR in the continuous period, 498 ($\pm$149) L/m^2·day for VFgravel and 576 ($\pm$193) L/m^2·day for VFmulch, were not statistically equal. During the pulse feeding period the HLRs of each reactor, 379 ($\pm$234) L/m^2·day and 313 ($\pm$124) L/m^2·day for the VFgravel and VFmulch, respectively were not significantly different either. The reason for such different HLRs in each period is that the influent volume required to achieve similar HLR to those of the continuous feeding was very high. Hence, the resulting HRT and the corresponding removals were notably low (data not shown). The results from the two continuous periods are considered together. The number of samples was 21 for the continuous period and 17 for the pulse feeding period. Table 4 provides the LR of both reactors and the removals achieved for each feeding mode. Figure 3 shows the concentrations of the influent (effluent of HF1) and those of the effluents of VFmulch2 and VFgravel2 vs. time for the continuous and pulse feeding periods.

In the case of BOD, it is not possible to compare the effect of the feeding mode because the resulting BOD LR for each period were very different (Table 4). Nevertheless, for the continuous feeding period the BOD LR of both reactors are comparable (VFgravel: 76 $\pm$ 48 g/m^2·day, VFmulch: 101 $\pm$ 59 g/m^2·day, $p = 0.1739$). The same applies for the pulse feeding period (VFgravel: 22 $\pm$ 26 g/m^2·day, VFmulch: 17 $\pm$ 16 g/m^2·day, $p = 0.729$). Hence, VFmulch (91%) was statistically better than VFgravel (78%) when the feeding was continuous ($p = 0.007$). This led to significantly lower average BOD effluent concentrations in VFmulch (11 mg/L) than in VFgravel (30 mg/L) for the continuous feeding (Table 4). During the pulse feeding VFmulch (95%) was also statistically better than VFgravel (70%) and the resulting average BOD concentration in VFmulch (3 mg/L) was significantly lower than that of VFgravel (14 mg/L) (Table 4). In the case of the continuous feeding of VFmulch (average LR: 101 g/m^2·day) and considering that 1 person equivalent (PE) corresponds to 60 g BOD/day, the surface area used was 0.6 m^2/PE.

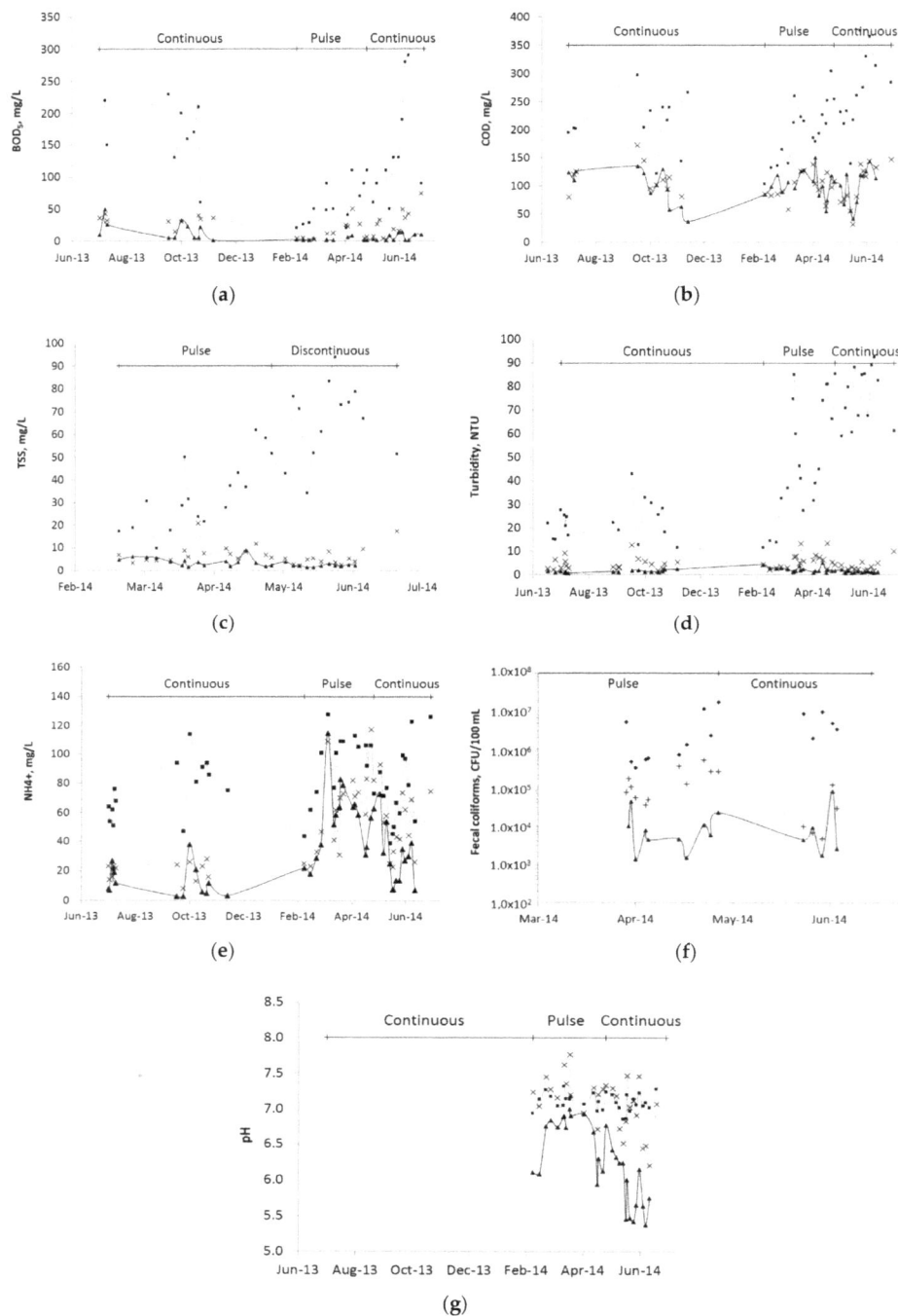

Figure 3. Values in the influent of the secondary constructed wetlands (-■-, dashed line) and effluents of VFmulch2 (-▲-, solid line) and VFgravel2 (-x-, dashed line) of: (**a**) BOD; (**b**) COD; (**c**) TSS; (**d**) turbidity; (**e**) NH$_4^+$; (**f**) fecal coliforms; (**g**) pH.

Table 4. Average LR, concentrations in the effluent (±standard deviation) and removals (in italics) for VFgravel2 and VFmulch2 in the continuous and pulse feeding periods.

LR * Effl. conc. *Removal*	VFgravel2		VFmulch2	
	Continuous	**Pulse**	**Continuous**	**Pulse**
BOD	76 (±48)	22 (±26)	101 (±59)	17 (±16)
	30 (±17)	14 (±14)	11 (±13)	3 (±2)
	78	*70*	*91*	*95*
COD	118 (±44)	81 (±63)	137 (±63)	66 (±39)
	106 (±34)	100 (±23)	99 (±32)	103 (±23)
	54	*47*	*57*	*44*
TSS	25 (±15)	15 (±15)	32 (±20)	11 (±10)
	6 (±4)	8 (±4)	2 (±1)	4 (±2)
	84	*73*	*93*	*84*
Turbidity	26 (±16)	21 (±21)	32 (±21)	17 (±14)
	4.1 (±2.6)	6 (±3)	1.3 (±0.5)	2 (±1)
	85	*85*	*95*	*92*
NH_4^+	39 (±16)	38 (±24)	43 (±21)	30 (±13)
	36 (±22)	64 (±28)	20 (±17)	55 (±25)
	53	*33*	*73*	*41*
PO_4^{3-}-P	6.2 (±1.96)	5.5 (±4.2)	7.2 (±2.9)	4.2 (±2.6)
	10 (±3.9)	10 (±2.9)	11.7 (±5.5)	10.4 (±3.6)
	15	*20*	*−3*	*16*
FC	5.3 (±8) × 10^{10}	2.4 (±5) × 10^{10}	10 (±17) × 10^{10}	1.8 (±3) × 10^{10}
	8.4 (±13) × 10^4	1.8 (±1) × 10^5	9.7 (±16) × 10^4	1.1 (±1.3) × 10^4
	98	*92.6*	*98*	*98.6*

Notes: * Units for removals, concentrations and LR are %, mg/L and g/m^2·day, with the exception of turbidity (NTU and NTUxL/m^2·day) and fecal coliforms (CFU/100 mL and CFU/m^2·day), respectively.

The average COD LR of the VFgravel during the continuous (118 g/m^2·day) and pulse (81 g/m^2·day) feeding modes were significantly different. The same results were obtained in the VFmulch. However, though the COD removals in the continuous mode seem to be greater (VFgravel: 54%, VFmulch: 57%) than those of the pulse feeding mode (VFgravel: 47%, VFmulch: 44%, Table 4), the difference was not significant. Consequently, the COD concentrations in the effluent of both VFs and feeding modes were similar. Thus, it can be concluded that the substrate (gravel vs. mulch) and the feeding mode (pulse vs. continuous) did not have any remarkable effect on COD removal. Nonetheless, most of the COD concentrations of the effluent of both VFs met the European legal limit of 125 mg/L (Figure 3).

Although the LR of TSS of VFgravel (25 g/m^2·day, Table 4) during the continuous feeding was significantly greater than that of the pulse mode (15 g/m^2·day), the removal of TSS (84%) was better (p = 0.0002) than that of the pulse one (73%). The same results were achieved with VFmulch, which obtained a significantly better (p = 0.0001) removal with continuous feeding (93%) than with the pulse one (84%) in spite of the fact that the average LR of TSS in continuous (32 g/m^2·day) was significantly greater than that of the pulse period (11 g/m^2·day). Consequently, the average TSS concentration in the effluent of VFmulch when fed in continuous (2 mg/L) was statistically lower than that of the pulse mode (4 mg/L). In the case of VFgravel, the concentration of TSS in the effluent was lower in the continuous mode (6 mg/L) but not significantly. In conclusion, the best TSS removals were achieved by VFmulch with the continuous feeding mode.

A similar result was obtained for turbidity. The LR of turbidity of VFgravel and VFmulch in both feeding modes were statistically similar. VFmulch improved significantly its performance (p = 0.00036) when fed in continuous (95%) compared with the pulse mode (92%). This was not the case for VFgravel as removals were the same in both feeding modes (85%). Besides, turbidity removal was significantly better with mulch independently of the feeding mode (p = 0.007). The average effluent turbidity values of the VFmulch were improved ($p = 8.9 \times 10^{-5}$) with the continuous regime (1.3 NTU) in comparison with the pulse one (2 NTU).

The average LR of ammonia of VFgravel and VFmulch were statistically similar in both feeding modes (Table 4). VFgravel improved ammonia removal significantly when fed continuously (53%) in comparison to that of the pulse feeding (33%). VFmulch also improved significantly with the continuous feeding (73%) compared with that of pulse one (41%, $p = 2.3 \times 10^{-5}$). Additionally, VFmulch was more efficient than VFgravel with both the continuous and pulse feeding modes. Between February and July 2014, the average pH of the influent of the VFs was 7.10 (± 0.12) and those of VFgravel and VFmulch were 7.09 (± 0.37) and 6.27 (± 0.37), respectively. The stronger acidification of the VFmulch effluent (Figure 3) is in agreement with its higher ammonia removal. Additionally, during the continuous feeding of both reactors their effluent pH were lower (VFgravel: 6.91, VFmulch: 5.86) than those of the pulse feeding period (VFgravel: 7.26, VFmulch: 6.61). These results suggest that the main ammonia removal mechanism was nitrification [35]. The efficiencies regarding COD and NH_4^+ removals of VFmulch with the continuous feeding (Table 4) are comparable to those achieved by Zhao et al. [18] in the treatment of anaerobic digested swine wastewater with a wood-chip-framework soil infiltrator. These authors used LR of COD and NH_4^+ LR of 26–118 g/m^2·day and 22–106 g/m^2·day and achieved removals of 67.5–48% and 82–78%, respectively.

Physical, chemical and biological factors are responsible for the removal of fecal bacteria and pathogens in CWs. Physical factors include mechanical filtration, sedimentation, and sorption to organic matter and the CW's substrate. Chemical factors comprise oxidation and exposure to biocides excreted by plants. Biological factors involve antimicrobial activity of root exudates, predation, retention in biofilms, natural die-off, etc. [36]. The LR of FC of VFgravel in the continuous mode (5.3×10^{10} CFU/m^2·day, Table 4) was significantly greater than that of the pulse one (2.4×10^{10} CFU/m^2·day). The same result was obtained with VFmulch. Yet, the removals of VFmulch in continuous and pulse feeding, and VFgravel in continuous (98.2%, 98.6% and 98%, respectively) were similar and superior to that of conventional VFgravel in the pulse feeding period (92.6%). Note that the lowest removal of FC coincided with those of TSS and ammonia (Table 4). This result suggests that both the mulch and the continuous feeding improved the effect of media filtration [37] and aeration [38]. Saeed and Sun [35] also observed improved removal of FC in wood mulch in comparison with gravel in bench-scale VFs. The authors attributed it to the effect of aerobic conditions on the promotion of the growth of heterotrophic protozoa and *E. coli* cell oxidation.

Regarding phosphate-P removal the differences shown by the VFs (Table 4) are not relevant. No significant differences were found between influent and effluent concentrations for both reactors fed in continuous nor in pulse. Removals were not significantly different with either the continuous or the pulse feedings. Thus, it can be concluded that the capacity of these reactors to remove dissolved phosphate-P was nil. Similarly to other forest and agricultural waste/by-products, the low efficiency of the palm mulch in the removal of the negatively charged phosphate ions can be attributed to the abundant availability of negatively charged functional groups (e.g., -OH, -COOH) and the absence of positively charged functional groups (e.g., -NH$_2$) on its surface [20]. Nevertheless, far from being a disadvantage the modest efficiency of CWs in the removal of N and P can be regarded as a way to decrease the demand for expensive inorganic fertilizers in agriculture [39]. For instance, García-Delgado et al. [40] saved considerable amounts of fertilizer (37% N, 66% P and 12% K) by applying treated urban wastewater in pepper cultivation. In fact, wastewater effluent reuse has been widely implemented in many countries with Israel and California (USA) leading wastewater reuse with 65–70% of the wastewater reused in agriculture [41].

These results show that palm mulch is a better substrate than gravel for VFs and that the continuous feeding mode can improve performance of VFs, particularly that of conventional (gravel-based, pulse fed) ones. Note that the goal of the pulse feeding of VFs is to improve the substrate aeration and the reactor efficiency. Nevertheless, the obtained results show that a continuous feeding can yield better results. Thus, it seems that the continuous feeding provided enough oxygen to the substrate. In fact, the pulse feeding requires adding larger volumes of water in shorter time periods to achieve the same HLR. This probably results in shorter HRTs and consequently worse performance. The combination of HF1 with VFmulch in the continuous feeding period provided remarkable removals of COD (82%), BOD (97%), TSS and turbidity (99%), ammonia (65%) and FC (99.8%) and nil of phosphate (-4%).

The better efficiency of the mulch with respect to gravel can be explained by considering the particular characteristics of the former. In this regard, one the most important features of the mulch is its small particle size and compressibility. A smaller particle size provides longer HRT and improved water distribution on the reactor surface, better retention of particles (TSS), micelles (turbidity) and bacteria, including the heterotrophic bacteria responsible for the degradation of dissolved BOD, nitrifying bacteria and fecal coliforms. In addition to this, Paing et al. indicated that the presence of a sludge layer on the surface of French VFs improved performance [11]. In the present case, the accumulation of sludge on the surface of VFmulch was more evident than in VFgravel. Saeed and Sun [35] found that eucalyptus wood mulch was a better substrate than gravel for VFs as improved removals of total nitrogen, organic matter and *E. coli* were obtained. The authors concluded that the higher void volume percentage of the organic substrate provided higher oxygen transfer efficiency. Another interesting feature of the mulch is its hygroscopicity which was determined to be about 10% in weight. A highly hygroscopic substrate would improve plant root growth, biofilm establishment and stability, providing longer HRT and consequently better treatment performance [42]. Moreover, the smaller particle size and hygroscopicity of the mulch can help to better distribute the influent inside the reactor, thus reducing the negative effects of shortcuts and preferential paths.

4. Conclusions

The first stage of a hybrid CW, a mulch-based HF, has shown a remarkable performance with no symptoms of clogging for 3 years.

Experiments with the hybrid CW second stage, the bench-scale VFs have shown that:

- palm mulch is a better substrate than gravel for VFs,
- with two vertical VFs in series the European legal limits regarding COD, BOD and TSS can be met even when high LR are applied,
- the continuous influent feeding mode significantly improved performance.

The high efficiency of the reactors studied makes them particularly adequate for places where palm mulch is available, evapotranspiration is high and the reclaimed water is intended to be reused for irrigation.

Acknowledgments: This work was partially funded by the Department of Chemistry of the University of Las Palmas de Gran Canaria. The English was revised by Colm Sullivan.

Author Contributions: The idea of this article was developed by José Alberto Herrera-Melián and Alejandro Borreguero-Fabelo. Sampling and analyses were performed by Alejandro Borreguero-Fabelo and Néstor Peñate-Castellano. The paper was written by José Alberto Herrera-Melián, Javier Araña and José Alejandro Ortega-Méndez.

Conflicts of Interest: The authors declare no conflict of interest.

References

1. García-Herrera, R.; Gallego, D.; Hernández, E.; Gimeno, L.; Ribera, P.; Calvo, N. Precipitation trends in the Canary Islands. *Int. J. Climatol.* **2003**, *23*, 235–241. [CrossRef]
2. Martín, I.; Betancort, J.R.; Pidre, J.R. Contribution of non-conventional technologies for sewage treatment to improve the quality of bathing waters (ICREW project). *Desalination* **2007**, *215*, 82–89. [CrossRef]
3. Rousseau, D.P.L.; Vanrolleghem, P.A.; De Pauw, N. Constructed wetlands in Flanders: A performance analysis. *Ecol. Eng.* **2004**, *23*, 151–163. [CrossRef]

4. Mander, Ü.; Dotro, G.; Ebie, Y.; Towprayoone, S.; Chiemchaisri, C.; Nogueira, S.; Jamsranjav, B.; Kasaka, K.; Truua, J.; Tournebize, J.; et al. Greenhouse gas emission in constructed wetlands for wastewater treatment: A review. *Ecol. Eng.* **2014**, *66*, 19–35. [CrossRef]

5. Dixon, A.; Simon, M.; Burkitt, T. Assessing the environmental impact of two options for small-scale wastewater treatment: Comparing a reedbed and an aerated biological filter using a life cycle approach. *Ecol. Eng.* **2003**, *20*, 297–308. [CrossRef]

6. Knowles, P.; Dotro, G.; Nivala, J.; García, J. Clogging in subsurface-flow treatment wetlands: Occurrence and contributing factors. *Ecol. Eng.* **2011**, *37*, 99–112. [CrossRef]

7. Zurita, F.; De Anda, J.; Belmont, M.A. Treatment of domestic wastewater and production of commercial flowers in vertical and horizontal subsurface-flow constructed wetlands. *Ecol. Eng.* **2009**, *35*, 861–869. [CrossRef]

8. Masi, F.; Martinuzzi, N. Constructed wetlands for the Mediterranean countries: Hybrid systems for water reuse and sustainable sanitation. *Desalination* **2007**, *215*, 44–55. [CrossRef]

9. Herrera-Melián, J.A.; Martín Rodríguez, A.J.; Araña, J.; González Díaz, O.; González Henríquez, J.J. Hybrid constructed wetlands for wastewater treatment and reuse in the Canary Islands. *Ecol. Eng.* **2010**, *36*, 891–899. [CrossRef]

10. Peñate, B.; Martel, G.; Vera, L.; Márquez, M.; Gutiérrez, J.; Moreno, E.; del Castillo, G.; Farrujia, I. El Agua en Canarias. Ed. Canarian Technological Institute, Department of Water. 2013. Available online: http://islhagua.org/c/document_library/get_file?p_l_id=23769&folderId=23758&name= DLFE-1002.pdf (accessed on 3 January 2018).

11. Paing, J.; Guilbert, A.; Gagnon, V.; Chazarenc, F. Effect of climate, wastewater composition, loading rates, system age and design on performances of French vertical flow constructed wetlands: A survey based on 169 full scale systems. *Ecol. Eng.* **2015**, *80*, 46–52. [CrossRef]

12. Fuchs, V.J.; Mihelcic, J.R.; Gierke, J.S. Life cycle assessment of vertical and horizontal flow constructed wetlands for wastewater treatment considering nitrogen and carbon greenhouse gas emissions. *Water Res.* **2011**, *45*, 2073–2081. [CrossRef] [PubMed]

13. Song, X.; Ding, Y.; Wang, Y.; Wang, W.; Wang, G.; Zhou, B. Comparative study of nitrogen removal and bio-film clogging for three filter media packing strategies in vertical flow constructed wetlands. *Ecol. Eng.* **2015**, *74*, 1–7. [CrossRef]

14. Vymazal, J. Removal of nutrients in various types of constructed wetlands. *Sci. Total Environ.* **2007**, *380*, 48–65. [CrossRef] [PubMed]

15. Tee, H.C.; Seng, C.E.; Noor, A.M.; Lim, P.E. Performance comparison of constructed wetlands with gravel- and rice husk-based media for phenol and nitrogen removal. *Sci. Total Environ.* **2009**, *407*, 3563–3571. [CrossRef] [PubMed]

16. Wang, R.Y.; Korboulewsky, N.; Prudent, P.; Domeizel, M.; Rolando, C.; Bonin, G. Feasibility of using an organic substrate in a wetland system treating sewage sludge: Impact of plant species. *Bioresour. Technol.* **2010**, *101*, 51–57. [CrossRef] [PubMed]

17. Gibert, O.; de Pablo, J.; Cortina, J.L.; Ayora, C. Chemical characterization of natural organic substrates for biological mitigation of acid mine drainage. *Water Res.* **2004**, *38*, 4186–4196. [CrossRef] [PubMed]

18. Zhao, B.; Li, J.; Leu, S. An innovative wood-chip-framework soil infiltrator for treating anaerobic digested swine wastewater and analysis of the microbial community. *Bioresour. Technol.* **2014**, *173*, 384–391. [CrossRef] [PubMed]

19. Bhatnagar, A.; Sillanpaa, M. Utilization of agro-industrial and municipal waste materials as potential adsorbents for water treatment—A review. *Chem. Eng. J.* **2010**, *157*, 277–296. [CrossRef]

20. Nguyen, T.A.H.; Ngo, H.H.; Guo, W.S.; Zhang, J.; Liang, S.; Lee, D.J.; Nguyen, P.D.; Bui, X.T. Modification of agricultural waste/by-products for enhanced phosphate removal and recovery: Potential and obstacles. *Bioresour. Technol.* **2014**, *169*, 750–762. [CrossRef] [PubMed]

21. Ribé, V.; Nehrenheim, E.; Odlare, M.; Gustavsson, L.; Berglind, R.; Forsberg, Å. Ecotoxicological assessment and evaluation of a pine bark biosorbent treatment of five landfill leachates. *Waste Manag.* **2012**, *32*, 1886–1894. [CrossRef] [PubMed]

22. Gao, R.Y.; Shao, L.; Li, J.S.; Guo, S.; Han, M.Y.; Meng, J.; Liu, J.B.; Xu, F.X.; Lin, C. Comparison of greenhouse gas emission accounting for a constructed wetland wastewater treatment system. *Ecol. Inform.* **2012**, *12*, 85–92. [CrossRef]

23. Sajdak, L.M.; Velazquez-Martí, B.; Lopez-Cortés, I. Quantitative and qualitative characteristics of biomass derived from pruning *Phoenix canariensis* hort. ex Chabaud. and *Phoenix dactilifera* L. *Renew. Energy* **2014**, *71*, 545–552. [CrossRef]

24. Directive, EU Urban Wastewater. Council directive 91/271/EEC of 21 May 1991, concerning urban waste water treatment (91/271/EEC). *Off. J. Eur. Communities* **1991**, 40–52. Available online: http://eur-lex.europa.eu/legal-content/EN/TXT/?uri=CELEX:31991L0271 (accessed on 3 January 2018).

25. Herrera-Melián, J.A.; González-Bordón, A.; Martín-González, M.A.; García-Jiménez, P.; Carrasco, M.; Araña, J. Palm tree mulch as substrate for primary treatment wetlands processing high strength urban wastewater. *J. Environ. Manag.* **2014**, *139*, 22–31. [CrossRef] [PubMed]

26. Santana Rodríguez, J.J.; Santana Hernández, F.J.; González González, J.E. The effect of environmental and meteorological variables on atmospheric corrosion of carbon steel, copper, zinc and aluminium in a limited geographic zone with different types of environment. *Corros. Sci.* **2003**, *45*, 799–815. [CrossRef]

27. American Public Health Association (APHA). *Standard Methods for the Examination of Water and Wastewater*, 21st ed.; American Public Health Association: Washington, DC, USA, 2005.

28. Albuquerque, A.; Arendacz, M.; Gajewska, M.; ObarskaPempkowiak, H.; Randerson, P.; Kowalik, P. Removal of organic matter and nitrogen in a horizontal subsurface flow (HSSF) constructed wetland under transient loads. *Water Sci. Technol.* **2009**, *60*, 1677–1682. [CrossRef] [PubMed]

29. Konnerup, D.; Koottatep, T.; Brix, H. Treatment of domestic wastewater in tropical, subsurface flow constructed wetlands planted with Canna and Heliconia. *Ecol. Eng.* **2009**, *35*, 248–257. [CrossRef]

30. Fan, J.; Wang, W.; Zhang, B.; Guo, Y.; Ngo, H.H.; Guo, W.; Zhang, J.; Wu, H. Nitrogen removal in intermittently aerated vertical flow constructed wetlands: Impact of influent COD/N ratios. *Bioresour. Technol.* **2013**, *143*, 461–466. [CrossRef] [PubMed]

31. Zhi, W.; Yuan, L.; Ji, G.; He, C. Enhanced long-term nitrogen removal and its quantitative molecular mechanism in tidal flow constructed wetlands. *Environ. Sci. Technol.* **2015**, *49*, 4575–4583. [CrossRef] [PubMed]

32. Kadlec, R.H.; Knight, R.L. *Treatment Wetlands*; CRC Press: Boca Raton, FL, USA, 1996; p. 893.

33. Sengupta, M.E.; Keraita, B.; Olsen, A.; Boateng, O.K.; Thamsborg, S.T.; Pálsdóttir, G.R.; Dalsgaard, A. Use of *Moringa oleifera* seed extracts to reduce helminth egg numbers and turbidity in irrigation water. *Water Res.* **2012**, *46*, 3646–3656. [CrossRef] [PubMed]

34. Zurita, F.; White, J.R. Comparative Study of Three Two-Stage Hybrid Ecological Wastewater Treatment Systems for Producing Reclaimed Water for Agricultural Reuse. *Water* **2014**, *6*, 213–228. [CrossRef]

35. Saeed, T.; Sun, G. Enhanced denitrification and organics removal in hybrid wetland columns: Comparative experiments. *Bioresour. Technol.* **2011**, *102*, 967–974. [CrossRef] [PubMed]

36. Wu, S.; Carvalho, P.N.; Müller, J.A.; Manoj, V.R.; Dong, R. Sanitation in constructed wetlands: A review on the removal of human pathogens and fecal indicators. *Sci. Total Environ.* **2016**, *541*, 8–22. [CrossRef] [PubMed]

37. Karim, M.R.; Manshadi, F.D.; Karpiscak, M.M.; Gerba, C.P. The persistance and removal of enteric pathogens in constructed wetlands. *Water Res.* **2004**, *38*, 1831–1837. [CrossRef] [PubMed]

38. Headley, T.; Nivala, J.; Kassa, K.; Olsson, L.; Wallace, S.; Brix, H.; van Afferden, M.; Müller, R. Escherichia coli removal and internal dynamics in subsurface flow ecotechnologies: Effects of design and plants. *Ecol. Eng.* **2013**, *61*, 564–574. [CrossRef]

39. Almuktar, A.; Scholz, M.; Al-Isawi, R.; Sani, A. Recycling of domestic wastewater treated by vertical-flow wetlands for irrigating Chillies and Sweet Peppers. *Agric. Water Manag.* **2015**, *149*, 1–22. [CrossRef]

40. García-Delgado, C.; Eymar, E.; Contreras, J.I.; Segura, M.L. Effects of fertigation with purified urban wastewater on soil and pepper plant (*Capsicum annuum* L.) production, fruit quality and pollutant contents. *Span. J. Agric. Res.* **2012**, *10*, 209–221. [CrossRef]

41. Kihila, J.; Mtei, K.M.; Njau, K.N. Wastewater treatment for reuse in urban agriculture; the case of Moshi Municipality, Tanzania. *Phys. Chem. Earth* **2014**, *72–75*, 104–110. [CrossRef]

42. Bruch, I.; Fritsche, J.; Bänninger, D.; Alewell, U.; Sendelov, M.; Hürlimann, H.; Hasselbach, R.; Alewell, C. Improving the treatment efficiency of constructed wetlands with zeolite-containing filter sands. *Bioresour. Technol.* **2011**, *102*, 937–941. [CrossRef] [PubMed]

water

MDPI

Article

High-Strength Domestic Wastewater Treatment and Reuse with Onsite Passive Methods

José de Anda [1,*], Alberto López-López [1,†], Edgardo Villegas-García [1] and Karla Valdivia-Aviña [2]

[1] Departamento de Tecnología Ambiental, Centro de Investigación y Asistencia en Tecnología y Diseño del Estado de Jalisco, A. C. Av. Normalistas 800, Colinas de la Normal, Guadalajara CP 44270, Jalisco, Mexico; evillegas@ciatej.mx
[2] Centro Universitario de Ciencias Biológicas y Agropecuarias, Universidad de Guadalajara, Camino Ramón Padilla Sánchez, Nextipac 2100, Zapopan, Jalisco, Mexico; kkk3_04@hotmail.com
* Correspondence: janda@ciatej.mx; Tel.: +52-33-3345-5200
† Died on 13 March 2017 in a car accident.

Received: 10 October 2017; Accepted: 19 December 2017; Published: 25 January 2018

Abstract: This paper describes the preliminary monitoring results of an onsite pilot wastewater treatment plant consisting of a septic tank, an anaerobic up-flow filter, and a horizontal subsurface flow wetland system planted with *Agapanthus africanus*. The system was designed to treat heavily polluted domestic wastewater produced in a research and development (R&D) center, reaching additional goals of zero energy consumption and eliminating the use of chemical additives. First water quality data shows that organic load in the treated sewage were removed achieving more than 95% efficiency. Nutrients were removed by almost 50%, and fecal and total coliform counts decreased by 99.96%. The results were compared to official Mexican regulations for wastewater discharged into lakes and reservoirs complied with all of them except for nutrients. In this pilot project, the resulting treated wastewater was directly reused for watering the green areas of the R&D center. The result was that the excess of nutrients improved the quality of the grass, avoiding the use of synthetic fertilizers, and created a wetland habitat for small wildlife species living in the area.

Keywords: water treatment; passive treatment systems; anaerobic processes; constructed wetlands; ornamental plants; treated wastewater reuse

1. Introduction

Wastewater treatment approaches vary from conventional centralized systems to entirely decentralized and clustered systems. The centralized systems, which are usually publicly owned, collect and treat large volumes of wastewater for entire large communities, thus requiring large pipes, major excavations, and manholes for access. While decentralized systems collect, treat, and reuse/dispose of treated wastewater on site or near the generation point, centralized systems often reuse/dispose of treated wastewater far from the generation point [1].

Maintenance and operation (M&O) costs associated with wastewater treatment include labor, energy, purchase of chemicals, and equipment replacement. Conventional centralized technologies normally require high amounts of energy due to the complexity of the processes which combine mechanical, chemical, and biological stages to remove contaminants in the sewage. Additionally, these systems also require further energy to treat and transport the produced biological sludge [2].

In developing countries, the treatment of domestic, commercial, or industrial wastewaters has become an important issue in recent years because of the increase of M&O costs involved in conventional wastewater treatment plants (WWTP) [3,4]. Thus, several WWTP facilities in developing countries have started to reduce their operation capacity, suspend operation, or end up being abandoned [4–6]. Therefore, decentralized wastewater treatment technologies based on anaerobic processes and constructed wetlands is attracting interest as a potential solution to reduce M&O costs [7–9].

This paper describes the performance of a functional pilot treatment process based on a combined anaerobic process and a horizontal subsurface flow wetland planted with the ornamental plant *Agapanthus africanus*, which uses zero energy consumption and no chemical additives. The treated wastewater comes from a food research and development (R&D) center and has high contents of organic pollutants and is rich in nitrogen because the sewages are mixed with nontoxic wastes produced in laboratories and pilot plants. As a result, the produced wastewater could be classified as high-strength domestic wastewater compared with common domestic wastewater according to the literature [10].

2. Background

The strategy of treating sewage by common and known aerobic processes has been shifted back to anaerobic processes in recent years with the advent of high rate anaerobic systems such as up-flow anaerobic sludge blanket reactors (UASB), anaerobic contact processes, anaerobic filters (AF), or fixed film reactors and fluidized bed reactors [11]. The high rate anaerobic processes have several advantages such as: low capital investment, lower M&O costs, energy recovery in the form of biogas, operational simplicity, and low production of digested sludge [11].

It is reported in warm tropical countries that, for domestic sewage, the UASB system is the best option for biological oxygen demand (BOD) removal due to the high attainable efficiencies and that a low BOD load, but the efficiencies of low BOD load removal could increase if the combination of a septic tank (ST) followed by an up-flow anaerobic filter (UAF) is used, as shown in Table 1 [12]. Therefore, most of the treatment systems based on anaerobic processes and constructed wetlands reported in the literature for warm tropical countries use the UASB as a preliminary treatment step prior to a constructed wetland [13–15]. Nevertheless, the use of up-flow anaerobic filters (UAF) is also widely used to treat municipal wastewater [16–19]. It was found few research works reporting use of a coupled ST with an UAF previous to the constructed wetland to improve the performance of the system [20,21]. Anaerobic ponds followed by constructed wetlands are also a convenient solution, especially for developing countries, due to their cost-effectiveness and high potential of removing different pollutants. However, these systems have to be installed far from residential areas due to the odor release, they need a larger surface area to construct them due to the higher residence time required, and algae could bloom in the ponds causing secondary pollution of the following stream [22–24].

Table 1. Removal efficiencies in anaerobic systems treating domestic sewage [12].

Anaerobic System	Effluent BOD [a] (mg/L)	BOD Removal Efficiency [a] (%)
Anaerobic pond	70–160	40–70
UASB reactor	60–120	55–75
Septic tank	80–150	35–60
Imhoff tank	80–150	35–60
Septic tank followed by anaerobic filter	40–60	75–85

[a] Ranges of effluent concentration and typical removal efficiencies based on Brazilian experience. Lower efficiency limits are usually associated with poorly operated systems.

Constructed wetlands with surface flow (SF CWs), horizontal sub-surface flow (HF CWs), vertical sub-surface flow (VF CWs), or hybrid systems have been used together with previous anaerobic and/or aerobic systems for wastewater treatment for at least 30 years [25,26]. By far the most frequently used plant around the globe to plant constructed wetlands with horizontal subsurface flow is *Phragmites australis* (Common reed). Species of the genera *Typha* (*latifolia*, *angustifolia*, *domingensis*, *orientalis*, and *glauca*) and *Scirpus* (e.g., *lacustris*, *validus*, *californicus*, and *acutus*) spp. are other commonly used species [27]. On the other hand, in many countries, and especially in the tropics and subtropics, local plants including ornamental species are used for HF CWs such as *Zantedeschia aethiopica* (giant white arum lily), *Strelitzia reginae* (crane flower, bird of paradise), *Anthurium andraenum* (flamingo flower), and *Agapanthus africanus* (agapanthus) [27]. Based on a search of the literature and in previous experience, it was concluded that it is possible to use ornamental plants in constructed wetlands without reducing the efficiency of the treatment system [28–31].

In the case of pilot experiments where a UASB with HF CW systems was used, the reviewed reports focus basically on estimating the performance of the HF CW system but water quality data of previous steps of the process are not included [32,33]. Thus, removal efficiencies in the reviewed HF CW systems planted with conventional or ornamental species achieves values up to 80% for COD, BOD_5, and total suspended solids (TSS) in most of the reviewed works [32,33]. The major removal mechanism for nitrogen in HF CWs is denitrification. Removal of ammonia is limited due to lack of oxygen in the filtration bed because of permanent waterlogged conditions [34]. The ammonia-N removal efficiency reported in the literature achieves values up to 65% and 45% for nitrate NO_3-N removal [34,35]. Phosphorus is removed primarily by ligand exchange reactions, where phosphate displaces water or hydroxyls from the surface of iron and aluminum hydrous oxides. Unless special materials are used, removal of P is usually lower to 50% in HF CWs [32–35].

Regarding the operation and maintenance (O&M) costs of tertiary treatment CWs with reuse purposes, they are lower than those of secondary treatment ones, not only because of the lower intensity of processes (lower loading rates) but also because of certain investment returns such as plant harvesting, aquaculture, production of ornamental plants, etc. [36].

3. Materials and Methods

The system was designed to treat wastewater from a food research and development (R&D) public institution located in the municipality of Zapopan, in the state of Jalisco, Mexico. The design of the present treatment system is based on experience gained in a demonstrative pilot plant installed previously in Chapala, Jalisco, Mexico [29]. In the sewage pipes, black and gray water are mixed together with discharges of non-hazardous liquid wastes generated in laboratories and food processing pilot plants. Because it is a functional pilot treatment plant discharging intermittently different type of wastes related to food industry, the content of solids, organic matter, and nutrients was not controlled at the entrance. The designed system consists basically of a septic tank (ST), an up-flow anaerobic filter (UAF), and a subsurface horizontal flow constructed wetland (HF CW) (see Figure 1).

It is convenient to mention that, due to failures in the connection of pluvial piping network during the construction of the pilot treatment plant, some rain water eventually entered into the piping system that conducts the sewage to the treatment plant. This situation created some efficiency problems for the system as will be discussed later.

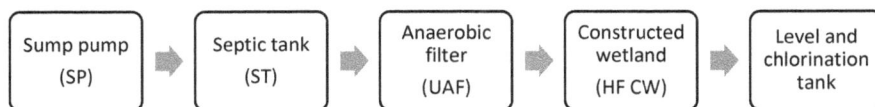

Sump pump (SP)	⇒	Septic tank (ST)	⇒	Anaerobic filter (UAF)	⇒	Constructed wetland (HF CW)	⇒	Level and chlorination tank

Figure 1. Block diagram describing the treatment process.

At the time when this work was carried out, about 150 people worked at this R&D center including researchers, administrative, and maintenance personnel, and students. All the sewage is directed to a sump pump (SP). Solar panels were installed to power the pump located in the pumping sump and in the receiving tank at the end of the system to irrigate the green areas of the R&D center. The average flow at the entrance was estimated to be 7.5 m^3/day (Q_i) based on the use and discharge of about 50 L per person per day. This amount was calculated from the monthly average consumption of water used directly in the buildings, laboratories, and pilot plants of the R&D center. Figure 1 displays the sequence of the treatment process.

The sump pump (SP) has a stainless-steel basket to trap coarse solids in the wastewater. The wastewater is pumped from the sump pump (SP) to a two-chamber septic tank (ST) as shown in Figure 2. The wastewater flows by gravity to an up-flow anaerobic filter (UAF) and then also by gravity to a horizontal subsurface flow constructed wetland (HF CW) as shown in Figure 3. The level tank (LT) at the end of the process controls the height of the water table in the constructed wetland. Immediately after the level tank, a disinfection system was installed which works with chlorination tablets. Finally, the treated water is stored in a plastic tank with a volume capacity of 10 m³ from which the treated water is pumped to irrigate the green areas of the R&D center. The pump installed for irrigation is also sun powered, making it a zero-energy consumption system.

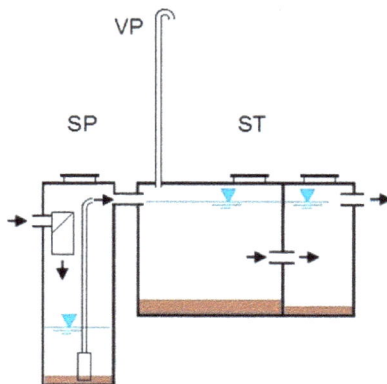

Figure 2. Wastewaters entering to the sump pump (SP). The waste waters are pumped to a septic tank (ST) which has a vent pipe (VP) to permit biogas generated during the anaerobic decomposition to exit.

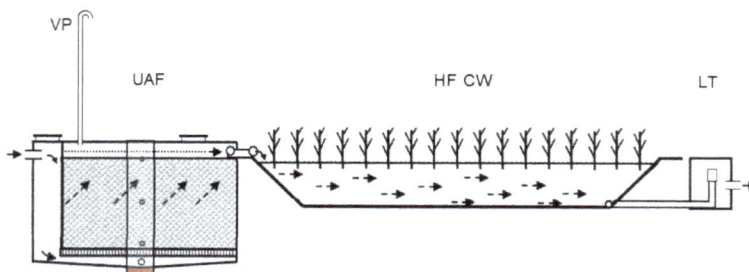

Figure 3. Wastewaters flowing to an up-flow anaerobic filter (UAF) and then to a horizontal subsurface constructed wetland (HF CW).

The septic tank (ST) is divided into two chambers: the first one is 11.3 m³ (V_1), the second chamber is 7.1 m³ (V_2). They are connected by two 3″ diameter pipes installed equidistant in the middle of both chambers and 1.22 m in height from the bottom. At the bottom of each chamber there is a sump that accumulates the biological sludge generated by the anaerobic bacteria. The amount of accumulated sludge in each chamber was partially removed every six months by using a vacuum pump to avoid the excess of solids in both chambers of the septic tank. Care was taken to leave part of the sludge at the bottom of each chamber to maintain a minimum of active methanogenic bacteria within the system. The removed sludge could be disposed of in a filtering bed. After a few days of sun exposure, the dried sludge could be used as fertilizer.

The up-flow anaerobic filter (UAF) is divided into two chambers, the function of the first one is to receive and distribute the wastewater from the bottom into a wider second chamber filled with a volcanic porous rock, known as lava rock, which is commonly called "tezontle" in Mexico [29]. The second chamber is the anaerobic up-flow chamber where the wastewater flows from the bottom to the top of the chamber through the porous media created with the use of tezontle. The fixed bed has a volume of 69.5 m^3 (V_3). The resulting porosity of the fixed bed is about 0.69. There are three homogeneous layers of volcanic rock settled in the filtering bed ranging from 3" average diameter at the bottom, 2" average diameter in the middle, and 1" diameter at the top. At the bottom of the anaerobic filter, a sump was constructed that collects the biological sludge. The sludge must be removed every six months by using a vacuum pump. The removed sludge can be disposed of in the same filtering bed used to treat the sludge of the septic tank (ST).

Tezontle was selected as filling material in the UAF since it is an inert volcanic material relatively abundant in the central portion of the country, it is neutral in pH, has a high porosity, is physically stable, does not contain nutrients, and is rich in minerals like calcium, iron, and zinc [33]. Two months after starting the operation, it was observed that a biofilm was created around the particles of tezontle and there was a constant bubbling due to the anaerobic digestion of organic matter.

The constructed wetland has a surface area of about 336 m^2 (A), and it is filled with 0.7 m of tezontle with an average diameter of 3/8". In the operation of the wetland, the hydraulic depth (δ) was 0.6 m. The total volume of the constructed wetland was 201.6 m^3 (V_4) and the resulting porosity of the filtering bed was about 0.68. The control level tank (LT) has a cubic shape and measures 0.40 m per side. The tube in the level tank (LT) was adjusted to control the water level in the constructed wetland 10 cm below the surface of the filling material. Tezontle was used as filling material in the HF CW because it is commonly used as substrate in the hydroponic production of ornamental commercial flowers and tomatoes in Mexico due to the richness of mineral content which is absorbed selectively by the plants [37,38].

The system was put into operation and after three months the constructed wetland was planted with African agapanthus (*Agapanthus africanus*) with a density of about three plants per square meter in a quincunx arrangement giving an approximated number of 1000 plants. It took about three months for plants to adapt to the system. After the level tank (LT), a disinfection system based on chlorination was installed to control the pathogens that could be present at the end of the treatment process.

After one year of a clear adaptation of the vegetation was observed, the process of monitoring water quality was initiated. Twelve water quality parameters where monitored every three months at the entrance and outlet of the system over a year period. All water quality parameters were determined by using accredited official Mexican norms that are in accordance with the standard methods for the examination of water and wastewater [39].

Water quality parameters were monitored at the entrance of the system, at the outlet of every treatment unit, and at the end of the treatment process after disinfection. Heavy metals were not monitored since potable water is used for the diverse services of the R&D center, and it was not contaminated with hazardous materials. The monitoring protocol was applied one year after its construction and operation of the treatment plant during the months of January, April, July, and October of the year 2016 and on January of 2017. The samples were taken only once in each reported month. A duplicate sample was taken at the beginning and at the end of every treatment unit. Samples were transported and analyzed in the laboratory following accredited standard methods for water examination [39]. The hydraulic residence time (HRT) was considered as constant along the system for purposes of analysis and discussion. The average use of water per month was calculated according with water consumption data of the R&D center finding slight variations in monthly water consumption. Pluvial water enters occasionally at the inlet of the system throughout the year. It was unavoidable that the precipitation fell over the surface of the wetland since it was constructed in an open area. The municipality of Zapopan is part of the metropolitan area of Guadalajara, it is located at 1548 m above sea level and its climate is sub-humid, with winters and dry and temperate springs. The average temperature is 23.5 °C, with a maximum of 35 °C and a minimum of 5.4 °C. The average annual rainfall is 906.1 mm, and it rains mainly between the months of June to October. During the rainfall period, the intensity of rainfall in Zapopan occurs mainly around midnight and in the early hours of the day. According with the nearest meteorological station located in the city of Guadalajara, the average monthly rainfall from January 2016 to January 2017 is shown in Figure 4.

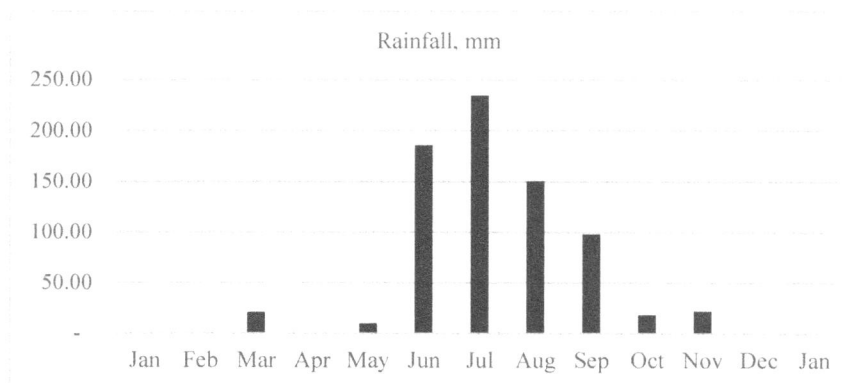

Figure 4. Amount of rainfall during the period of January 2016 to January 2017 in the meteorological station located in Guadalajara.

4. Results

Measured water quality parameters are shown in Tables 2–6. According to Table 2 the mean fats, oil, and grease (FOG) during the monitoring period was 37.04 ± 8.97 mg L^{-1}, settleable solids (SS) was 2.18 ± 2.37 mg L^{-1}, total suspended solids (TSS) was 204.50 ± 89.19 mg L^{-1}, biological oxygen demand (BOD) was 505.1 ± 202.0 mg L^{-1}, chemical oxygen demand (COD) was 987.9 ± 295.4 mg L^{-1}, and the total phosphorus (TP) was 11.98 ± 2.08 mg L^{-1}. A high concentration of total nitrogen (TN) was also found of 196.56 ± 91.13 mg L^{-1}. Presence of pathogens increased markedly during the month of July due to non-desirable input of rain water to the inlet of the system which mixed with wastewater in the sump pump. The temperature was not reported since the samples were taken according to the environmental conditions at midday which was an average of 23.5 °C with a maximum of 35 °C. In general, the range of measured values in the water quality parameters overpass by far the values reported in the literature corresponding to domestic wastewater and it is closer to the so named 'high-strength domestic wastewater' [10]. The reason of high standard deviation in BOD, COD, and TSS as well as in TN is because some of the activities of the R&D center generates sewages of different content in solid and organic matter depending on the type of raw material processed in the laboratories and pilot plants. The R&D areas of the center are food technology, industrial biotechnology and plant biotechnology. The activities of the plant biotechnology department in particular use nitrogen rich compounds in the laboratory essays.

It is noteworthy that pathogens indicators increase abnormally in July at the entrance of the system (see Table 2). This phenomenon was probably due to the entrance of pluvial water to the system which dragged out part of the settled solids in the sump pump. Therefore, during this month, also at the outlet of the septic tank (ST) and in the up-flow anaerobic filter (UAF) an abnormal increase in the number of the pathogens indictors was measured (Tables 3 and 4). This increase in the count of pathogen indicators was already observed at the outlet of the constructed wetland (HF CW) (Table 5).

Table 2. Water quality parameters at the entrance of the septic tank (ST) (SD = standard deviation).

Water Quality Parameter	Unit	2016				2017	Average and SD
		Jan	Apr	Jul	Oct	Jan	
pH	-	8.5	7.5	6.3	7.0	8.4	7.54 ± 0.93
Floating material	-	Presence	Presence	Presence	Presence	Presence	Presence
Turbidity	NTU	158.0	87.3	88.5	196.0	223.0	150.56 ± 61.69
Fats, oil, and grease	mg L^{-1}	40.9	44.9	38.8	21.6	39.0	37.04 ± 8.97
Settleable solids	mg L^{-1}	3.0	0.8	0.5	0.6	6.0	2.18 ± 2.37
Total Suspended Solids	mg L^{-1}	112.5	157.5	152.5	290.0	310.0	204.50 ± 89.19
Biological Oxygen Demand	mg L^{-1}	854.0	426.0	329.5	458.0	458.0	505.10 ± 202.03
Chemical Oxygen Demand	mg L^{-1}	1453.0	863.5	666.5	1061	895.5	987.90 ± 295.37
Total Nitrogen	mg L^{-1}	121.0	295.1	296.2	122.0	148.5	196.56 ± 91.13
Total Phosphorus	mg L^{-1}	14.2	13.9	12.0	10.0	9.8	11.98 ± 2.08
Total Coliforms ($\times 10^5$)	CFU/100 mL	920	210	1600	920	920	914.00 ± 491.51
Fecal Coliforms ($\times 10^5$)	CFU/100 mL	540	820	1600	540	540	808.00 ± 459.04
E. coli ($\times 10^5$)	CFU/100 mL	240	820	1600	540	540	748.00 ± 518.57
Residual chlorine	mg L^{-1}	<0.1	<0.1	<0.1	<0.1	<0.1	CELL
Color	Pt-Co.	312	739	376	959	1574	792.00 ± 511.29

Table 3. Water quality parameters at the outlet of the septic tank (SD = standard deviation).

Water Quality Parameter	Unit	2016				2017	Average and SD
		Jan	Apr	Jul	Oct	Jan	
pH	-	7.1	6.9	6.7	7.1	7.1	6.98 ± 0.18
Floating material	-	Absence	Absence	Absence	Absence	Absence	CELL
Turbidity	NTU	53.0	53.0	75.8	116.0	147.0	88.96 ± 41.41
Fats, oil, and grease	mg L^{-1}	1.9	5.8	27.6	5.56	19.4	12.05 ± 10.95
Settleable solids	mg L^{-1}	0.3	0.4	0.3	0.2	1.5	0.54 ± 0.54
Total Suspended Solids	mg L^{-1}	45.0	90.0	45.0	165.5	110.0	91.10 ± 50.36
Biological Oxygen Demand	mg L^{-1}	157.0	136.5	235.5	276.0	177.5	196.50 ± 57.80
Chemical Oxygen Demand	mg L^{-1}	458.0	554.0	476.5	534.0	465.5	497.60 ± 43.44
Total Nitrogen	mg L^{-1}	111.5	505.7	270.8	120.7	140.0	229.74 ± 167.16
Total Phosphorus	mg L^{-1}	12.4	12.5	11.4	9.4	6.8	10.50 ± 2.41
Total Coliforms ($\times 10^5$)	CFU/100 mL	35	49	220	79	170	110.60 ± 80.63
Fecal Coliforms ($\times 10^5$)	CFU/100 mL	3.1	49	47	33	170	60.42 ± 63.94
E. coli ($\times 10^5$)	CFU/100 mL	3.1	49	47	33	170	60.42 ± 63.94
Residual chlorine	mg L^{-1}	<0.1	<0.1	<0.1	<0.1	<0.1	CELL
Color	Pt-Co.	302	499	354	631	1075	572.20 ± 309.13

Table 4. Water quality parameters at the outlet of the up-flow anaerobic filter (UAF) (SD = standard deviation).

Water Quality Parameter	Unit	2016				2017	Average and SD
		Jan	Apr	Jul	Oct	Jan	
pH	-	7.1	7.1	6.8	7.1	7.4	7.10 ± 0.21
Floating material	-	Absence	Absence	Absence	Absence	Absence	-
Turbidity	NTU	32.8	32.8	14.5	3.7	13.5	19.46 ± 12.89
Fats, oil, and grease	mg L^{-1}	0.9	1.87	4.2	1.1	0.8	1.77 ± 1.42
Settleable solids	mg L^{-1}	0	0	0	0	0	-
Total Suspended Solids	mg L^{-1}	22.5	17.5	12.5	27.5	21.5	20.30 ± 5.63
Biological Oxygen Demand	mg L^{-1}	35.0	29.4	31.5	19.0	25.5	28.08 ± 6.13
Chemical Oxygen Demand	mg L^{-1}	239	245.5	52.5	56.5	126.5	144.00 ± 94.42
Total Nitrogen	mg L^{-1}	120.5	261	230.1	145.3	144.0	180.10 ± 61.46
Total Phosphorus	mg L^{-1}	12.0	7.0	9.1	6.9	4.3	7.86 ± 2.87
Total Coliforms ($\times 10^5$)	CFU/100 mL	7.9	2.6	540	1.3	6.4	4.55 ± 3.11
Fecal Coliforms ($\times 10^5$)	CFU/100 mL	4.3	2.6	350	0.2	4.3	2.85 ± 1.94
E. coli ($\times 10^5$)	CFU/100 mL	4.3	2.6	350	0.2	4.3	2.85 ± 1.94
Residual chlorine	mg L^{-1}	<0.1	<0.1	<0.1	<0.1	<0.1	-
Color	Pt-Co.	207	126	97	82	193	152.00 ± 58.54

Table 5. Water quality parameters at the outlet of constructed wetland (SD = standard deviation).

Water Quality Parameter	Unit	2016				2017	Average and SD
		Jan	Apr	Jul	Oct	Jan	
pH	-	7.0	7.0	6.9	7.1	7.4	7.08 ± 0.19
Floating material	-	Absence	Absence	Absence	Absence	Absence	-
Turbidity	NTU	1.7	1.7	4.2	3.6	2.5	2.74 ± 1.13
Fats, oil, and grease	mg L^{-1}	0.1	0.1	3.6	1.1	0.6	1.10 ± 1.46
Settleable solids	mg L^{-1}	0	0	0	0	0	-
Total Suspended Solids	mg L^{-1}	12.5	5.5	4.5	13.0	10.0	9.10 ± 3.93
Biological Oxygen Demand	mg L^{-1}	15	10.3	8	8.0	3.0	8.86 ± 4.35
Chemical Oxygen Demand	mg L^{-1}	36.7	20.7	32.7	21.2	25.8	27.42 ± 7.08
Total Nitrogen	mg L^{-1}	98.8	123.1	100.3	105.0	75.5	100.54 ± 17.02
Total Phosphorus	mg L^{-1}	11	6.1	4.4	6.0	3.3	6.16 ± 2.95
Total Coliforms ($\times 10^5$)	CFU/100 mL	0.017	0.068	0.79	0.078	1.3	0.37 ± 0.62
Fecal Coliforms ($\times 10^5$)	CFU/100 mL	0.0078	0.018	0.49	0.002	1.3	0.33 ± 0.65
E. coli ($\times 10^5$)	CFU/100 mL	0	0.018	0.49	0.002	1.3	0.33 ± 0.65
Residual chlorine	mg L^{-1}	<0.1	<0.1	<0.1	<0.1	<0.1	-
Color	Pt-Co.	36	45	19	44	40	41.25 ± 4.11

Table 6. Overall efficiency of the system in the pollutants removal and compliance with the regulations.

Water Quality Parameter	Unit	Inlet to the Septic Tank	Outlet of the HF CW	Removal Efficiency	After Chlorination	NOM-001/003 [a]	Compliance with the Regulation
Temperature	°C	-	-	-	-	40	√ [c]
pH	-	7.54	7.08	-	7.02	5–10	√
Floating material	-	Presence	Absence	-	Absence	Absence	√
Turbidity	NTU	150.56	2.74	98.2%	1.78	NS [b]	
Fats, oil, and grease	mg L^{-1}	37.04	1.10	97.0%	0.98	15	√
Settleable solids	mg L^{-1}	2.18	-	100.0%	-	1	√
Total Suspended Solids	mg L^{-1}	204.50	9.10	95.6%	5.90	30	√
Biological Oxygen Demand	mg L^{-1}	505.10	8.86	98.2%	8.70	30	√
Chemical Oxygen Demand	mg L^{-1}	987.90	27.42	97.2%	23.88	NS	
Total Nitrogen	mg L^{-1}	196.56	100.54	48.9%	103.06	15	NA [d]
Total Phosphorus	mg L^{-1}	11.98	6.16	48.6%	5.32	5	NA
Total Coliforms ($\times 10^5$)	CFU/100 mL	914.00	0.37	99.96%	-	NS	
Fecal Coliforms ($\times 10^5$)	CFU/100 mL	808.00	0.33	99.96%	-	0.01	PA [e]
E. coli ($\times 10^5$)	CFU/100 mL	748.00	0.33	100.0%	-	NS	
Residual chlorine	mg L^{-1}	<0.1	<0.1	-	0.34	NS	
Color	Pt-Co.	792.00	41.25	94.8%	28.20	NS	

[a] Maximum permitted water quality values for treated water discharge to natural lakes or reservoirs which later are used for urban public uses, measured as monthly average. [b] NS = Not specified by the Mexican regulation. [c] "√" Means in accordance with the regulation. [d] "NA" Means not in accordance with the regulation. [e] "PA" Means partially in accordance with the regulations since fecal coliforms were controlled after the HF CW by chlorination.

4.1. Septic Tank

By applying Equation (1) we have an estimation of the residence time (τ) in the septic tank of about 2.45 days. In Equation (1), V_1 is the volume of the first chamber of the septic tank, V_2 is the volume of the second chamber, and Q_i is the inlet flow. The residence time could diminish eventually during the rainy season due to the increase of flow at the inlet of the system.

$$\tau = \frac{(V_1 + V_2)}{Q_i} \tag{1}$$

Comparing the data of the water quality parameters from Tables 2 and 3, the septic tank could remove 67.5% of FOG, 75.2% of the SS, 55.5% of TSS, 61.1% of the BOD$_5$, 49.6% of the COD, and 12.4% of TP. On the other hand, TN increased 16.9%. In a conventional septic tank, organic nitrogen in household wastes is transformed into ammonia products under the anaerobic conditions of the septic tank (ammonification). Some of the organic nitrogen, however, is not degraded and becomes part of the sludge at the bottom of the septic tank [40]. However, the abnormal increase of measured total nitrogen concentration could be due to an introduction of excess of nitrogen-rich compounds during some tests carried out in the R&D facilities. On the other hand, reduction of pathogens was very effective in the first stage of the treatment process achieving an average of 90.6% removal of all measured pathogens.

4.2. Anaerobic Filter

To estimate the residence time of the up-flow anaerobic filter (UAF) we used Equation (2) where V_3 is the total volume of the packed bed in the filter (69.5 m^3) and ε is the average porosity of the filling material (0.69). By applying Equation (2) we calculated a residence time in UAF of about 6.4 days. The UAF considered a high hydraulic residence time to ensure the maximum removal of the organic matter before it arrives to the wetland.

$$\tau = \frac{V_3 \, \varepsilon}{Q_i} \tag{2}$$

Comparing water quality conditions among the measured outflow of septic tank (Table 3) and UAF (Table 4) it was concluded that this treatment unit reduced 85.3% of the FOG, 100% of the SS, 77.7% of the TSS, 85.7% of the BOD, 71.1% of COD, 21.6% of TN, and 25.1% of TP. In this second treatment step, the pathogens indicators were reduced an average of 95.6%. Due to the abnormal increase of pathogens indicators explained before, to calculate the efficiency of the UAF regarding bacteria removal, the measured pathogens indicators during July were not included, otherwise the efficiency of bacteria removal in this unit falls to negative values (Table 4).

4.3. Constructed Wetland

To estimate the residence time of the horizontal flow constructed wetland (HF CW), we used Equation (3) where V_4 is the total volume of the constructed wetland (201.6 m^3), ε_w is the average porosity of the filling material in the wetland (0.68), and E is the estimated loss of water due to evaporation and evapotranspiration through the plants in the wetland according to the average local climatic conditions. Based on the findings of Headley et al. [41], measuring the rate of evapotranspiration from subsurface horizontal flow wetlands planted with *Phragmites australis* in sub-tropical environment, an average loss of 10% of water by evapotranspiration from soil and plants was considered. By applying Equation (3) we got a residence time in the HF CW of 11.75 days.

$$\tau = \frac{V_4 \, \varepsilon_w}{Q_i} (1 - E) \tag{3}$$

Comparing water quality conditions in the measured outflow of the UAF (Table 4) and in the HF CW (Table 5) it was observed that this unit could reduce 38.0% of the FOG, 55.2% of the TSS, 68.4% of the BOD, 81.0% of COD, 44.2% of TN, and 21.6% of TP. In the third treatment step, the pathogen indicators were reduced by an average of 88.5%. As it was explained before, the increase of pathogens indicators in July happened because of the rainfall that entered the inlet of the system which removed part of the pathogens settled in the septic tank. If we leave out the values measured in July, the pathogen removal efficiency of the system increases to 90.0%, fulfilling the requirements of the official Mexican regulations (NOM-001-SEMARNAT-1996) without using chlorination [42].

4.4. Outlet of the Desinfection Stage

The final process of the treatment system is the disinfection which works with chlorination tablets and is installed immediately after the level tank (LT) (see Figure 3). The disinfection practically reduces the pathogen indicators to zero, fulfilling the Mexican official regulations [42]. In the case of discharges to surface waterbodies, the upper limit for FC is 1000 CFU/100 mL, and for the reuse of treated wastewater the upper limit is 240 CFU/100 mL in the case of direct contact with persons and 1000 for indirect or occasional contact with persons. Basically, we observe the same removal efficiencies obtained at the outflow of the constructed wetland for FOG, TSS, BOD, and COD. Nutrients remain practically the same but there is a reduction to zero of each of the pathogen indicators.

5. Discussion

The passive treatment system was designed and constructed during the period of 2014–2015 to support the treatment needs of an R&D center producing high-strength domestic wastewater. An amount of close to 120,000 USD was invested in the construction of this system. After almost three years of operating the system, the M&O costs of the system were estimated to be around 700 USD per year. The actual savings in water for irrigation were estimated to be close to 2000 USD per year. Additionally, during the period of study the system discharges consistently complied almost with all parameters observed by the official Mexican standards NOM-001-SEMARNAT-1997 and NOM-003-SEMARNAT-1997 [42]. These norms establish the limits of wastewater parameters in treated wastewater discharged into surface water bodies and those reused for public services respectively [42]. Table 6 shows the average water quality parameters reached at the end of the process after the chlorination stage and the estimated overall efficiency of the system.

As explained throughout this paper, the system was designed to treat highly polluted domestic wastewater. Therefore, the actual hydraulic retention times (HRT) in each stage of the treatment process are relatively high compared to those recommended in the literature for septic tanks of one to three days [40,43], 2 to 96 h for UAF units treating domestic or rural sewages [16–19], 10 to 20 days for high rate anaerobic digesters treating high-strength wastewater [44], and the range of two to seven days in horizontal subsurface constructed wetlands [45]. In the pilot treatment plant, it was established a HRT for the septic tank of 2.45 days which is in accordance with the literature. For the UAF, a HRT of 6.4 days was established which is 1.6 times higher than the best HRT suggested for UAF systems but in the middle of those suggested to treat high-strength wastewater. Finally, the HF CW was designed with a HRT of 11.75 days which is double too high to that suggested by the literature to treat domestic wastewater. The content of total nitrogen in domestic wastewater is in the range of 20 mg L^{-1} to 50 mg L^{-1} and the content of total phosphorus is in the range of 5 mg L^{-1} to 15 mg L^{-1} [10]. As shown in Table 2, total nitrogen at the inlet of the system is three to six times higher and total phosphorus is in the upper limit. It was expected that the longer the residence time designed for the HF CW system could capture most of the nitrogen and phosphorus entering to the system.

According with the results shown in Table 7, the system efficiently removes most of the contaminants entering to the system except for nutrients. Phosphorus concentration was reduced almost to the limits permitted by the Mexican regulations for treated wastewater discharged to surface waterbodies [42]. Nitrogen was reduced only to almost 50% but the end concentration is far from the permitted concentration discharged to surface waterbodies [42].

Table 7. Water quality parameters at the outlet of the disinfection system (SD = standard deviation).

| Water Quality Parameter | Unit | 2016 | | | | 2017 | Average and SD |
		Jan	Apr	Jul	Oct	Jan	
pH	-	7.3	7.0	6.5	7.2	7.2	7.02 ± 0.31
Floating material	-	Absence	Absence	Absence	Absence	Absence	-
Turbidity	NTU	1.70	1.70	4.2	3.6	2.5	1.78 ± 0.22
Fats, oil, and grease	mg L^{-1}	0.1	1.02	2.9	0.8	0.1	0.98 ± 1.15
Settleable solids	mg L^{-1}	0	0	0	0	0	-
Total Suspended Solids	mg L^{-1}	9.0	4.5	2.5	12.0	1.5	5.90 ± 4.46
Biological Oxygen Demand	mg L^{-1}	15	10.3	8	7.2	3.0	8.70 ± 4.40
Chemical Oxygen Demand	mg L^{-1}	36.1	18.62	23.6	16.3	24.8	23.88 ± 7.67
Total Nitrogen	mg L^{-1}	83.5	140.9	116.8	99.6	74.5	103.06 ± 26.60
Total Phosphorus	mg L^{-1}	11.0	5.6	4	4	2	5.32 ± 3.42
Total Coliforms ($\times 10^5$)	CFU/100 mL	0	0	0	0	0	-
Fecal Coliforms ($\times 10^5$)	CFU/100 mL	0	0	0	0	0	-
E. coli ($\times 10^5$)	CFU/100 mL	0	0	0	0	0	-
Residual chlorine	mg L^{-1}	0.3	0.3	0.4	0.2	0.5	0.34 ± 0.11
Color	Pt-Co.	20	41	16	38	26	28.20 ± 10.96

On the other hand, overabundance or deficiency of available nitrogen are both problematic for grass plants. Excessive levels of nitrogen: (1) stimulate rapid shoot growth while slowing down root growth and increasing the need for more frequent mowing; (2) deplete the plant's carbohydrate reserves more rapidly, which in turn can result in less stress tolerance and slower recovery from any injury to the plant; (3) result in thinner, more succulent leaf tissue, which increases moisture loss and therefore creates a greater need for water; (4) can predispose the plant to greater insect and disease problems; (5) contribute to more rapid and excessive thatch development; (6) leach through the soil beyond the root system, potentially polluting groundwater resources when not used by the grass plant [46]. To date, the grass of the R&D Center shows healthy development in both rainy and dry season. However, it will be necessary to monitor the subsurface land to measure if the excess of nitrogen is leaching and polluting the underground.

Later monitoring data shows that the system still works satisfactorily removing chemical organic matter, even with COD loads of 2560 mg L^{-1} with an efficiency removal of 93.9%. In counterpart, the efficiency in nutrients removal barely achieved 50% for both nitrogen and phosphorus. Since the treated wastewater is directly reused for irrigation, the excess of nutrients results in benefits to the green areas of the R&D center, because the use of synthetic fertilizers is not required. The quantification of the number of plants produced per square meter per month in the HF CW is still pending but preliminary counts establish that about 50% of the planted wetland (about 500 plants) produce at least one lateral bud per month.

6. Conclusions

The elevated organic content of high-strength domestic wastewater makes aerobic treatment systems uneconomical. High-strength domestic wastewater was preferably treated anaerobically, thus providing a potential for energy generation while producing low surplus sludge [44]. Additionally, in the tested pilot plant project it was possible to produce successfully *Agapanthus africanus* as an ornamental plant.

As explained by previous authors, the efficiency in the removal of pollutants from high-strength domestic wastewater by using anaerobic processes are mainly controlled by the hydraulic residence time selected in the design of the anaerobic units and in the HF CW [44,45]. The studied wastewater treatment process efficiently reduced the FOG, TSS, BOD, and COD loads from the R&D center sewage, meeting the water quality standards requested by Mexican regulations [42]. The present results are in accordance with the experience of similar treatment systems reported for tropical climates [12]. Through this work, it is possible to extend the principles of BOD and COD removal reported for domestic wastewater to high-strength domestic wastewater [44].

Discharged Total-N concentration was close to seven times higher compared with the requested official standard which controls the treated wastewater discharged to surface water bodies [42]. Most of the total-N concentration was removed in the HF CW and the results show that, despite the longer hydraulic residence time, it was not sufficient to remove it to values below 50%. In counterpart, total-P was very close to fulfil the limits of the official regulations [42]. Since treated wastewaters were used for irrigation, chlorination was necessary at the end of the treatment process to meet environmental regulations regarding fecal coliforms [42]. Total-N removal was safely solved by the reuse of treated wastewater in grass irrigation, but in case of discharges to surface waterbodies it will be necessary to use combined aerobic and anaerobic processes to improve the denitrification process [47].

Nowadays, the pilot treatment system still works close to the facilities where R&D activities are regularly carried out without releasing offensive odors. The community at the R&D center enjoys an environmentally friendly area because they preserve green areas all year long, a nice view was created specially during the flowering time of the African agapanthus, and a habitat was created within the constructed wetland, where several species of birds, lizards, butterflies, and bees are frequent visitors to this artificial ecosystem.

Acknowledgments: The authors acknowledge the "Centro de Investigación y Asistencia en Tecnología y Diseño del Estado de Jalisco, A. C." (CIATEJ) for financing the monitoring work during the period of study. We particularly acknowledge the work done by the team of the Department of Analytical Chemistry at CIATEJ. We are also very thankful for Dana Erickson's help in reviewing the grammar of this work.

Author Contributions: José de Anda and Alberto López-López are the authors of the technology described in this work. Edgardo Villegas-García implemented the sampling protocols and followed up with the analytical procedures during the period of study.

Conflicts of Interest: The authors declare no conflict of interest. The technology involved in this paper was already protected by "Centro de Investigación y Asistencia en Tecnología y Diseño del Estado de Jalisco, A. C." under the law of the Mexican Institute of Industrial Protection, number MX/a/2010/014332.

References

1. Massoud, M.A.; Tarhini, A.; Nasr, J.A. Decentralized approaches to wastewater treatment and management: Applicability in developing countries. *J. Environ. Manag.* **2009**, *90*, 652–659. [CrossRef] [PubMed]
2. Muga, H.E.; Mihelcic, J.R. Sustainability of wastewater treatment technologies. *J. Environ. Manag.* **2008**, *88*, 437–447. [CrossRef] [PubMed]
3. Libralato, G.; Ghirardini, A.V.; Avezzù, F. To centralise or to decentralise: An overview of the most recent trends in wastewater treatment management. *J. Environ. Manag.* **2012**, *94*, 61–68. [CrossRef] [PubMed]
4. Noyola, A.; Morgan-Sagastume, J.M.; Güereca, L.P. *Selección de Tecnologías Para el Tratamiento de Aguas Residuales Municipales. Guía de Apoyo Para Ciudades Pequeñas y Medianas*; Universidad Nacional Autónoma de México, Instituto de Ingeniería: Mexico City, México, 2013; p. 140. ISBN 978-607-02-4822-1. (In Spanish)
5. Abdel-Halim, W.; Weichgrebe, D.; Rosenwinkel, K.-H.; Verink, J. Sustainable sewage treatment and re-use in developing countries. In Proceedings of the Twelfth International Water Technology Conference, IWTC12 2008, Alexandria, Egypt, 1 January 2008; pp. 1397–1409. Available online: http://www.iwtc.info/2008_pdf/15-2.PDF (accessed on 20 August 2017).
6. De Anda, J.; Shear, H. Searching a sustainable model to manage and treat wastewater in Jalisco, Mexico. *Int. J. Dev. Sustain.* **2017**, *5*, 278–294.
7. Sing, N.K.; Kazami, A.A.; Starkl, M. A review on full-scale decentralized wastewater treatment systems: Techno-economical approach. *Water Sci. Technol.* **2015**, *71*, 468–474. [CrossRef] [PubMed]
8. Wua, H.; Zhang, J.; Ngo, H.H.; Guo, W.; Hub, Z.; Liang, S.; Fan, J.; Liu, H. A review on the sustainability of constructed wetlands for wastewater treatment: Design and operation. *Bioresour. Technol.* **2015**, *175*, 594–601. [CrossRef] [PubMed]
9. ElZein, Z.; Abdou, A.; Abd ElGawad, I. Constructed Wetlands as a Sustainable Wastewater Treatment Method in Communities. *Procedia Environ. Sci.* **2016**, *34*, 605–617. [CrossRef]
10. Tchobanoglous, G.; Stensel, H.D.; Tsuchihashi, R.; Burton, F.L.; Abu-Orf, M.; Bowden, G.; Pfang, W. *Wastewater Engineering: Treatment and Reuse*, 5th ed.; McGraw-Hill: New York, NY, USA, 2014; p. 2018.
11. Khan, A.A.; Gaur, R.Z.; Kazmi, A.A.; Lew, B. *Sustainable Post Treatment Options of Anaerobic Effluent, Biodegradation—Engineering and Technology*; Chamy, R., Ed.; InTech: Rijeka, Croatia, 2013. [CrossRef]
12. Chernicharo, C.A.L. Post-treatment options for the anaerobic treatment of domestic wastewater. *Rev. Environ. Sci. Bio/Technol.* **2006**, *5*, 73–92. [CrossRef]
13. Kaseva, M.E. Performance of a sub-surface flow constructed wetland in polishing pre-treated wastewater—A tropical case study. *Water Res.* **2004**, *38*, 681–687. [CrossRef] [PubMed]
14. Mbuligwe, S.E. Comparative effectiveness of engineered wetland systems in the treatment of anaerobically pre-treated domestic wastewater. *Ecol. Eng.* **2004**, *23*, 269–284. [CrossRef]
15. Hamouri, B.E.; Nazih, J.; Lahjouj, J. Subsurface-horizontal flow constructed wetland for sewage treatment under Moroccan climate conditions. *Desalination* **2007**, *215*, 153–158. [CrossRef]
16. Przywara, L.; Mrowiec, B.; Suschka, J. The Application of Anaerobic Filter for Municipal Wastewater Treatment. *Chem. Pap.* **2000**, *54*, 159–164.
17. Bodík, I.; Herdová, B.; Drtil, M. The use of up-flow anaerobic filter and AnSBR for wastewater treatment at ambient temperature. *Water Res.* **2002**, *36*, 1084–1088. [CrossRef]
18. Manariotis, I.D.; Grigoropoulos, S.G. Municipal-Wastewater Treatment Using Up-flow-Anaerobic Filters. *Water Environ. Res.* **2006**, *78*, 233–242. [CrossRef] [PubMed]
19. Ladu, J.L.C.; LÜ, X.-W. Effects of hydraulic retention time, sewage temperature and effluent recycling on efficiency of up-flow anaerobic filter reactor in treating rural domestic sewage. *Int. J. Waste Resour.* **2016**, *6* (Suppl. S3). [CrossRef]

20. Villegas-Gómez, J.D.; Jhonniers-Guerrero, E.; Castaño-Rojas, J.M.; Paredes-Cuervo, D. Septic Tank (ST)-Up Flow Anaerobic Filter (UAF)-Subsurface Flow Constructed Wetland (SSF-CW) systems aimed at wastewater treatment in small localities in Colombia. *Rev. Téc. Fac. Ing. Univ. Zulia* **2006**, *29*, 269–281.

21. Nguyen, A.V.; Pham, N.T.; Nguyen, T.H.; Morel, A.; Tonderski, K. Improved septic tank with constructed wetland, a promising decentralized wastewater treatment alternative in Vietnam. In Proceedings of the Paper XI-RCS-07-30 NOWRA 16th Annual Technical Education Conference & Exposition, Baltimore, Maryland, 10–14 March 2007.

22. Kadlec, R.H. Pond and wetland treatment. *Water Sci. Technol.* **2003**, *48*, 1–8. [PubMed]

23. Senzia, M.A.; Mashauri, D.A.; Mayo, A.W. Suitability of constructed wetlands and waste stabilisation ponds in wastewater treatment: Nitrogen transformation and removal. *Phys. Chem. Earth* **2003**, *28*, 1117–1124. [CrossRef]

24. Peng, J.-F.; Wang, B.-Z.; Wang, L. Multi-stage ponds-wetlands ecosystem for effective wastewater treatment. *J. Zhejiang Univ. Sci. B* **2005**, *6*, 346–352. [CrossRef] [PubMed]

25. Brix, H. Use of constructed wetlands in water pollution control: Historical development, present status, and future perspectives. *Water Sci. Technol.* **1994**, *30*, 209–223.

26. Vymazal, J.; Brix, H.; Cooper, P.F.; Perfler, R.; Laber, J. Removal mechanisms and types of constructed wetlands. In *Constructed Wetlands for Wastewater Treatment in Europe*; Vymazal, J., Brix, H., Cooper, P.F., Green, M.B., Eds.; Backhuys Publishers: Leiden, The Netherlands, 1998; pp. 17–66, ISBN-10 9073348722.

27. Vymazal, J. Plants used in constructed wetlands with horizontal subsurface flow: A review. *Hydrobiologia* **2011**, *674*, 133–156. [CrossRef]

28. Belmont, M.A.; Metcalfe, C.D. Feasibility of using ornamental plants (*Zantedeschia aethiopica*) in subsurface flow treatment wetlands to remove nitrogen, chemical oxygen demand and nonylphenol ethoxylate surfactants—A laboratory-scale study. *Ecol. Eng.* **2003**, *21*, 233–247. [CrossRef]

29. Merino-Solís, M.L.; Villegas, E.; de Anda, J.; López-López, A. The effect of the hydraulic retention time on the performance of an ecological wastewater treatment system: An anaerobic filter with a constructed wetland. *Water* **2015**, *7*, 1149–1163. [CrossRef]

30. Zurita, F.; De Anda, J.; Belmont, M.A. Treatment of domestic wastewater and production of commercial flowers in vertical and horizontal subsurface-flow constructed wetlands. *Ecol. Eng.* **2009**, *35*, 861–869. [CrossRef]

31. Konnerup, D.; Koottatep, T.; Brix, H. Treatment of domestic wastewater in tropical, subsurface flow constructed wetlands planted with Canna and Heliconia. *Ecol. Eng.* **2009**, *35*, 248–257. [CrossRef]

32. Vymazal, J.; Kröpfelová, L. Types of Wastewater Treated in HF Constructed Wetlands. In *Wastewater Treatment in Constructed Wetlands with Horizontal Sub-Surface Flow, Environmental Pollution*; Vymazal, J., Kröpfelová, L., Eds.; Springer: Dordrecht, The Netherlands, 2008; Volume 14, pp. 323–354. [CrossRef]

33. Vymazal, J. The use constructed wetlands with horizontal sub-surface flow for various types of wastewater. *Ecol. Eng.* **2009**, *35*, 1–17. [CrossRef]

34. Vymazal, J. Removal of nutrients in various types of constructed wetlands. *Sci. Total Environ.* **2007**, *380*, 48–65. [CrossRef] [PubMed]

35. Mekonnen, A.; Leta, S.; Njau, K.N. Wastewater treatment performance efficiency of constructed wetlands in African countries: A review. *Water Sci. Technol.* **2015**, *71*, 1. [CrossRef] [PubMed]

36. Rousseau, D.P.L.; Lesage, E.; Story, A.; Vanrolleghem, P.A.; De Pauw, N. Constructed wetlands for water reclamation. *Desalination* **2008**, *218*, 181–189. [CrossRef]

37. Trejo-Téllez, L.I.; Ramírez-Martínez, M.; Gómez-Merino, F.C.; García-Albarado, J.C.; Baca-Castillo, G.A.; Tejeda-Sartorius, O. Physical and chemical evaluation of volcanic rocks and its use for tulip production. *Rev. Mex. Cienc. Agrícolas* **2013**, *5*, 863–876.

38. Gayosso-Rodríguez, S.; Borges-Gómez, L.; Villanueva-Couoh, E.; Estrada-Botello, M.A.; Garruña-Hernández, R. Substrates for Growing Flowers. *Agrociencia* **2016**, *50*, 617–631.

39. Rice, E.W.; Baird, R.B.; Eaton, A.D. (Eds.) *Standard Methods for the Examination of Water and Wastewater*, 23rd ed.; American Public Health Works Association: Washington, DC, USA; American Water Works Association: Denver, CO, USA; Water Environment Federation: Alexandria, VA, USA, 2017; ISBN(s) 9781625762405.

40. Bedinger, M.S.; Fleming, J.S.; Johnson, A.I. (Eds.) *Site Characterization and Design of On-Site Septic Systems*; STP1324; ASTM International: West Conshohocken, PA, USA, 1997.

41. Headley, T.R.; Davison, L.; Huett, D.O.; Müller, R. Evapotranspiration from subsurface horizontal flow wetlands planted with *Phragmites australis* in sub-tropical Australia. *Water Res.* **2012**, *46*, 345–354. [CrossRef] [PubMed]

42. Secretaría de Medio Ambiente y Recursos Naturales (SEMARNAT). Normas Oficiales Mexicanas NOM-001-SEMARNAT-1996, NOM-002-SEMARNAT-1996 NOM-003-SEMARNAT-1997. Secretaría de Medio Ambiente y Recursos Naturales (SEMARNAT) 2007, Comisión Nacional del Agua (CONAGUA), México, 1997. p. 65. Available online: http://www.conagua.gob.mx/CONAGUA07/Publicaciones/Publicaciones/SGAA-15-13.pdf (accessed on 13 September 2017). (In Spanish)

43. Nasr, F.A.; Mikhaeil, B. Treatment of domestic wastewater using modified septic tank. *Desalination Water Treat.* **2014**, *56*, 2073–2081. [CrossRef]

44. Hamza, R.A.; Iorhemen, O.T.; Joo Hwa Tay, J.H. Advances in biological systems for the treatment of high-strength wastewater. *J. Water Process Eng.* **2016**, *10*, 128–142. [CrossRef]

45. Rousseau, D.P.L.; Vanrolleghem, P.A.; De Pauw, N. Model-based design of horizontal subsurface flow constructed treatment wetlands: A review. *Water Res.* **2004**, *38*, 1484–1493. [CrossRef] [PubMed]

46. Mugaas, B. The Good, Bad and Interesting Roles of Nitrogen (N) and Nitrogen Fertilizers in Home Lawn Care—Part 2 of a 3 Part Series on Understanding and Using Home Lawn Fertilizers. University of Minnesota Extension. Available online: http://blog-yard-garden-news.extension.umn.edu/2011/03/the-good-bad-and-interesting-roles-of.html (accessed on 31 March 2011).

47. Washington State Department of Health (WSDH). *Nitrogen Reducing Technologies for Onsite Wastewater Treatment Systems. Wastewater Management Program*; Division of Environmental Health, Washington State Department of Health: Olympia, WA, USA, 2005; p. 14. Available online: http://www.doh.wa.gov/portals/1/Documents/Pubs/337-093.pdf (accessed on 12 September 2017).

Article

Treatment of Dairy Wastewater by Oxygen Injection: Occurrence and Removal Efficiency of a Benzotriazole Based Anticorrosive

Santiago Martín-Rilo [1], Ricardo N. Coimbra [1], Carla Escapa [1] and Marta Otero [2,*]

[1] Department of Applied Chemistry and Physics and IMARENABIO (Institute of Environment, Natural Resources and Biodiversity), Universidad de León, Campus de Vegazana, 24071 León, Spain; smrilo69@gmail.com (S.M.-R.); ricardo.decoimbra@unileon.es (R.N.C.); carla.escapa@unileon.es (C.E.)

[2] Department of Environment and Planning and CESAM (Centre for Environmental and Marine Studies), University of Aveiro, Campus de Santiago, 3810-193 Aveiro, Portugal

* Correspondence: marta.otero@ua.pt

Received: 12 December 2017; Accepted: 30 January 2018; Published: 6 February 2018

Abstract: Benzotriazole is used as corrosion inhibitor in many industrial sectors, such as the dairy industry. Due to its widespread use in various applications and everyday consumer products, this chemical easily reaches the aquatic environment, where it may have deleterious effects. In fact, benzotriazole has been included among the so-called emerging contaminants. In this work, the occurrence and fate of a benzotriazole based anticorrosive (BTA-A) during wastewater treatment in a dairy industry has been assessed. At this dairy, a new system for wastewater treatment based on the injection of pure oxygen was recently started. This system has been proved to be efficient, economic and able to stably operate under a wide range of chemical oxygen demand and total suspended solids inputs. Then, after detecting the presence of BTA-A in the effluent of the wastewater treatment plant, it was aimed to optimize oxygen injection for the removal of this anticorrosive together with the regulated parameters. The performance of the system was evaluated at a real scale during a month period, during which the mean removal performance of the oxygen injection based treatment was 91%, 90% and 99% for chemical oxygen demand, total suspended solids and BTA-A, respectively.

Keywords: food industry; anticorrosive agent; benzotriazole; emerging contaminant; oxygen injection

1. Introduction

Benzotriazole (BTA), whose chemical structure is shown in Figure 1, is a heterocyclic compound containing three nitrogen atoms. This aromatic compound is colorless and polar and has been widely utilized in several fields such as plastics, coatings, dyes, and sunscreen. BTA has also been extensively used as metal corrosion inhibitor in a wide range of industrial applications. It is characterized by posing high water solubility (28 g/L), low vapor pressure and low octanol water distribution coefficient (log Kow: 1.23) [1].

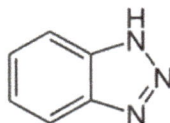

Figure 1. Chemical structure of benzotriazole (BTA, $C_6H_5N_3$).

BTA concentrations above 0.97 mg/L have been shown to pose chronic adverse effects to *Daphnia galeata* [2] and concentrations above 40 mg/L have revealed toxic effects in Microtox® tests [3]. Therefore, BTA has been classified as a toxic compound to aquatic organisms that can cause long-term adversary effects in the aquatic environment [1].

Apart from its toxic properties, due to its persistence and bioaccumulation, BTA cannot be totally but only partially removed from wastewater by conventional treatment processes [4]. In fact, such conventional processes and wastewater treatment plants are not designed for the removal of unregulated contaminants such as BTA. Consequently, and due to its widespread applications, BTA has become a ubiquitous contaminant in the aquatic environment, having been classified as an emerging contaminant [5]. The definition of emerging contaminants, which include an extensive and expanding spectrum of compounds, is still under discussion, but it may be said that they are compounds that are not currently covered by existing water-quality regulations, have not been studied before, and are thought to be potential threats to environmental ecosystems and human health and safety [6]. In fact, BTA has recently been detected in water supplies around the world, which has called the attention of many environmental researchers [7].

In the industry, water has numerous uses—heating, cooling, washing, cleanup, etc.—but has traditionally been over-used due to its low cost. Nevertheless, actual increasing environmental regulations, concerns around human and ecological health, and consumer expectations of high environmental performance have placed water conservation onto the agenda of the process industry [8]. Due to the advances on water/wastewater treatment technologies, a variety of options is actually available to provide a high standard of wastewater treatment. These technologies include advanced oxidation technologies (AOTs), which have been widely investigated for the treatment of industrial wastewaters, particularly where the source waters contain high concentration of ambiguous, refractory and recalcitrant chemical compounds such as aromatics, pesticides, pharmaceuticals and personal care products, drugs and endocrine disruptors [9–11]. Despite their efficiency, the implementation of AOTs is not always economically affordable for local industries. Alternatively, simple oxidation processes, which can be applied straightforward at low investment costs, may, in some cases, enable quality requirements to be met and matched to specific end-uses. In fact, for any industry, decisions on wastewater treatment require the analysis of economic criteria combined with the associated environmental issues [12].

In a previous work [13], data on the start-up of a new system for dairy wastewater treatment based in the injection of pure oxygen in the homogenization tank of a traditional physicochemical treatment were presented. It was concluded that this system was able to stably operate under a wide variety of both input chemical oxygen demand (COD) and total suspended solids (TSS). Furthermore, compared with the previous physicochemical system and also with a conventional biological treatment [14,15], it was proved to be more efficient and cheap. In this work, given the use of a benzotriazole based anticorrosive (BTA-A) agent for the refrigeration towers, the main aim was to determine the occurrence and removal of this BTA-A from wastewater at the dairy treatment plant. To the best of our knowledge, few publications deal with the removal of BTA or BTA products during wastewater treatment and there are not published results on dairy wastewater.

2. Materials and Methods

2.1. Dairy Wastewater Treatment and Anticorrosive Agent Utilization

The wastewater treatment plant under study was implemented in Lácteos Ibéricos, which is a dairy and juice factory in Northwest Spain. This factory processes 2,700,000 L of milk/week and 900,000 L of juice/week. Of total production, 72% consists of dairy products and the remaining 28% of juices and nectars. Such a production involves the generation of around 1000 m^3/day of wastewater with 61.1% of total COD corresponding to milk fat and 38.9% to juices and nectars. In the industry, wastewater must be treated before being discharged in the local sewage treatment plant (STP) under the obligatory accomplishment of tabulated limits (TSS, COD and biological oxygen demand after five days (BOD_5)) and a tax payment.

Figure 2 shows the current layout of the wastewater treatment plant at Lácteos Ibéricos. From the factory, wastewater is pumped to two homogenization rafts of 110 m^3 each, where the wastewater is neutralized under CO_2 injection (hydraulic retention time (HRT) = 30 min). Neutralized wastewater is then pumped to a 360 m^3 raft. Next, gravity passed to another raft of 110 m^3, pure O_2 (99.9999%, 6.5 bar) being injected in these rafts (total HRT = 352.5 min). Injected O_2 is purchased from Praxair (Madrid, Spain) at a price of 0.5 €/m^3. Subsequently, wastewater is pumped to the dissolved air flotation tank (DAF), where a coagulation–flocculation treatment is applied (HRT = 15 min). Finally, the treated wastewater exits the manifold and is discharged to the municipal STP while the extracted sludge is dried by means of a horizontal decanter. The clean water extracted from the mud in the decanter as well as the excess of sludge are sent back to head of the plant (homogenization rafts).

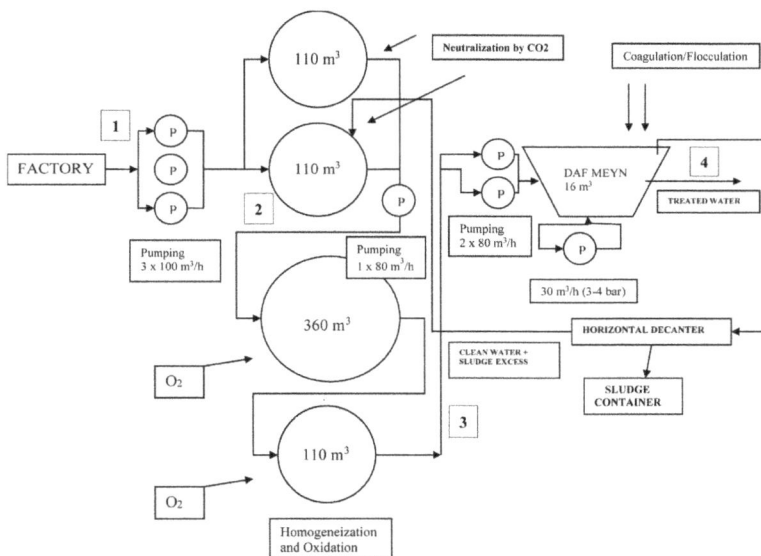

Figure 2. Dairy wastewater treatment plant layout. Sampling points considered for this work were: 1.—entrance of the plant (inlet); 2.—after neutralization under CO_2; 3.—after O_2 injection; and 4.—after the dissolved air flotation (DAF) tank, which corresponds to the effluent of the plant being discharged to the municipal sewage treatment plant (outlet).

At the dairy, a BTA-A, namely 3D TRASAR® 3DT265, from Nalco Water (Oviedo, Spain), is employed to avoid corrosion of the refrigeration towers. For this purpose, BTA-A is supplied by a peristaltic pump in the water feeding these refrigeration towers. The concentration of BTA-A in this water is weekly determined by a Nalco TRASAR Pen Fluorometer and dosage is adjusted in order to keep it in the range 90–100 mg/L, which guarantees that the desired purpose is fulfilled. Benzotriazole (BTA) is part of the composition of the referred anticorrosive agent (1.13 g/mL, 0.5–1.5% *w/w* BTA), which is widely used in different types of industries.

Actually, 11,750 L of BTA-A are used at the Lácteos Ibéricos dairy on a yearly basis. Due to the cleaning operations on the refrigeration towers, this anticorrosive agent will end in the dairy wastewater. Therefore, considering a yearly production of 365,000 m^3 of wastewater at the dairy, a concentration of around 36.4 mg/L of BTA-A in the wastewater entering the treatment plant at the dairy may be expected, which means about 0.55–0.18 mg/L of BTA.

2.2. Laboratory-Scale Experiments

Three wastewater samples (2 L) were collected just at the entrance of the treatment plant (pointed as 1 in Figure 2). Each sample was divided into four 0.5 L sub-samples. One of them was used as the control, in the absence of oxygen injection. The other three sub-samples were treated by the injection of O_2 (99.9999%, 6.5 bar), each at a different flow rate, namely 15, 20 and 25 m^3/h. During such an oxygen injection treatment at laboratory, aliquots were withdrawn throughout time so to quantify TSS and COD in water, which was done according to standard methods [16]. In addition, the concentration of BTA-A in aliquots was determined by a Nalco TRASAR Pen Fluorometer. The oxygen injection treatment was maintained until the stabilization of the above parameters in wastewater samples.

The experimental results on the removal of TSS, COD and BTA-A were described by a pseudo first-order kinetic model (Equation (1)). Fittings to this equation were obtained by GraphPad Prism 7 (GraphPad Software Inc., San Diego, CA, USA):

$$\%_{\text{removal}} = \%_{\text{max}} \left(1 - e^{-k_1 t}\right), \tag{1}$$

where $\%_{\text{removal}}$ is the percentage of TSS, COD or BTA-A removed at a certain time under O_2 injection, $\%_{\text{max}}$ is the maximum percentage of removal that is expected to attain, k_1 (min^{-1}) is the pseudo-first order rate constant and t (min) is the time under a certain flow rate of O_2.

The differences among the three O_2 flow rates in terms of kinetic parameters were tested using the non-parametric Kruskal–Wallis test. When the Kruskal–Wallis test pointed to significant differences ($p < 0.05$), the differences between the three combinations of flows were tested using the posthoc Dunn's test.

2.3. Plant Study at a Real Scale

Wastewater was sampled from Monday to Friday during four consecutive weeks at the following points of the wastewater treatment plant (Figure 2): 1.—at the entrance of the plant (inlet); 2.—after neutralization under CO_2; 3.—after O_2 injection; and 4.—after the DAF, which corresponds to the effluent of the plant being discharged to the municipal STP (outlet). At each point, three grab samples (0.5 L each) were collected daily, one every eight hours. Sampling was made to coincide with the start of each work shift so to avoid disturbing the work routine at the dairy. The following parameters were analysed according to standard methods [16]: temperature, pH, dissolved oxygen, TSS, COD. Furthermore, the BTA-A concentration in the samples was determined by a Nalco TRASAR Pen Fluorometer. Considering the flow stability of the influent and effluent in the plant (1000 m^3/day) and that three samples were collected daily, the removal efficiency (%) of TSS, COD and BTA-A was calculated on the basis of their respective average daily concentrations by applying error calculation rules.

3. Results and Discussion

3.1. Laboratory-Scale Experiments

Control experiments allowed to verify that TSS, COD and BTA-A concentration remained the same during the duration of the laboratory-scale experiments. Regarding the laboratory experiments under oxidation, the percentages of TSS, COD and BTA-A removal throughout time under different O_2 flow rates are represented in Figure 3 together with the corresponding fittings to the pseudo first-order kinetic model (Equation (1)). The fitted parameters of the kinetic model (k_1 and $\%_{\text{max}}$) are depicted in Table 1, where the goodness of the fittings is represented by the determination coefficient (R^2) and the deviation (S_{xy}).

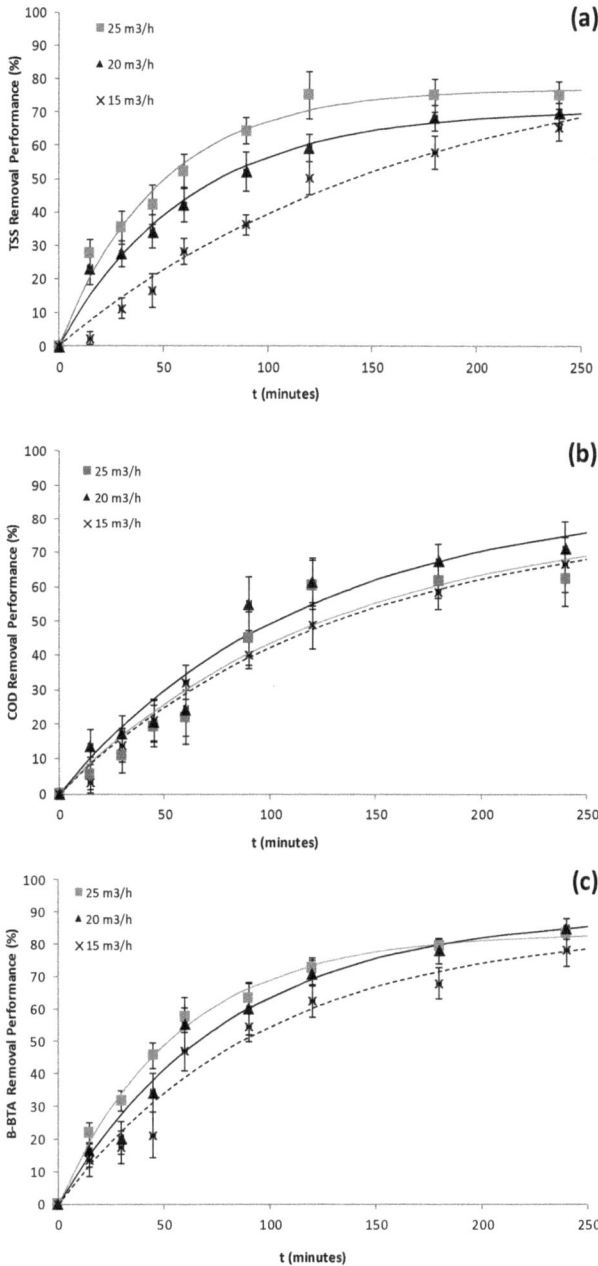

Figure 3. Laboratory study investigating the effect of different O_2 flow rates on the removal of: (**a**) total suspended solids (TSS); (**b**) chemical oxygen demand (COD); and (**c**) benzotriazole based anticorrosive agent (BTA-A). Note: Error bars stand for standard deviation values (*n* = 3). Experimental results are shown together with fittings to the pseudo first-order kinetic model (grey continuous line, black continuous line and black discontinuous line represent fittings to removal under 25, 20 and 15 m^3/h of O_2, respectively).

Table 1. Parameters obtained from the fittings of experimental results on the removal of total suspended solids(TSS), chemical oxygen demand (COD) and benzotriazole based anticorrosive (BTA-A) under different O_2 flow rates to the pseudo first-order kinetic model.

Contaminant Removed	O_2 Flowrate	Parameters of the Pseudo First-Order Kinetic Model		Goodness of Fitting	
	(m^3/h)	k_1 (min^{-1})	$\%_{max}$	R^2	S_{xy}
	25	0.0207 ± 0.0021 [a]	77.24 ± 2.76 [a]	0.9818	3.74
TSS	20	0.0158 ± 0.0017 [b]	71.02 ± 3.05 [a]	0.9814	3.37
	15	0.0057 ± 0.0013 [c]	80.05 ± 12.65 [a]	0.9818	3.50
	25	0.0077 ± 0.0026 [a]	80.72 ± 15.07 [a]	0.9394	6.73
COD	20	0.0085 ± 0.0023 [a]	85.93 ± 12.47 [a]	0.9491	6.46
	15	0.0074 ± 0.0010 [a]	80.62 ± 6.27 [a]	0.9894	2.62
	25	0.0175 ± 0.0010 [a]	83.40 ± 1.82 [a]	0.9947	2.19
BTA-A	20	0.0124 ± 0.0017 [b]	85.59 ± 5.45 [a]	0.9789	4.67
	15	0.0102 ± 0.0021 [b]	85.15 ± 8.39 [a]	0.9640	5.62

Note: Significantly different parameters ($p < 0.05$) are marked with different superscripts (a), (b), (c), while the same superscript is used in the case of no significant differences ($p > 0.05$). Those parameters for which significantly different values have been determined are in bold.

As may be seen in Figure 3a, under the flow rates tested, TSS are mostly removed during the first 120 min under oxygen injection. Moreover, from 180 to 240 min, there is not a noteworthy increase of the removal percentage. Parameters in Table 1 show that the largest O_2 flow rate (25 m^3/h) allowed for a significantly larger kinetic constant for the removal of TSS, followed by 20 and 15 m^3/h. In any case, the $\%_{max}$ attained under these flow rates were not significantly different. With respect to the COD, removal curves in Figure 3b are similar to those of TSS (Figure 3a), COD being removed mainly during the first 120 min and with no remarkable increments between 180 and 240 min. Under the considered O_2 flow rates, neither the kinetic constant nor the maximum percentage of COD removal were significantly different according to the data displayed in Table 1. Finally, results on BTA-A removal, which are represented in Figure 3c, show that elimination is near 85% at the end of 240 min under oxygenation. As for TSS and COD, BTA-A is mainly removed during the first 120 min under O_2 injection and irrelevant increments on the BTA-A removal were observed for oxygenation times longer than 180 min. As for the parameters in Table 1, the k_1 under 25 m^3/h was significantly higher than that under 20 or 15 m^3/h. However, the $\%_{max}$ attained under these flow rates were not significantly different.

On the basis of the above results, a flow rate of 20 m^3/h was selected to be injected into the dairy wastewater treatment plant for the subsequent study. Although it is true that a slightly faster TSS and BTA-A removal was attained under 25 m^3/h, the maximum TSS, COD and BTA-A removals attained under the O_2 flow rates considered here were not significantly different. Furthermore, at a real scale, if the plant works 720 h/month, using 25 m^3/h instead of 20 m^3/h would represent an extra expense of 1800 €/month, which is not worth taking into account the results here obtained at the laboratory-scale experiments.

3.2. Plant Study at a Real Scale

Figure 4 represents the average inlet concentration (Figure 4a) and removal efficiency (Figure 4b) of the wastewater treatment plant for TSS, COD and BTA-A. Results on the inlet TSS and COD in Figure 4a are within the range of values determined for these parameters in a previous study [13]. Regarding BTA-A inlet concentration, values are all between 30 and 35 mg/L, which is a quite narrow range as compared with TSS and COD. While TSS and COD contents in wastewater are related to production fluctuations at the dairy, the inlet BTA concentration is more stable since it is due to its dosed use as anticorrosive agent. In fact, the mean BTA-A concentration at the entrance of the wastewater treatment plant is 33 mg/L, which is close to that referred above as the expected concentration of BTA-A (36.4 mg/L) on the basis of its yearly consumption.

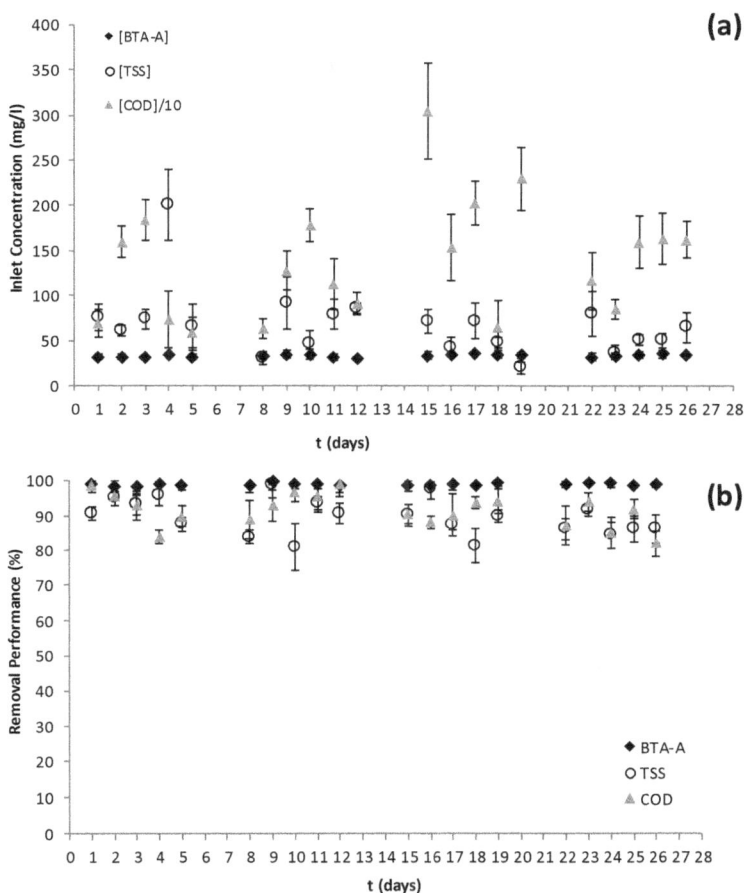

Figure 4. Results at a real plant scale on: (**a**) the inlet concentration of total suspended solids (TSS), chemical oxygen demand (COD) and benzotriazole based anticorrosive agent (BTA-A) concentration; and (**b**) the efficiency performance of the dairy wastewater treatment plant on the removal of TSS, COD and BTA-A. Note: Error bars stand for standard deviation values ($n = 4$).

Figure 4b represents the average daily removal of TSS, COD and BTA in the dairy wastewater treatment plant throughout the sampling period. As may be seen, the percentage of TSS average removal remained between 81% and 98%. With respect to COD, average removal percentages between 82% and 99% were observed. These efficiencies are mostly higher than those obtained when the oxygen injection system was started [13]. It should be highlighted that, as said above, effluent from the dairy wastewater treatment plant is discharged to the municipal sewage system, which involves the payment of a tax on the basis of the load and volume discharged as described by the following equation:

$$\text{Tax } (\text{€}/\text{m}^3) = P \times K, \tag{2}$$

where P is a fixed coefficient (currently, $P = 0.5 \text{ €}/\text{m}^3$) and K ($1 < K < 5$) is a correction factor. This factor (K) is a function of the contamination index (I) of the effluent to be discharged in the municipal STP system.

The calculation of I is as follows:

$$I = COD + 1.65 \, BOD_5 + 1.10 \, TSS, \tag{3}$$

where BOD_5, COD and TSS are expressed in kg/m^3. At the dairy industry considered here, BOD_5 is not regularly determined but estimated as $BOD_5 = 0.65 \, COD$, thereby I is calculated as indicated next:

$$I = 2.3 \, COD + 1.10 \, TSS. \tag{4}$$

The value of I is periodically determined for a composed sample and it may be verified by the STP at any moment. Then, the K coefficient is the value of I rounded to units (units will be left the same if the tenth value is less than 5, but units will be increased by one if the tenth value is five or more). The minimum applicable value of K is 1 and the maximum value is 5, which will be applied when the calculated I is 5 or larger. For the dairy, it is a priority that the STP tax is as low as possible. Therefore, it must be ensured that wastewater treatment implemented at the dairy allows for a $K = 1$ in order to minimize the STP tax to be paid. In the present study, the mean outlet TSS and COD concentrations within the sampling period were 6.5 and 126 mg/L, respectively, which guarantees that the correction factor (K) that is applied for the calculation of the municipal STP tax is equal to 1.

Regarding the removal of BTA-A, percentages above 98% have been always attained throughout the four weeks considered period. It may be observed that the removal of BTA-A is larger and remains mostly unchanged as compared with the removal of TSS and COD. These facts must be associated to the more stable inlet BTA-A concentrations, which do not depend on the dairy production. On the other hand, comparing results in Figure 3 with those in Figure 4, it is evident that removal efficiencies at a real scale were higher than those observed at a laboratory-scale. This must be related to the existence of three consecutive steps at the dairy wastewater treatment plant, namely CO_2 injection, O_2 injection and coagulation–flocculation, while, at a laboratory scale, only O_2 injection was applied.

A main question about BTA-A removal in the dairy wastewater treatment plant is the contribution of each treatment stage. In order to assess this issue, the concentration of BTA-A in the dairy wastewater and the removal at each stage of the treatment plant is represented in Figure 5. Figure 5a allows for comparing the daily average concentrations of BTA-A at the entrance of the wastewater treatment plant (sampling point (1)), after neutralization by CO_2 injection (sampling point (2)), after O_2 injection (sampling point (3)) and after the DAF (sampling point (4)). As it was highlighted above, the mean BTA-A concentration at the entrance of the wastewater treatment plant throughout the period under study is 33 mg/L, with values between 26 and 40 mg/L. Meanwhile, the mean BTA-A concentration after CO_2 injection is 19 mg/L, with values between 13 and 23 mg/L. After O_2 injection, the mean BTA-A concentration is 8 mg/L, varying between 6 and 12 mg/L. Finally, after the DAF, BTA-A daily average concentrations are between 0.86 and 0.06 mg/L with a mean of 0.35 mg/L. Therefore, BTA-A concentration progressively decreases along with subsequent treatments at the dairy wastewater treatment plant. In addition, it may be seen that, from one stage to the subsequent one, BTA-A concentration values get more stable within the sampling period with a decreasing daily standard deviation.

Accumulative percentages of the BTA-A removal in the dairy wastewater treatment plant throughout the period under study are shown in Figure 5b for each treatment stage. Throughout the whole period, the average removal of BTA-A under CO_2 injection (sampling point (2)) remains between 33% and 56%, with a mean of 44%. Then, after the O_2 injection stage (sampling point (3)), a mean removal of 74% is achieved within the whole sampling period. Finally, after the DAF (sampling point (4)), the mean performance of the plant on the elimination of BTA-A is 99% within the sampling period, with average values ranging between 98.3% and 99.7%. Then, the individual weight of each stage for this overall removal may be calculated as 44%, 30% and 25% for CO_2 injection, O_2 injection and coagulation–flocculation, respectively. Therefore, the importance of each stage on the overall removal of BTA-A from wastewater is progressively decreasing. However, when considering the BTA-A concentration entering at each stage, the mean removal performance of the coagulation–flocculation

DAF stage is 95.6%. This is a remarkable efficiency as compared with the previous treatments, namely CO_2 and O_2 injection, which respective mean efficiencies were of 44% and 56% relatively to the corresponding BTA-A input. Polymeric cationic flocculants are used at the dairy wastewater treatment plant; thus, given the low Kow of BTA [1], either a combination of charge neutralization and bridging or bridging, but not sorption, must be the main mechanisms for BTA-A removal at the DAF. On the other hand, as compared with the DAF treatment, large daily deviations in the performance of the CO_2 and, especially, of the O_2 injection stages may be observed in Figure 5b. In addition, for these stages, a certain variation from day to day between average performance values may be observed. On the contrary, the global removal shows relative small variations within and inter days throughout the period here considered. This is quite relevant, since the dairy wastewater treatment plant gets to ensure a short range of low BTA-A concentrations in the effluent. Even so, considering the BTA content of the BTA-A used here, the effluent from the dairy wastewater treatment plant must have a BTA concentration of 0.003–0.013 mg/L.

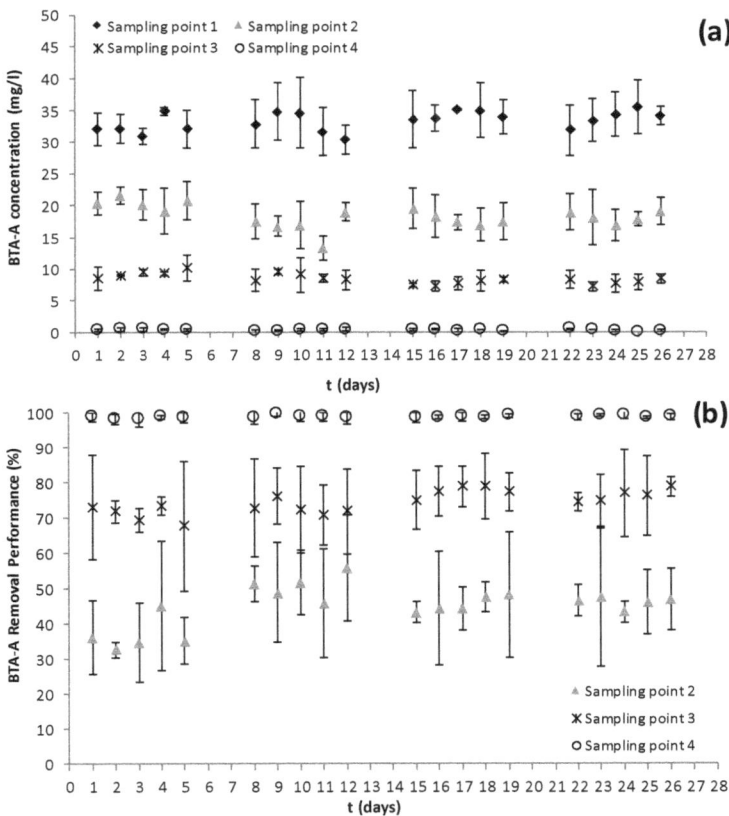

Figure 5. Results on (**a**) the benzotriazole based anticorrosive (BTA-A) concentration; and (**b**) cumulative BTA-A removal percentage at the different stages of the dairy wastewater treatment plant, namely at the sampling point 2.—after neutralization under CO_2; sampling point 3.—after O_2 injection; and sampling point 4.—after the DAF, which corresponds to the effluent of the plant being discharged to the municipal sewage treatment plant (outlet). Cumulative removal percentages were calculated respect the entering BTA-A concentration at the sampling point 1.—entrance of the plant (inlet). Note: Error bars stand for standard deviation values ($n = 3$).

To our best knowledge, there are not published results on the occurrence of benzotriazole or benzotriazole based anticorrosives in dairy wastewaters, neither on their removal. Several authors have studied the removal of BTA during activated sludge (AS) batch experiments [17–19]. However, information on the removal of BTA during wastewater treatment is scarce in the literature. In the case of municipal wastewaters, Voutsa et al. [1] made a deep study on the occurrence and fate of benzotriazoles in wastewater treatment plants (WWTPs) and presented data on the concentrations of BTA in samples of primary and secondary effluents from 24 different WWPTs in Switzerland. The median outlet concentrations of BTA were 18 and 10 μg/L, respectively [1], which are similar to the outlet BTA concentration at the dairy wastewater treatment plant throughout the sampling period carried out in the present work. Voutsa et al. [1] highlighted that the elimination of BTA in WWTPs was relatively low, with maximum values of 62% obtained from 10 WWTPs in the Glatt Valley catchment. More recently, it was highlighted that benzotriazoles, which are polar and poorly degradable pollutants, are insufficiently removed by biological treatment at conventional wastewater treatment plants as concluded from the reported BTA removals (29–58%) in wastewater treatment plants [20]. Even under higher removals (around 75% for BTA), it has been pointed out that large average daily loads are discharged via treated wastewater to the aquatic environment [21].

Compared with the above values, the performance of the oxygen injection based dairy wastewater treatment plant here studied is quite satisfactory. Likewise, it must be considered that the treated effluent from the dairy is discharged to the local STP [13]), where BTA-A may be further removed. In any case, efficient wastewater treatment processes and zero discharge treatment units are desirable for dairies sustainability [22]. Additional in-plant treatments could be implemented in the dairy wastewater treatment plant to improve the quality of the outlet effluent. In this sense, hybrid membrane processes combining powdered activated carbon (PAC) adsorption with ultrafiltration (UF), which have been successfully used for the removal of BTA from wastewater [23] could be implemented after the DAF. Future studies must be carried out on the implementation of such treatments, and on the direct analysis of BTA and its transformation products in the dairy wastewater.

4. Conclusions

A benzotriazole based anticorrosive agent (BTA-A) is used at the dairy for inhibiting corrosion of the refrigeration towers. Then, water used for cleaning operations of these towers is treated at the dairy wastewater treatment plant where a new oxygen injection system was recently established. It was verified in this work that BTA-A is present in the dairy wastewater, concentration values between 26 and 40 mg/L having been determined during a four-week sampling period. After a laboratory scale study, a 20 m^3/h O_2 (99.9999%, 6.5 bar) injection flow rate was set up at the dairy wastewater treatment plant in order to remove total suspended solids (TSS), chemical oxygen demand (COD) and BTA-A and to minimize oxygen associated costs. During the sampling period, the removal of TSS and COD at the dairy wastewater treatment plant remained between 81–98% and 82–99%, respectively, with mean outlet concentrations of 6.5 mg/L TSS and 126 mg/L COD. This operation efficiency guaranteed the minimization of the tax to be paid to the municipal sewage treatment plant (STP) where the treated wastewater from the dairy is discharged. Regarding BTA-A, removal at the dairy wastewater treatment plant was always above 97%, progressive elimination occurring at the three treatment stages, namely neutralization under CO_2 injection, oxidation under O_2 injection and coagulation–flocculation at the dissolved air flotation tank (DAF). No published results have been found on BTA or BTA based anticorrosive agents in dairy wastewater, but, compared with data on conventional STPs, the removal efficiency of the wastewater treatment plant here considered is quite satisfactory.

Acknowledgments: The *Fundação para a Ciência e a Tecnologia* (FCT, Lisboa, Portugal) is acknowledged for support through the FCT Investigator Program (IF/00314/2015). Thanks are due for the financial support to CESAM (UID/AMB/50017)through national funds (*Programa de Investimento e Despesas de Desenvolvimento da Administração Central*, PIDDAC) by FCT/*Ministério da Ciência, Tecnologia e Ensino Superior* (MCTES), and through co-funding (POCI-01-0145-FEDER-007638) by the *Fundo Europeu de Desenvolvimento Regional* (FEDER), within the PT2020 Partnership Agreement and Compete 2020 (*Programa Operacional Competitividade e Internacionalização*, POCI).

Author Contributions: Santiago Martín-Rilo, Ricardo N. Coimbra and Marta Otero conceived the work and designed the experiments; Santiago Martín-Rilo performed the sampling and chemical analysis; Ricardo N. Coimbra, Carla Escapa and Marta Otero analysed the results and wrote the manuscript; and Marta Otero supervised this study. The final version was approved by all authors.

Conflicts of Interest: The authors declare no conflict of interest. Authors also declare that the founding agents did not participate in the design of the study; in the sample collection or sample analysis; in the interpretation of data; in the writing of the manuscript; and in publishing results.

References

1. Voutsa, D.; Hartmann, P.; Schaffner, C.; Giger, W. Benzotriazoles, alkylphenols and bisphenol A in municipal wastewaters and in the Glatt River, Switzerland. *Environ. Sci. Pollut. Res.* **2006**, *13*, 333–341. [CrossRef]
2. Seeland, A.; Oetken, M.; Kiss, A.; Fries, E.; Oehlmann, J. Acute and chronic toxicity of benzotriazoles to aquatic organisms. *Environ. Sci. Pollut. Res.* **2012**, *19*, 1781–1790. [CrossRef] [PubMed]
3. Pedrazzani, R.; Ceretti, E.; Zerbini, I.; Casale, R.; Gozio, E.; Bertanza, G.; Gelatti, U.; Donato, F.; Feretti, D. Biodegradability, toxicity and mutagenicity of detergents: Integrated experimental evaluations. *Ecotoxicol. Environ. Saf.* **2012**, *84*, 274–281. [CrossRef] [PubMed]
4. Ye, J.; Zhou, P.; Chen, Y.; Ou, H.; Liu, J.; Li, C.; Li, Q. Degradation of 1H-benzotriazole using ultraviolet activating persulfate: Mechanisms, products and toxicological analysis. *Chem. Eng. J.* **2018**, *334*, 1493–1501. [CrossRef]
5. Xu, J.; Li, L.; Guo, C.; Zhang, Y.; Wang, S. Removal of benzotriazole from solution by BiOBr photocatalysis under simulated solar irradiation. *Chem. Eng. J.* **2013**, *221*, 230–237. [CrossRef]
6. Farré, M.L.; Pérez, S.; Kantiani, L.; Barceló, D. Fate and toxicity of emerging pollutants, their metabolites and transformation products in the aquatic environment. *TrAC Trends Anal. Chem.* **2008**, *27*, 991–1007. [CrossRef]
7. Alotaibi, M.D.; McKinley, A.J.; Patterson, B.M.; Reeder, A.Y. Benzotriazoles in the aquatic environment: A review of their occurrence, toxicity, degradation and analysis. *Water Air Soil Pollut.* **2015**, *226*, 226. [CrossRef]
8. Barrington, D.J.; Prior, A.; Ho, G. The role of water auditing in achieving water conservation in the process industry. *J. Clean. Prod.* **2013**, *52*, 356–361. [CrossRef]
9. Asghar, A.; Raman, A.A.A.; Daud, W.M.A.W. Advanced oxidation processes for in-situ production of hydrogen peroxide/hydroxyl radical for textile wastewater treatment: A review. *J. Clean. Prod.* **2015**, *87*, 826–838. [CrossRef]
10. Chatzisymeon, E.; Foteinis, S.; Mantzavinos, D.; Tsoutsos, T. Life cycle assessment of advanced oxidation processes for olive mill wastewater treatment. *J. Clean. Prod.* **2013**, *54*, 229–234. [CrossRef]
11. Chong, M.N.; Sharma, A.K.; Burn, S.; Saint, C.P. Feasibility study on the application of advanced oxidation technologies for decentralised wastewater treatment. *J. Clean. Prod.* **2012**, *35*, 230–238. [CrossRef]
12. Rodríguez, R.; Espada, J.J.; Gallardo, M.; Molina, R.; López-Muñoz, M.J. Life cycle assessment and techno-economic evaluation of alternatives for the treatment of wastewater in a chrome-plating industry. *J. Clean. Prod.* **2018**, *172*, 2351–2362. [CrossRef]
13. Martín-Rilo, S.; Coimbra, R.N.; Martín-Villacorta, J.; Otero, M. Treatment of dairy industry wastewater by oxygen injection: Performance and outlay parameters from the full scale implementation. *J. Clean. Prod.* **2015**, *86*, 15–23. [CrossRef]
14. Asplund, S. The Biogas Production Plant at Umeå Dairy. Evaluation of Design and Start Up. Master's Thesis, Linköpings Universitet, Sweden, 2005. Available online: http://www.diva-portal.org/smash/get/diva2:21340/FULLTEXT01.pdf (accessed on 15 November 2017).
15. Passeggi, M.; López, I.; Borzacconi, L. Modified UASB reactor for dairy industry wastewater: Performance indicators and comparison with the traditional approach. *J. Clean. Prod.* **2012**, *26*, 90–94. [CrossRef]
16. Eaton, A.D.; Clesceri, L.S.; Greenberg, A.E. *Standard Methods for the Examination of Water and Wastewater*, 19th ed.; American Public Health Association (APHA): Washington, DC, USA, 1995; ISBN 0-87553-223-3.
17. Herzog, B.; Lemmer, H.; Huber, B.; Horn, H.; Müller, E. Xenobiotic benzotriazoles-biodegradation under meso- and oligotrophic conditions as well as denitrifying, sulfate-reducing, and anaerobic conditions. *Environ. Sci. Pollut. Res.* **2014**, *21*, 2795–2804. [CrossRef] [PubMed]
18. Liu, Y.; Ying, G.; Shareef, A.; Kookana, R.S. Biodegradation of three selected benzotriazoles under aerobic and anaerobic conditions. *Water Res.* **2011**, *45*, 5005–5014. [CrossRef] [PubMed]

19. Mazioti, A.A.; Stasinakis, A.S.; Pantazi, Y.; Andersen, H.R. Biodegradation of benzotriazoles and hydroxy-benzothiazole in wastewater by activated sludge and moving bed biofilm reactor systems. *Bioresour. Technol.* **2015**, *192*, 627–635. [CrossRef] [PubMed]

20. Reemtsma, T.; Miehe, U.; Duennbier, U.; Jekel, M. Polar pollutants in municipal wastewater and the water cycle: Occurrence and removal of benzotriazoles. *Water Res.* **2010**, *44*, 596–604. [CrossRef] [PubMed]

21. Stasinakis, A.S.; Thomaidis, N.S.; Arvaniti, O.S.; Asimakopoulos, A.G.; Samaras, V.G.; Ajibola, A.; Mamais, D.; Lekkas, T.D. Contribution of primary and secondary treatment on the removal of benzothiazoles, benzotriazoles, endocrine disruptors, pharmaceuticals and perfluorinated compounds in a sewage treatment plant. *Sci. Total Environ.* **2013**, *463–464*, 1067–1075. [CrossRef] [PubMed]

22. Tiwari, S.; Behera, C.R.; Srinivasan, B. Simulation and experimental studies to enhance water reuse and reclamation in India's largest dairy industry. *J. Environ. Eng. Chem.* **2016**, *4*, 601–616. [CrossRef]

23. Löwenberg, J.; Zenker, A.; Baggenstos, M.; Koch, G.; Kazner, C.; Wintgens, T. Comparison of two PAC/UF processes for the removal of micropollutants from wastewater treatment plant effluent: Process performance and removal efficiency. *Water Res.* **2014**, *56*, 26–36. [CrossRef] [PubMed]

MDPI

St. Alban-Anlage 66

4052 Basel

Switzerland

Tel. +41 61 683 77 34

Fax +41 61 302 89 18

www.mdpi.com

Water Editorial Office

E-mail: water@mdpi.com

www.mdpi.com/journal/water

www.ingramcontent.com/pod-product-compliance
Lightning Source LLC
Chambersburg PA
CBHW051844210326
41597CB00033B/5767